Nanocrystals - Synthesis, Characterization and Applications

Editor

Hua Fen Han

scitus
academics

Nanocrystals - Synthesis, Characterization and Applications
Edited by **Hua Fen Han**

ISBN: 978-1-68117-527-0
Library of Congress Control Number: 2016942927

© 2017 by
SCITUS Academics LLC,
www.scitusacademics.com
Box No. 4766, 616 Corporate Way,
Suite 2, Valley Cottage,
NY 10989

This book contains information obtained from highly regarded resources. Copyright for individual articles remains with the authors as indicated. All chapters are distributed under the terms of the Creative Commons Attribution License, which permits unrestricted use, distribution, and reproduction in any medium, provided the original author and source are credited.

Notice

Reasonable efforts have been made to publish reliable data and views articulated in the chapters are those of the individual contributors, and not necessarily those of the editors or publishers. Editors or publishers are not responsible for the accuracy of the information in the published chapters or consequences of their use. The publisher believes no responsibility for any damage or grievance to the persons or property arising out of the use of any materials, instructions, methods or thoughts in the book. The editors and the publisher have attempted to trace the copyright holders of all material reproduced in this publication and apologize to copyright holders if permission has not been obtained. If any copyright holder has not been acknowledged, please write to us so we may rectify.

Printed in United States of America on Acid Free Paper

Preface

Nanocrystals are molecular-sized solids formed with a repeating, 3D pattern of atoms or molecules with an equal distance between each part. Nanocrystals are aggregates of anywhere from a few hundred to tens of thousands of atoms that combine into a crystalline form of matter known as a 'cluster'. Typically around 10nm in diameter, nanocrystals are larger than molecules but smaller than bulk solids and therefore frequently exhibit physical and chemical properties somewhere in-between. Nanocrystals are believed to have potential in optical electronics because of their ability to change the wavelength of light. Nanocrystal" is expected to lead to the creation of new materials with revolutionary properties and functions. It will open up fresh possibilities for the solution to the environmental problems and energy problems. During the last decade, nanocrystals have been widely adopted in various electronic and optoelectronic applications. They provide alternative options in terms of ease of processing, low cost, better flexibility, and superior electronic/optoelectronic properties. By taking advantage of solution-processing, self-assembly, and surface engineering, nanocrystals could serve as new building blocks for low-cost manufacturing of flexible and large area devices. Tunable electronic structures combined with small exciton binding energy, high luminescence efficiency, and low thermal conductivity make nanocrystals extremely attractive for FET, memory device, solar cell, solid-state lighting/display, photodetector, and lasing applications.

Nanocrystals - Synthesis, Characterization and Applications provides a comprehensive overview on nanocrystals processing and characterization of their structural, optical, electronic, magnetic and mechanical properties. Various methods for nanocrystals synthesis are discussed in the book.

Table of Contents

Chapter 1

Characterization of Nanocrystals Using Spectroscopic Ellipsometry

Peter Petrik[1]

[1] Institute for Technical Physics and Materials Science (MFA), Research Centre for Natural Sciences, Budapest, Hungary

1. INTRODUCTION

First applications of ellipsometry to the measurement of poly- and nanocrystalline thin films date back to many decades. The most significant step towards the ellipsometric investigation of composite thin films was the realization of the first spectroscopic ellipsometers in the '70s [3, 4, 8], which allowed the measurement of the dielectric function, the imaginary part of which is directly related to the joint density of electronics states sensitively depending upon the changes of the crystal structure. The first models were based on the effective medium approach using constituents of known dielectric functions [5], whereas the volume fraction of the components can be related to the crystal properties of the thin films. This approach is popular ever since, based on its robustness.

The effective medium methods have been followed by a range of different analytical models based on the parameterization of the dielectric function. These models allow the determination of the material properties also in cases when the material cannot be considered as a homogeneous mixture of phases with known dielectric function. These models can also be used for small grains that show size effects (and hence a modified electronic structure and dielectric function), i.e. for grains that can not be modeled by bulk references.

Additional to the nanocrystal properties, the ellipsometric approach allows the sensitive characterization of further layer characteristics like the interface quality (e.g. nanoroughness at the layer boundaries), the lateral or vertical inhomogeneity or the thicknesses in multi-layer structures.

2. BASICS OF ELLIPSOMETRY

If polarized light will be reflected on the boundary of two media, the state of polarization of the reflected beam will be elliptical, circular, or linear depending on the properties of the sample. In most cases, the reflected light is elliptically polarized, that's why the method is called ellipsometry. Ellipsometry directly measures the change of polarization caused by the reflection, i.e. the complex reflectance ratio defined by

$$\rho = \frac{\chi_r}{\chi_i} = \frac{r_p}{r_s} = \tan\Psi e^{i\Delta}, \qquad (1)$$

where $\chi r = Er{,}p/Er{,}s$ and $\chi i = Ei{,}p/Ei{,}s$ (*r*: reflected, *i*: incident; *p*: parallel to the plane of incidence; *s*: perpendicular to the plane of incidence) are the states of polarization with $E = E_0 e^{i(\omega t+\delta)} e^{-i\omega \frac{N}{c} x}$ describing the electromagnetic plane wave (ω: angular frequency; t: time; δ: phase; *N*: complex refractive index; *c*: speed of light; *x*: position), *rp* and *rs* are the reflection coefficients (the ratio of reflected and incident *E* values) of light polarized parallel and perpendicular to the plane of incidence, respectively, while Ψ and Δ are the ellipsometric angles [15, 17]. The latter are the raw data of an ellipsometric measurement usually plotted as a function of wavelength, angle of incidence, or time.

The sample properties like the layer thickness or the refractive index are determined using optical models, in which one assumes a layer structure and the refractive index of each layer at each wavelength. Using such optical models, the Ψ and Δ values can be calculated (for each wavelength and angle of incidence) and compared with the measured Ψ and Δ spectra. Next, the parameters of the optical model (e.g. thickness, refractive index, or parameters of the dispersion) are fitted using optimization algorithms like the Levenberg-Marquardt method in order to minimize the difference between the measured and calculated Ψ and Δ values.

3. MEASURABLE NANOCRYSTAL PROPERTIES

Ellipsometry can measure nanocrystals in a thin film form. The optical model consists of a substrate with usually known optical properties (e.g. single-crystalline silicon or glass) and one or more thin films containing the nanocrystals. The information needed to calculate the ellipsometric angles Ψ and Δ (that in turn can be compared with the measured angles) are the angle of incidence, the thickness(es) of the layer(s) and the complex refractive indices of each medium. The square of the refractive index is the dielectric function, the imaginary part ($\int_2$) of which is directly proportional to the joint density of electronic states of the measured crystalline or partly crystalline material. Compared to the typical sensitivities ($< 10^{-3}$ in $\int$) the difference between the dielectric functions of the amorphous and single-crystalline phases are huge, as shown in Fig. 1. It is clearly seen that the dielectric functions are largely different depending on the crystallinity of silicon ranging from single-crystalline through nanocrystalline to amorphous. Note that the sensitivity is especially high around the critical point energies, which appear as peaks in $\int_2$. For example, the difference of the three cases in $\int_2$ is larger than 10 at the critical point energy of 4.2 eV.

The key of measuring nanocrystals is to relate the dielectric function to nanocrystal properties. The technique that is most widely used for more than 30 years is the effective medium theory [5, 10]. In this theory it is assumed that the material is a mixture of phases large enough to retain their bulk-like properties, but smaller than the wavelength of the probing light, so that scattering can be avoided. In case of a mixture of two components with dielectric functions of $\int a$ and $\int b$ the effective dielectric function ($\int$) can be determined using effective medium theory, the most general form of which is the self-consistent Bruggeman effective medium approximation (B-EMA):

$$0 = f_a \frac{\epsilon_a - \epsilon}{\epsilon_a + 2\epsilon} + f_b \frac{\epsilon_b - \epsilon}{\epsilon_b + 2\epsilon}. \qquad (2)$$

The fit parameters of such models are the volume fractions *fa* and *fb* of components *a* and *b*, respectively ($\int a$ and $\int b$ are known, $\int$ is calculated assuming *fa* and *fb* values). In turn, the change of the volume fractions can be related to the morphological changes of the material.

As an example of using B-EMA, lets consider an intensively investigated nanocrystalline material, namely the nanocrystalline silicon prepared by a range of deposition techniques from low temperature chemical vapor deposition to magnetron sputtering. A typical optical model for the measurement of this material uses components of single-crystalline silicon (c-

Si, dielectric function from Ref. [16]), amorphous silicon (a-Si, dielectric function from Ref. [6]), nanocrystalline silicon (nc-Si, dielectric function from Ref. [18]), and voids. Some optical models for large-grain (Model 1) and small-grain (Model 2) nanocrystalline silicon layers deposited on oxidized single-crystalline silicon wafers are shown in Fig. 2 (see also Ref. [25]). The surface nanoroughness is also considered as a homogeneous layer [9], the thickness of which correlates with the root mean square roughness [22]. There are more sophisticated theories to analyze the surface roughness from the polarized optical response [14], but the robustness of the above effective medium approach makes it a popular, most widely used method. Note however, that the accurate, validated determination of the root mean square roughness is a problem even by scanning probe methods, because the obtained numerical value may strongly depend on the measurement configuration (e.g. the window size or the quality of the tip, [22]).

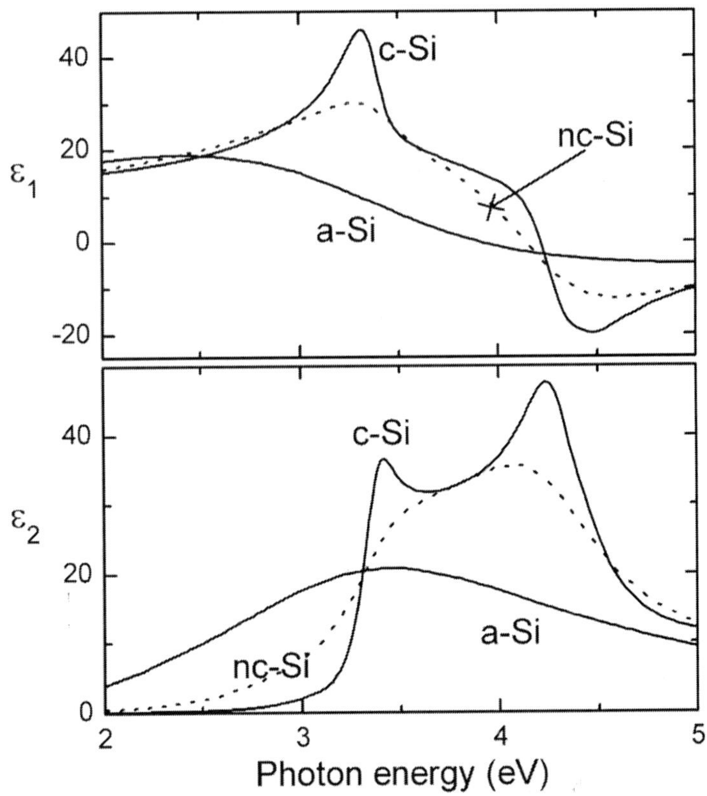

Figure 1. Real (ʃ₁) and imaginary (ʃ₂ parts of the dielectric function of single-crystalline (c-Si), amorphous (a-Si), and nanocrystalline (nc-Si) silicon.

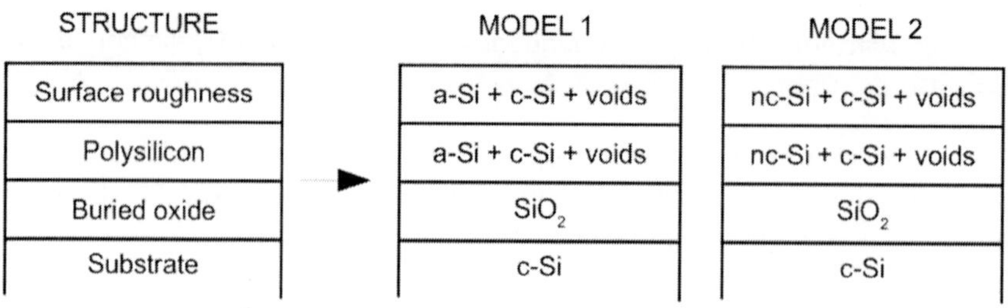

Figure 2. Some B-EMA models for nanocrystalline Si on c-Si with different expected grain sizes. Models 1 and 2 can be used for large and small grain sizes, respectively.

In Fig. 2 both the surface nanoroughness and the bulk layer properties are described by a combination of a-Si, c-Si, and voids or nc-Si, c-Si, and voids. The voids ratio serves as a density correction in the bulk layer, and it describes the character of the surface nanoroughness in the roughness layer. In many cases, the void ratio is fixed at 50 % due to a possible correlation with the thickness of the surface roughness layer. The model also contains the buried thermal oxide. The thickness of this oxide layer can also be determined in one fitting step together with the volume fractions of the components in the roughness and bulk nanocrystalline silicon layer, as well as their layer thicknesses.

For layers with smaller nanocrystals (several times 10 nanometers or below) the use of the nc-Si reference dielectric function [18] provides a better fit than using the conventional components of c-Si, a-Si, and voids. The volume fraction of the nc-Si component in the model is an indication of the crystallinity (Fig. 3). The higher the volume fraction of the nc-Si component, the smaller the grain size [25]. The sensitivity is revealed to be high, however, quantitative measurements verified by reference methods were not performed. Micrographs of transmission electron microscopy were available, but the technique of a reliable grain size analysis using image processing is missing.

There are a range of methods for the analytical representation of the dielectric function of semiconductors. A general empirical approach is to use a combination of Lorentz oscillators. This method allows the determination of the critical point energies and the layer thickness [13].

The critical point features can also be characterized by calculating the derivative of the dielectric function [7]. Using this approach the broadening of the critical point features can especially well be characterized. In turn, the grain size can indirectly be measured, because the broadening is proportional to the

electron scattering at the grain boundaries, i.e. the amount of grain boundaries is measured.

The generalized oscillator model applies a standard analytical equation for all types of critical points [11]. The dimensionality of the critical point can be adjusted by a parameter of the model.

Another general approach for the parameterization of the critical point features of the dielectric function is the generalized critical point model suggested by B. Johs et al. [19]. In this model the dielectric function around the critical point is described by four Gaussian-broadened polynomials. A detailed description can be found in Ref. [12].

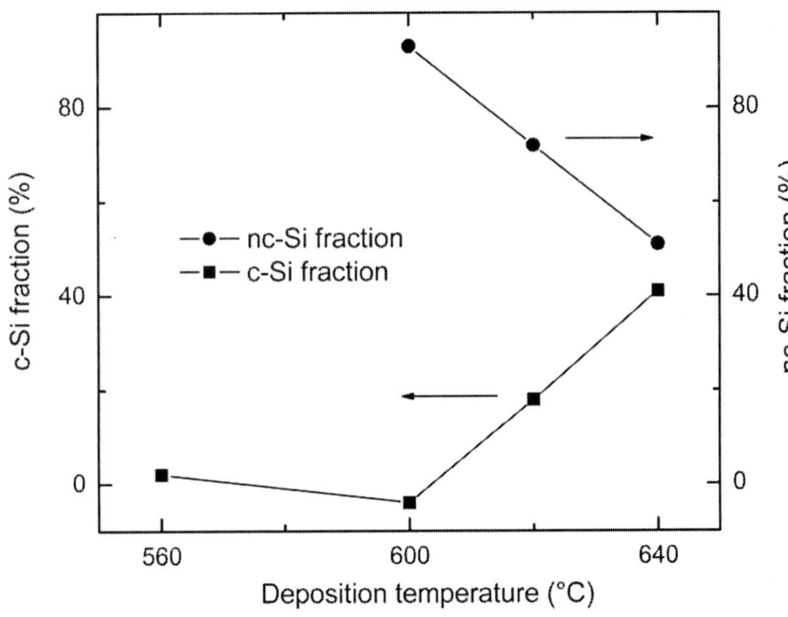

Figure 3. c-Si and nc-Si volume fractions as functions of the deposition temperature. The grain size increases with increasing temperature. Reprinted with permission from Journal of Applied Physics 87, P. Petrik, Ellipsometric study of polycrystalline silicon films prepared by low-pressure chemical vapor deposition, 1734 (2000). Copyright 2000, American Institute of Physics.

A special parameterization has been developed by S. Adachi. This model dielectric function describes each critical point with a specific analytical

equation depending on the type of the critical point [1,26]. As a result, one has a set of Kramers-Kroenig-consistent analytical equations, with each equation describing an oscillator. The superposition of all the fitted oscillators results in the final dielectric function as shown in Fig. 4. It is also shown that the effective medium approximation using the combination of c-Si and a-Si as described above doesn't allow an acceptable fit (short-dashed line). Although the number of parameters to fit is large, there are a lot of possibilities to couple or fix non-sensitive parameters [23].

The most important properties of the nanocrystal layers that can be measured by ellipsometry are the layer thickness and the grain size (Table 1). The crystallinity can be defined for ellipsometry as the "position" of the dielectric function between that of the single-crystalline and the amorphous ones. In terms of effective medium theory this can be defined as the c-Si to a-Si ratio of the model components. The density is also a parameter that can be expressed compared to a reference value. Staying at the above example the model "c-Si + a-Si + void" compares the density to the purely "c-Si + a-Si" case.

Additional to the above parameters a range of thin film-related parameters can be determined. Most important may be the vertical inhomogeneity and the interface quality. Using ellipsometry it can sensitively be checked whether the layer is of optical quality.

4. APPLICATIONS

The range of ellipsometric applications for nanocrystals is very large. The most important applications comprise semiconductor nanocrystals, with silicon still being the number one.

In this section some relevant applications known to the author are selected, with no claim to completeness.

The correlation between the grain size and the broadening of the dielectric function at the E_1 critical point was revealed by H. V. Nguyen and R. W. Collins in an in situ study during plasma enhanced chemical vapor deposition [21]. The grain size was determined from the layer thickness assuming a three-dimensional, isolated particle growth. It was shown that the theoretically predicted relationship that the broadening is proportional to $1/d$, where d denotes the grain size, holds for the investigated nanocrystals (Fig. 5). Furthermore, the extrapolation of $d \rightarrow \infty$ results in the expected broadening value of the single-crystalline silicon, which suggests that this behavior is valid for smaller crystals as well. It is important to note that the investigated

thickness range of 20-25 nm was large enough to neglect the shift of the critical point energy due to the quantum confinement.

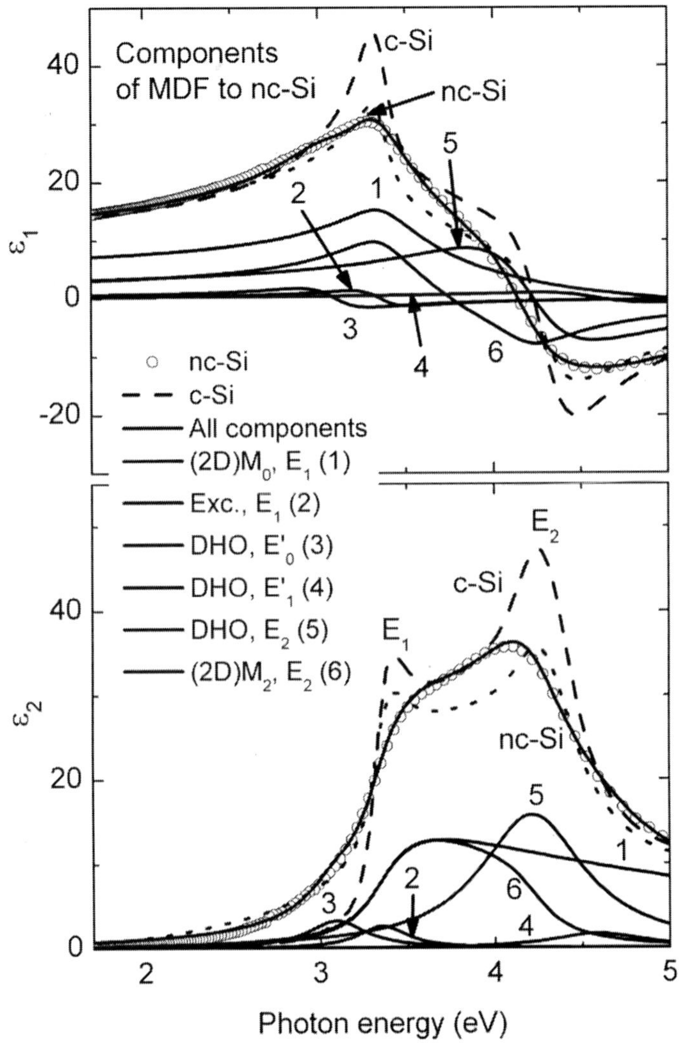

Figure 4. Oscillators of the model dielectric function of S. Adachi (numbered solid lines) fitted to the reference dielectric function of a fine-grained polycrystalline silicon (nc-Si). The long-dashed line is the single-crystalline silicon (c-Si) reference. The short-dashed line shows the fit using the effective medium approximation described above with the components of c-Si and a-Si.

Table 1. Properties of nanocrystal layers that can be measured by ellipsometry.

Property	Notes
Grain size	Sub-nanometer sensitivity Indirect: "calibration" needed
Crystallinity	Amount of grain boundaries In case of one-phase layers indicative to grain size
Layer thickness	Sub-nanometer sensitivity Also in multi-layers
Density	Compared to the reference materials used in the model
Inhomogeneity	Both vertical and lateral
Interface quality	Whether one needs an additional transition layer to achieve a good fit
Surface nanoroughness	As a homogeneous roughness layer or using more sophisticated methods
Shape of crystals	In principle possible using anisotropy and B-EMA screening

G. F. Feng and R. Zallen [13] used Raman spectroscopy to verify the linearity between broadening and reciprocal grain size (Fig. 6) on nanocrystals created by ion implantation in GaAs in the size range of 5-50 nm. A good linearity was found for all investigated critical point energies. This example also proves that the sensitivity is high enough, and the quantitative determination of the grain size is possible by validation using a reference method.

A further example of a nanocrystalline structure is porous silicon prepared by electrochemical etching. After etching, the sample has a well-defined nanostructure depending on the etching conditions. The size of the remaining nanocrystals strongly depends on the doping level of the substrate. In the study of P. Petrik et al. [23] substrates with different B-doping were used to achieve substrate resistivities from 0.01 to 0.09 Ωcm and nanocrystal sizes from 3 to 12 nm. In case of porous silicon the use of proper optical models is of crucial importance, because the layers are usually vertically non-uniform. In this study the dielectric function of the porous silicon layers were analyzed using the model dielectric function of S. Adachi described above. The fitted broadening parameters of the model are plotted as a function of the grain size in Fig. 7. Similar to the above cases, a more or less systematic behavior was found also in this study, especially for the E_1 critical point (Γ_1). The reason is that the porous silicon layers are vertically inhomogeneous, and the information depth

(in c-Si) at the E_1 and E_2 critical point energies are about 30 and 10 nm, respectively. So Γ_2 is very sensitive to surface deteriorations. The dependence on the layer thickness is also due to the vertical gradient in the nanostructure.

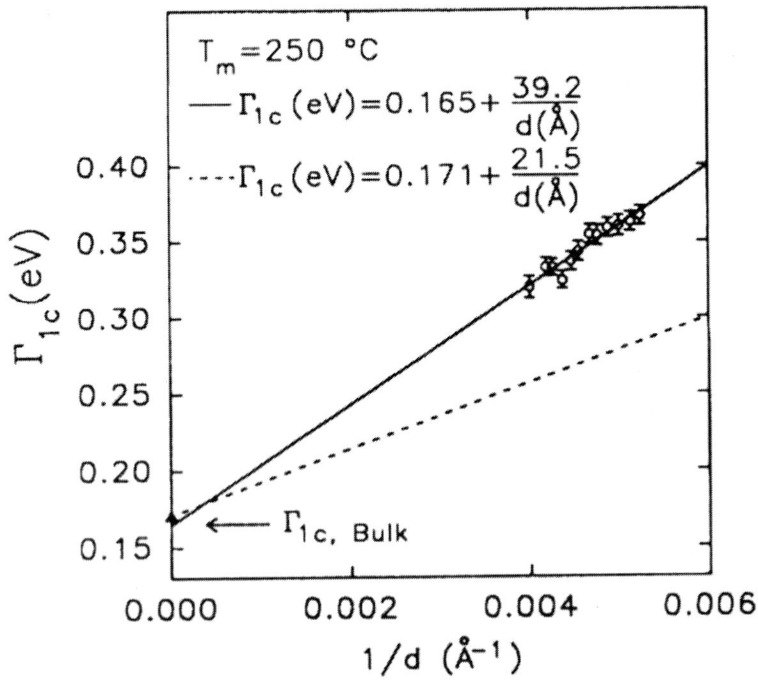

Figure 5. Broadening of the dielectric function at the E_1 critical point energy for grain sizes of d measured during layer growth. Reprinted from Physical Review B 47, H. V. Nguyen and R. V. Collins, Finite-size effects on the optical functions of silicon microcrystallites: A real-time spectroscopic ellipsometry study, 1911 (1993). Copyright 1993, American Physical Society.

Chemical vapor deposition combined with oxidation was used to prepare nanocrystal layers with well-defined grain sizes [20]. The thin nanocrystal films were deposited on quarz substrates, and the nanocrystal size was set by the layer thickness and the time of the oxidation process, with nanocrystal sizes verified by electron microscopy and X-ray diffraction. For the analysis of the nanocrystal layers the above mentioned effective medium approximation could successfully be applied using a combination of nc-Si, c-Si, and SiO_2 reference dielectric functions [2]. It was shown that the nanocrystallinity defined by $f_{nc\text{-}Si}/(f_{c\text{-}Si}+f_{nc\text{-}Si})$, where f denotes the fitted volume fractions of the

components (see equation 2), correlates well with the nanocrystal size determined by the layer thickness (see Fig. 8). This example also shows that the nanocrystal size can sensitively (though indirectly) be measured by this optical technique.

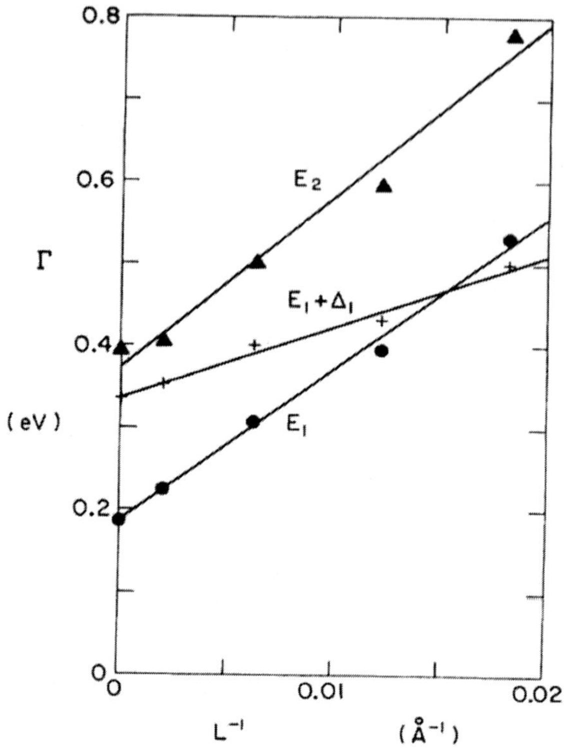

Figure 6. Broadening of the dielectric function at different critical points as a function of the reciprocal crystallite size in ion implanted GaAs. Reprinted from Physical Review B 40, G. F. Feng and R. Zallen, Optical properties of ion-impanted GaAs: The observation of finite-size effects in GaAs microcrystals, 1064 (1989). Copyright 1989, American Physical Society.B

The importance of an optical technique is not only the speed and sensitivity, but also the in situ capability. Using special beam-guiding techniques the ellipsometric measurement can even be made in a vertical furnace through the base plate (i.e. without modifying the furnace walls) [24]. The performance of current computers allows a real time evaluation of the process, providing the layer thicknesses and the information on the morphology. Fig. 9 shows the results of an in situ measurement made in a vertical furnace during

crystallization of a deposited amorphous layer. The thickness of the amorphous silicon layer was about 40 nm, deposited on a thermally oxidized (10 nm oxide) silicon wafer. Accordingly, the optical model consists of a single-crystalline silicon substrate, a buried SiO_2layer (with a thermal SiO_2 reference dielectric function), an initially amorphous silicon layer described by the a-Si reference, and an SiO_2 surface oxide layer. The surface roughness could be neglected for the deposited amorphous silicon layer. During annealing crystals are formed in the amorphous silicon layer. This can be followed by a simple effective medium combination of components a-Si and c-Si, as described above. The increasing volume fraction of c-Si represents the process of crystallization during annealing (see Fig. 9).

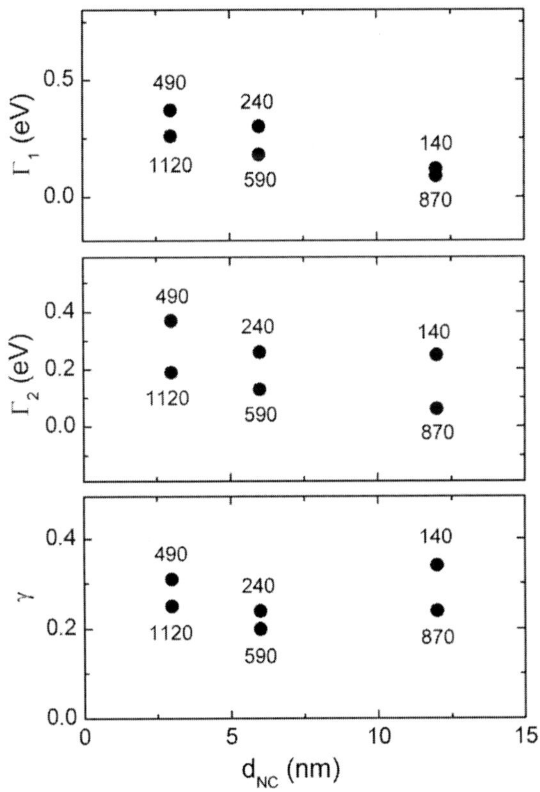

Figure 7. Broadening values measured using the model dielectric function of S. Adachi for different grain sizes (*dNC*). The numbers at the measured points show the thicknesses of the measured layers. Reprinted from Journal of Applied Physics 105, P. Petrik et al., Nanocrystal characterization by ellipsometry in porous silicon using model dielectric function, 024908 (2009). Copyright 2009, American Institute of Physics.

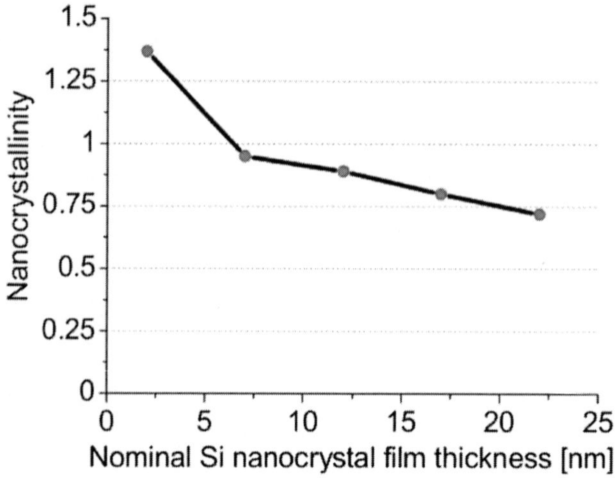

Figure 8. Nanocrystallinity as a function of the thickness of nanocrystal films prepared by chemical vapor deposition and oxidation of quartz substrates. Reprinted from Mater. Res. Soc. Symp. Proc. 1321, E. Agocs et al., Optical characterization using ellipsometry of si nanocrystal thin layers embedded in silicon oxide, DOI: 10.1557/opl.2011.949. Copyright 2011, Cambridge University Press.

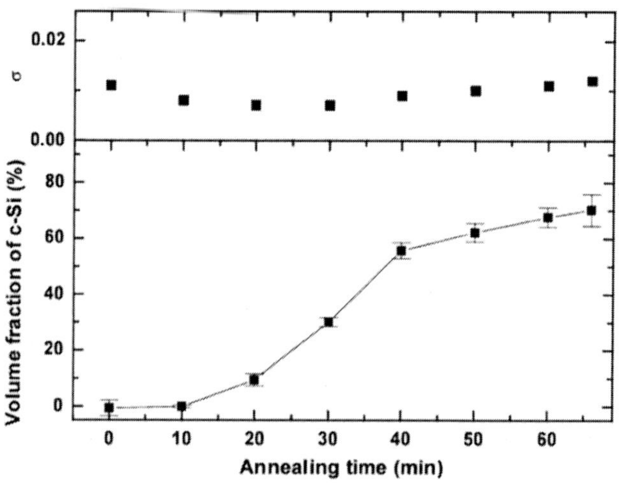

Figure 9. Fitted volume fraction of crystalline silicon in the effective medium model as a function of annealing time. Reprinted from Thin Solid Films 364, P. Petrik et al, In situ spectroscopic ellipsometry for the characterization of polysilicon formation inside a vertical furnace, 150 (2000). Copyright 2000, with permission from Elsevier.

5. CONCLUSIONS

The capability of measuring the change of phase by the reflection of polarized light on a surface or layer structure allows a typical sensitivity of less than one nanometer for the layer thickness and 10^{-3} - 10^{-4} for the refractive index. These capabilities together with a spectroscopic use in the photon energy range of the characteristic interband transitions makes the method applicable for an accurate determination of nanocrystal properties especially for Si, but also for wide band gap semiconductors using an extension into the deep UV spectral range. The sensitivity is high down to the smallest nanocrystal sizes. The requirement for an accurate nanocrystal measurement is a layered form with optical quality interfaces. The thickness, homogeneity and interface qualities of the layers can be measured directly, whereas the properties related to the nanocrystal structure (like the crystallinity, the nanocrystal size or the density of the layer) can be obtained indirectly using proper optical models and "calibration" by reference methods.

REFERENCES

1. S. Adachi, 1988 Model dielectric constants of si and gePhys. Rev. B 38: 12966 EOF12976 EOF

2. E. Agocs, P. Petrik, M. Fried, A. G. Nassiopoulou, 2011Optical characterization using ellipsometry of si nanocrystal thin layers embedded in silicon oxideMater. Res. Soc. Symp. Proc. 1321

3. D. E. Aspnes, 1975Extended spectroscopy with high-resolution scanning ellipsometryPhysical Review B4008 EOF4011 EOF

4. D. E. Aspnes, [1981]. Studies of surface, thin film and interface properties by automatic spectroscopic ellipsometry, SPIE Proc. 276: 312.

5. D. E. Aspnes, 1982Optical properties of thin filmsThin Solid Films249 EOF262 EOF

6. D. E. Aspnes, 1985The accurate determination of optical properties by ellipsometry, in E. D. Palik (ed.), Handbook of Optical Constants of Solids, Academic Press, New York.

7. D. E. Aspnes, S. Kelso, C. Olson, D. Lynch, 1982Phys. Rev. Lett. 48: 1863.

8. D. E. Aspnes, A. A. Studna, 1978Methods for drift stabilization and photomiltiplier linearization for photometric ellipsometers and polarimeters, Review of Scientific Instruments 49: 291.

9. D. E. Aspnes, J. B. Theeten, F. Hottier, 1979Investigation of effective-medium models of microscopic surface roughness by spectroscopic ellipsometryPhys. Rev. B 20: 3292 EOF3302 EOF

10. D. A. G. Bruggeman, 1935Berechnung verschiedener physikalischer konstanten von heterogenen substanzen i. dielektrizitätskonstanten und leitfähigkeiten der mischkörper aus isotropen substanzen, Ann. Phys. (Lepzig) 24: 636.

11. M. Cardona, 1969Modulation spectroscopy,in F. Seitz, D. Turnbull & H. Ehrenreich (eds), Solid State Physics, Academic Press, New York.

12. R. W. Collins, A. S. Ferlauto, 2005Optical physics of materials, in E. G. Irene & H. G. Tomkins (eds), Handbook of ellipsometry, William Andrew, Norwich, NY.

13. G. F. Feng, R. Zallen, 1989Optical properties of ion-implanted gaas: The observation of finite-size effects in gaas microcrystalsPhysical Review B1064 EOF1073 EOF

14. D. Franta, I. Ohlidal, D. Necas, 2008Optical quantities of rough films calculated by rayleigh-rice theoryPhysica Status Solidi C 5: 1395 EOF1398 EOF

15. H. Fujiwara, [2007]. Spectroscopic Ellipsometry, Wiley, Chichester.

16. C. M. Herzinger, B. Johs, W. A. Mc Gahan, J. A. Woollam, W. Paulson, 1998Ellipsometric determination of optical constants for silicon and thermally grown silicon dioxide via a multi-sample, multi-wavelengt, multi-angle investigation,J. Appl. Phys. 83: 3323 EOF

17. E. A. Irene, H. G. Tompkins, 2005Sio2 films, in E. G. Irene & H. G. Tomkins (eds), Handbook of ellipsometry, William Andrew, Norwich, NY.

18. G. E. Jellison, Jr , M. F. Chisholm, S. M. Gorbatkin, 1993Optical functions of chemical vapor deposited thin-film silicon determined by spectroscopic ellipsometryAppl. Phys. Lett. 62: 348.

19. B. Johs, C. M. Herzinger, J. H. Dinan, A. Cornfeld, J. D. Benson, 1998Thin Solid Films 313: 137.

20. C. B. Lioutas, N. Vouroutzus, I. Tsaioussis, N. Frangis, S. Gardelis, A. G. Nassiopoulou, 2008Columnar growth of ultra-thin nanocrystalline si films on quartz by low pressure chemical vapor deposition: accurate control of vertical sizePhysica Status Solidi a 11: 2615 EOF2620 EOF

21. H. V. Nguyen, R. W. Collins, 1993Finite-size effects on the optical functions of silicon microcrystallites: A real-time spectroscopic ellipsometry studyPhysical Review B1911 EOF1917 EOF

22. P. Petrik, L. P. Biró, M. Fried, T. Lohner, R. Berger, C. Schneider, J. Gyulai, H. Ryssel, 1998Surface roughness measurement on polysilicon produced by low pressure chemical vapor deposition using spectroscopic ellipsometry and atomic force microscopy, Thin Solid Films 315: 186.

23. P. Petrik, M. Fried, E. Vazsonyi, P. Basa, T. Lohner, P. Kozma, Z. Makkai, 2009Nanocrystal characterization by ellipsometry in porous silicon using model dielectric functionJ. Appl. Phys. 105: 024908.

24. P. Petrik, W. Lehnert, C. Schneider, M. Fried, T. Lohner, J. Gyulai, H. Ryssel, 2000In situ spectroscopic ellipsometry for the characterization of polysilicon formation inside a vertical furnaceThin Solid Films150 EOF155 EOF

25. P. Petrik, T. Lohner, M. Fried, L. P. Biro, N. Q. Khanh, J. Gyulai, W. Lehnert, C. Schneider, H. Ryssel, 2000Ellipsometric study of polycrystalline silicon films prepared by low pressure chemical vapor deposition, J. Appl. Phys. 87: 1734 EOF

26. K. Tsunoda, S. Adachi, M. Takahashi, 2002Spectroscopic ellipsometry study of ion-implanted si(100) wafers, J. Appl. Phys. 91: 2936.

Chapter 2

Semiconductor Nanocrystals

Anurag Srivastava[1] and Neha Tyagi[1]

[1] Advance Material Research Group, Computational Nanoscience and Technology Laboratory, ABV- Indian Institute of Information Technology and Management, Gwalior (M.P.), India

1. INTRODUCTION

In terms of continuous miniaturization of electronic devices nanotechnology is a ray of hope for semiconductor industries due to peculiar properties of nano-materials that changes significantly the efficiency of the devices. Research in nanoscale materials get started because of the unique properties that are obtain at this scale, by changing the shape or size of these materials. At nanoscale, the behavior of materials drastically get changed and hence their properties. Particularly in semiconductors, it results due to the motion of electron to a length scale that is equal or smaller to the length scale of the electron Bohr radius that is generally a few nanometers. Continuous efforts are being made to explore the new physical properties of materials and to engineer them to fit for various technological applications. Scientific community has paid much attention to study various aspects of variety of nanostructured materials like; fullerenes, nanotubes (NT), nanoribbons (NR), nanowires (NW) and most recently, the nanocrystals (NCs). As the name indicates, NCs are the structures exhibiting crystalline structure but with one, two or all the three dimensions within the range of 1-100 nm. NCs enjoys advancement of crystalline periodicity at nano regime (10^{-9} m) and often possesses new properties, most of the time reverse from those of the equivalent bulk materials due to their large surface to volume ratio, quantum size effect, and confinement effects [1-

3]. Quantum confinement effect is the main phenomena that often been observed in NC, which deals with the spatial inclusion of the electronic charge carriers within the NC. The quantum size effect is analogous to the well known quantum mechanical problem i.e when a particle is place in a box, where the energy partition between the levels raises as the dimensions of the box reduces. Due to the same, we can observe an increment in the energy band gap of the semiconductor as the size of the crystal reduces. Nanostructuration is not confined to human imagination only; in fact it's an important choice of nature. The word nanotechnology is not new for the nature, where it has been realized for over billions of years which can be for instance in clays, circumstellar, dusts or many biological systems too. One of the most interesting examples of nature inspired nanotechnology is the water proof leaves of lotus owing to which a lotus can sustain in water for a very long time. Recently, it has been verified that lotus leaves has a special coating of nanoparticles which behaves as water repellant, this water repellant property has motivated number of researchers to perform research in this area and some of them even succeeded to manufacture hydrophobic coatings [4]. Interaction at the nanoscale are usually governed by the fact that the characteristic nano-length becomes comparable with other critical lengths of the system as mean free paths, scattering or coherence lengths. At the same time, confinement translates into a reorganization of the electronic density of states towards more distinct states. These two consequences of nanostructuration and their combination lead to the so called quantum related effects which determines most of the peculiar properties of nano-materials. A major corollary of nano confinement is the elevated ratio of the number of surface to volume atoms, which becomes another key point for the understanding of these extraordinary properties. In small NCs the ratio of the number of surface to the total number of atoms is less than one, for which it scales as ~1/(Particle diameter)^3 and exactly one in nanotubes or fullerenes, whereas in bulk crystal it goes to 10^{-20} or less.

NCs are very tiny crystals that results when a single crystal experience a solid to solid phase transition. Generally, NCs can be prepared by a huge number of atoms presented on the surface and no interior defects, unlike bulk materials. We all know that under the application of pressure abrupt change in the arrangement of the atoms i.e. structural transformation from one phase to another has been observed in crystals. The understanding of pressure-induced phase transition in various nanomaterials plays an important role in probing the properties of new materials. The consequences of finite crystal size on the structural phase transition and bulk modulus as well as compressibility has widely been demonstrated to get better understanding towards the stability, electronic, mechanical as well as other properties of nano-materials. Difference in surface energies of the two crystal phases is mainly responsible for the change in transition pressure in case of nano-materials. In fact the general rule

states that smaller the size of the crystal, higher the transition pressure has been verified by several systems [5,6]. A report shows that NC shape can be easily observed using transmission electron microscopy at atmospheric pressure [7].

2. APPLICATIONS

The enormous potential of exploring the new sciences and technology at nanoscale, may impact on industrial productivity, realized all over the world to develop the new materials for variety of newer applications. NCs possess variety of applications in the field of electronics, opto-electronics and photonics, and also several biological (medical) treatments following are the few important applications of semiconductor NCs.

2.1. Nanocrystals Leds

In the last two decades very basic and more efficient white light-emitting diodes (LEDs) have been formed by replacing phosphors (a colour converter) with CdSe based NCs that are integrated directly into the p-n junction [8]. CdSe NCs/nanorods emit linearly polarized light along the crystallites axis and the degree of polarization depends upon the surface to volume ratio of the NCs [9]. A hydride structure comprises of single layer of CdSe NCs accumulate at the top of an InGaN/GaN quantum well (QW) has been utilized as LED which provides ~10% color conversion efficiency. An inverted LED design is shown in Fig. 1(a), where an InGaN QW grow on the top of a thick p-doped GaN barrier and complete the structure with a thin n-type GaN cap layer (normally a QW is grown on top of the n-type layer). Electrical contacts have been deposited by electron beam evaporation using an inter-digitated mask that comprised of several different device-mesa and contact geometries, as shown in Fig.1(b)[10].

The conjugated polymers and InAs based NCs have been utilized to produce near infrared LEDs, where the emission can be tuned from the range of 1 to 1.3μm that efficiently covers the short-wavelength telecommunications [11]. Semiconductor NCs when excited electrically using GaN injection layer can be used as color selectable chromophores, shows high emission efficiencies with excellent photo stability and chemical flexibility. In a report by Mueller et al. [12] multicolour LEDs have been demonstrated where semiconductors NCs are integrated into a p-n junction made by GaN injection layers. Si NCs LEDs have been fabricated using ion implantation and observed the enhancement in LEDs optical output power (~20 nW), with typical power efficiencies ranging from 1

to 10% [13]. Light emitting transistors integrated with individual CdSe NCs shows occurrence of Coulomb blockade at low bias voltage and low temperature, which signifies that electrons pass through the NC by single electron tunneling. Fig. 2(a), shows a single nanorod contacted by two Au electrodes with about ~30 nm separation. The representative current-voltage and electroluminescence (EL) data collected from a device "D1" is illustrated in Fig. 2(b). A nonlinear increase in current (*I*) at high bias voltage (*V*) clearly reveals that the device exhibits a low-bias conductance gap. Once *V* reaches a threshold (V_{th}) the device starts emitting light as shown by EL intensity measurement [14].

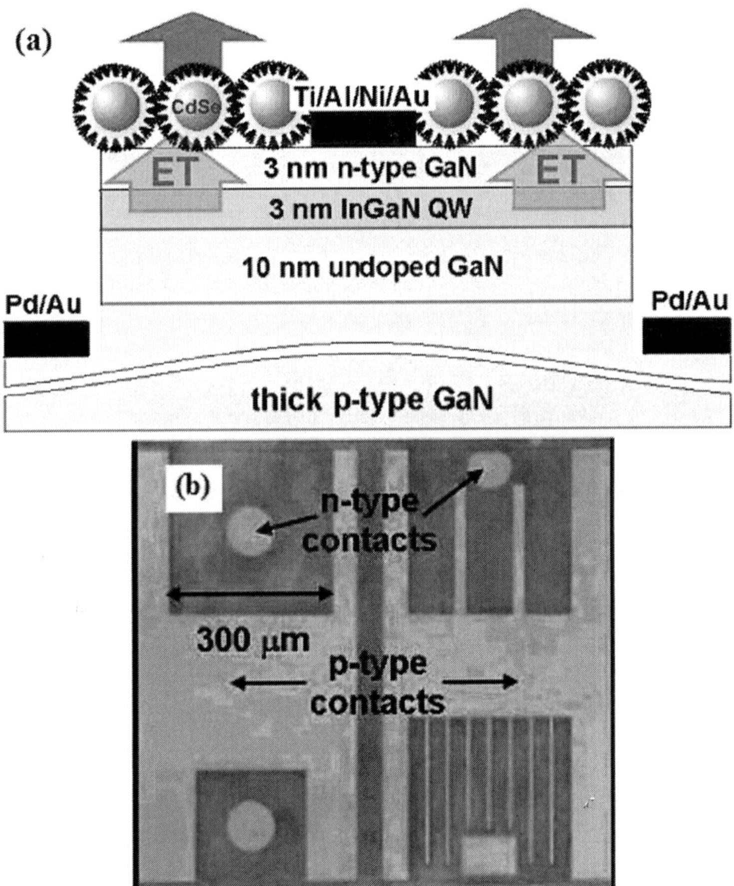

Figure 1. a) Schematic of the nonradiative energy transfer (ET) LED structure. (b) Contact geometry of real device Ref.[**10**]

2.2. Nanocrystal Memories

Floating gate structure is firstly invented by Sze and Kahnd in 1967, which was used to construct flash memories [15] (Fig.3a). In 1995, IBM proposed first discrete NC memory and in early 2000's scientists have considered NCs as the capable candidate that can solve the current scaling problem (Fig.3b). The metal-oxide-semiconductor (MOS) memory structures based on Si NCs have potential applications in flash memory, reason being Si NCs are implanted as charge storage nodes in an oxide layer between the control gate and tunneling layer that reduces the difficulty of charge loss which can generally seen in conventional flash memories. Si NC provides faster write/remove speeds, small injection oxides, little operating voltages and better stamina [16]. Literature confirms that the information is stored in the NCs by injecting charges, due to which a transistor needs much voltage for turning it ON, known as program operation illustrated in Fig.3(c). If we apply read voltage (V_{read}) to the gate between the program and erase operations to read the corresponding drain current (I_d) then we obtain the memory status given by values 0 and 1 [17]. Generally the quantum dot/NCs flash memories have used Si NCs as a substitute to floating gate layer, however many results demonstrates the superiority of Ge based NC memories over those based on Si, because Ge NCs provides large nonvolatile charge conservation time because of their small band gap.

NCs may work as memory cells, where the conventional poly-silicon floating gate is replaced by an array of Si NCs, as the single cell and cell array of 1 Mb and 10 k have been realized by using a conventional 0.15 μm FLASH technology [18,19]. Ge NCs have been fabricated by pulsed laser deposition for their use in floating gate memory, such memories shows excellent charge preservation characteristic. [20].

2.3. Nanocrystal Photocharge Generators

The PbSe NCs represent a fascinating system due to the easiness of knowing the quantum modulated optical nature in the infrared range with a Bohr exciton radius of PbSe around 46 nm, where the quantum confinement effect come into view at comparatively large particle size. As we know that PbSe bulk crystal has rocksalt type phase and a direct band gap semiconductor (E_g = 0.28 eV) whereas the PbSe NC show well defined band-edge excitonic transition, tuned from 0.9 to 2.0 eV [21], leads to an efficient photocharge generators for communicating the IR wavelengths [22]. Highly efficient photo-detectors based on composites of the semiconducting polymer and PbSe NCs have been prepared and the outer quantum efficiency in these devices is larger than one

[23]. Ge NCs photo-detector possesses powerful optical absorption and its photocurrent response has been measured within the wavelength range of 1.3 to1.55 μm with a low dark current of 61.4 nA along with a a photocurrent responsivity of 56 mA/W at the 5 V reverse bias [24].

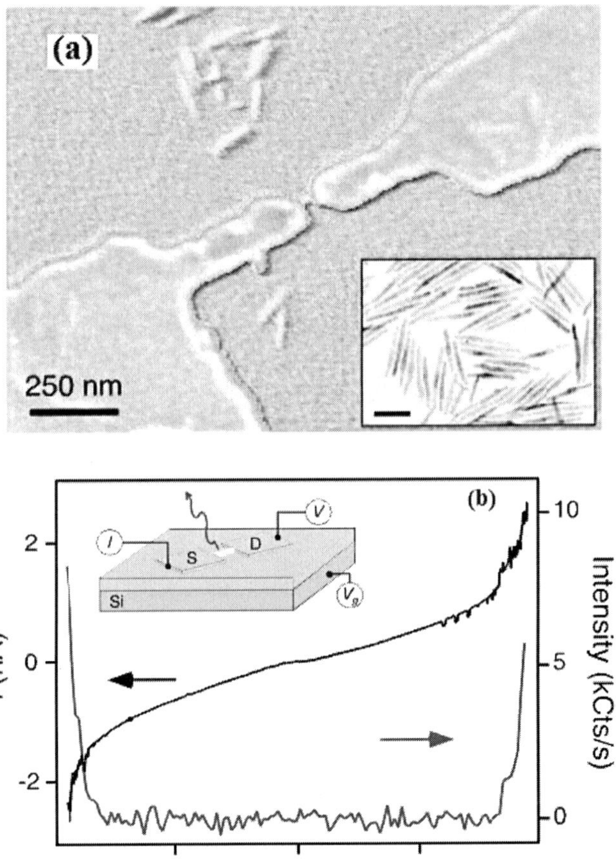

Figure 2. a) SEM image (false color) of CdSe nanorod transistor. The inset shows a TEM image of the CdSe nanorods (scale bar is 50nm) (b) Current (black) and concurrently calculated EL intensity (red) plotted against bias voltage obtained from device. Inset illustrates the transistor model, with source (S) and drain (D) electrodes connecting the nanorod with a back gate is shown [**14**].

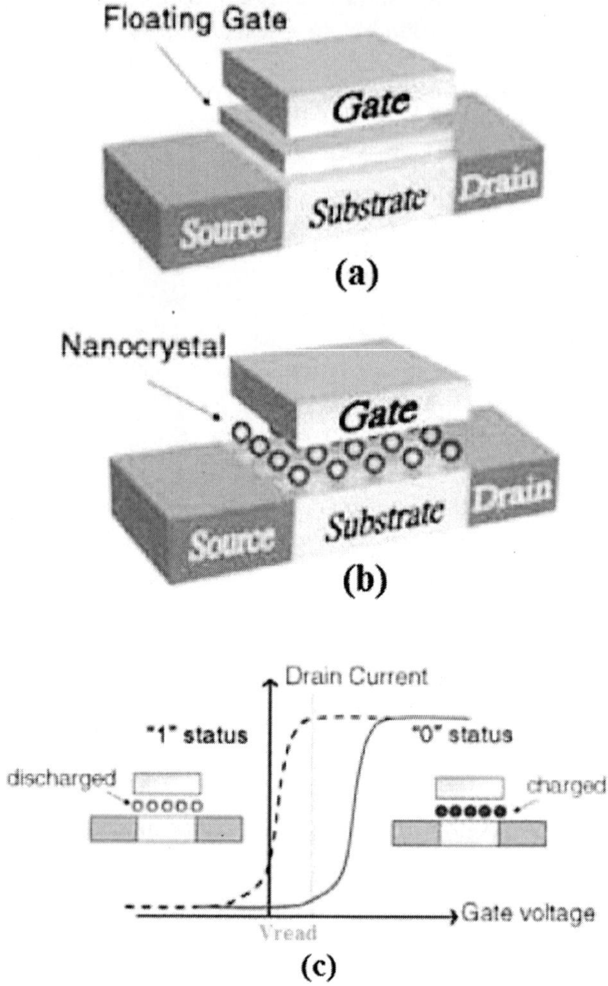

Figure 3. a) Floating gate non volatile memory structure. (b) NC non volatile memory structure. (c) Program and erase mode of the NC memory device Ref.[**17**].

2.4. Nanocrystal Solar Cells

An InAs/GaAs quantum dot solar cell has been synthesized and observed superior photocurrent without reduction of open circuit voltage compared to a solar cell without quantum dots. These solar cells have a light absorption range,

which is extended up to 1.3 μm and confirms a trade-off between open circuit voltage and quantum dot ground state energy [25]. An ultrathin solar cell composed of CdSe and CdTe NCs of size 40nm shows 3% power conversion efficiency and the device remain stable in air [26] (Fig.4).

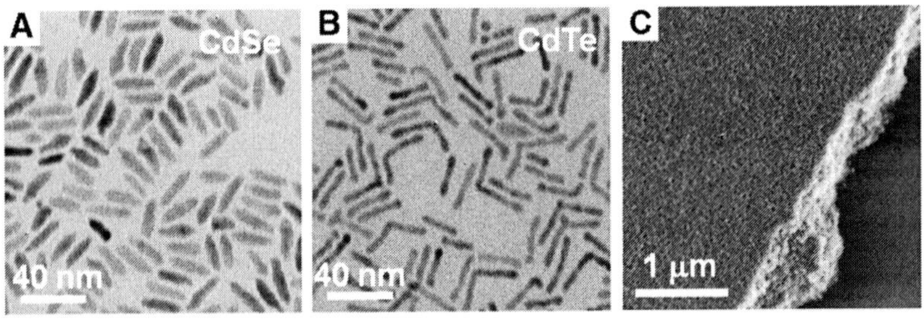

Figure 4. Transmission electron micrographs of (A) CdSe and (B) CdTe NCs (C) Spin-cast film of colloidal NCs imaged by scanning electron micrography is uniform and without defect; the film edge of this ~100 nm film is shown for difference with the Si substrate Ref[**26**].

An induce electric field is large enough when two different molecules attached to a single NC and the field significantly changes the electronic as well as optoelectronic properties of the NC. By the help of mixed ligand, induced electric field significantly increases the efficiency of charge generation in CdSe NC solar cells, which enhanced the complete cell efficiency [27]. Si NC have quantum confinement property and so it can be used as photovoltaic materials since its ability to collect the photo-generated current through efficient electronic transport and for the development of Si NC based solar cells the exciton dissociation is a recent challenge [28].

2.5. Nanocrystals In Health

Enzymes, membranes, nucleic acids, etc. are the elementary functional units made up of complex nanoscale particles in biological systems. In the era days of miniaturization where metals, semiconductors and magnets through which we construct optical and electrical sensors, can now be prepared on the scale of individual biological macromolecules that will have large impact on forthcoming medical treatments. In drug delivery and clinical applications, nanotechnology is one of the key factors for modern drug therapy. Due to the

simplicity in preparation and general utility NCs are the new carrier-free colloidal drugs delivery systems having a particle size of 100-1000 nm, these NCs have been thought of as a practical drug delivery approach to develop the poorly soluble drugs. Literature shows that the number of drugs coming directly from synthesis is nowadays poorly soluble and the drugs which are very less soluble in water combine very poorly with bioavailability. If we do not get any way to improve drug solubility then it will be very difficult to get absorbed from the gastrointestinal zone in the bloodstream and then to reach at the proper site for required action [29,30]. Recently intense research is being made on the colloidal quantum dots due to their stable light emitting nature, which can be broadly tuned by varying the size of the NC. Lots of efforts have been devoted in last two years on the development of large range of methods for bio conjugating colloidal NCs [31] because of their diverse applications such as in vivo imaging [32], cell tracking [33], DNA detection [34] etc. While studying the optical properties of NCs, it has been noticed that NCs fluorescence wavelength sturdily depends on their size as a result of which NCs photo bleach property get reduced. These NCs are fascinating fluorescence probes for various types of labelling research like cellular structures labeling, tracking the path and absorption of NCs by living cells [35].

3. SYNTHESIS & CHARACTERIZATION

In general the performance of the material depend on its properties whereas the properties depend on the crystal geometry, constituents, defects and interfaces that are definitely prohibited by thermodynamics and kinetics of the fabrication process. In 1980's when the first theoretical explanation was proposed for colloidal spherical NCs by Brus [36], together with advances in the synthetic procedures [37] lead to a rapid increase in research in the field of nano sized materials. Following are the popular methods developed for the fabrication and characterizations of NCs:

Solution-phase methods are being used for preparing CdS NCs with a wide variety of morphologies, starting with oleylamine as capping agent and by varying the reaction conditions [38]. Low temperature inverse micelle **solvo-thermal route** method has been used to synthesis large amount of single crystalline Ge NCs and X-ray diffraction measurement illustrates that these NCs are composed of pure cubic Ge structure. The size and morphology of the NCs are observed by transmission electron microscopy (TEM). In Fig.5(a) a TEM image of Ge NCs is illustrated and Fig.5(b) shows selected area electron area diffraction (SAED) sample for the NC of 25 nm size and displays the typical spot pattern of a single crystal domain and also the lattice planes of Ge with

[111] hexagonal symmetry. High resolution TEM (HRTEM) images of high quality Ge NCs are shown in Fig. 5(c) and (d) [39].

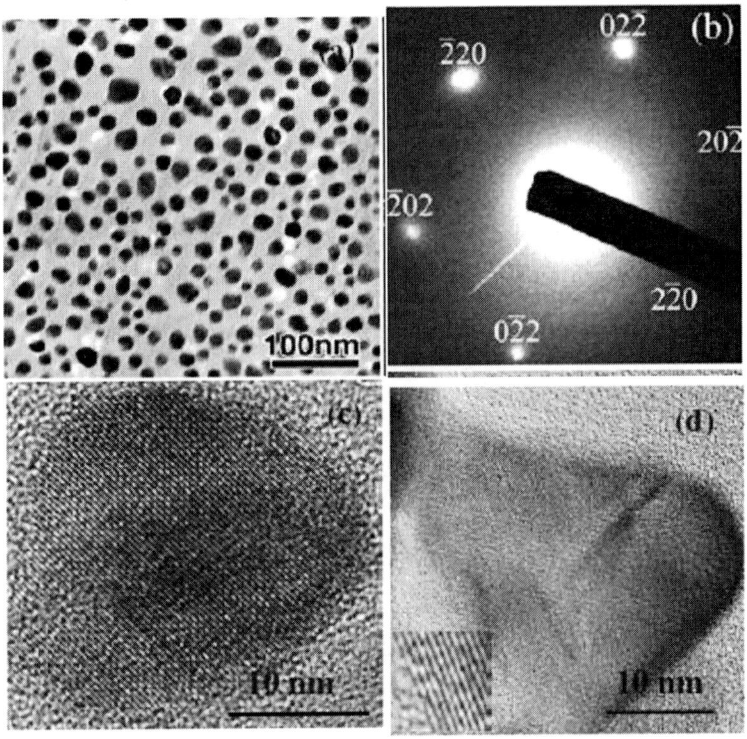

Figure 5. a) As-prepared Ge NCs TEM image, (b) SAED sample of Ge NC with 25nm size; HRTEM images of spherical (c) and triangular(d) Ge NCs Ref.[**39**]

Few **non-hydrolytic methods** have also been developed for the synthesis of ZnO NCs with controlled size, shape and surface-defect at low temperature. But such methods need very cautious control of conditions and carefully engineered precursor.

Direct liquid phase precipitation is a new process proposed recently for NC fabrications, however slight hard from the previous complex fabrication methods in superficial and efficient way (at room temperature). This process produced perfect ZnO NCs of different diameter (5 to 12 nm) and of various shapes [40].

Benzene thermal method at low temperature and low pressure (0.12–0.14 MPa) has been used for the synthesis of AlN NCs. Owing to the catalytic effect of AlN nanocrystals, ~8% benzene was converted at 150°C into several polymers, like cyclohexylbenzene [41].

Sol-gel technique can be used for the fabrication of amorphous or crystalline materials from a liquid phase at low temperature. This technique employed for the production of laser dyes, enzymes, nano sized semiconductors and metal nanoparticles. AlP NCs embedded in silica glasses have been prepared by sol gel process, where the preparation of gels take place by the complex solution hydrolysis and after about 10hours heating in an air atmosphere to form gel glass, again get heated in the presence of H_2N_2 gas and react with Al(III), finally a cubic AlP crystallites are produced [42]. Similar technique has been used in producing InAs and InP microcrystallites embedded in SiO_2 gel glasses [43,44]. GaAs NCs have been synthesized by an **electrochemical route** method from the acidic solutions of metallic gallium and arsenic oxide [45]. The growth of 1D nanostructures have traditionally been associated with such highly technological processes/analysis tools as **vapor deposition**, **lithography**, **thermally assisted electromigration** method [46], **annealing process** [47], scanning tunneling microscope (STM), atomic force microscope (AFM) [48]. By using size selective **precipitation techniques**, cadmium chalcogenide NCs have been successfully obtained [49]. The most common semiconductor materials grown by precipitation technique are CdTe, CdSe, CdS, and ZnS of which CdTe has special technological importance because it is the only known II-VI material that can form conventional p-n junctions [50]. For the synthesis of III-V compound semiconductors on the semiconductor substrate variety of methods have been developed starting from **metaorganic chemical vapour phase epitaxy**(MOVPE), **molecular beam epitaxy** (MBE) [51], various **inorganic techniques** [52], **porous glass**[53] to photo **chemical vapour deposition** [54].

In **chemical vapour deposition (**CVD), the vaporized precursors are introduced into a CVD reactor and adsorbed onto a substance held at high temperature. These adsorbed molecules will either thermally crumble or react with other gases to form crystals. The CVD process has been organised in three steps (i) mass transport of reactants to the growth surface through a boundary layer by diffusion (ii) chemical reactions on the growth surface, and (iii) removal of the gas-phase reaction by products from the growth surface. NCs have been characterized by X-ray diffraction, scanning transmission electron microscopy (STEM), optical absorption spectroscopy and fluorescence spectroscopy.

Keeping in view of all the techniques discussed for the synthesis and characterization of NCs we can state that in current scenario it has been easier task to fabricate a NC of desired morphology.

4. PROPERTIES

4.1. Structural Properties

The structural properties of nanocrystals have been performed using various experimental as well as theoretical approaches. In general, a structural property means the analysis of stability of the crystal structure and then the evaluation of ground state parameters such as bond length, bond angle, lattice parameter etc. For example a periodic cluster approach with an atomistic pair potential has been employed to simulate AlN NC and predictes that the NCs displays graphitic-like layers and non-buckled wurtzite structure. The variation of surface tension, *c/a* ratio, and equation of state for AlN NC with respect to both the pressure and size has also been examined. The study also reveals that the bulk modulus of the wurtzite phase of NC is smaller (although decreasing with size) than its bulk counterpart [55]. Recently, Franceschetti [56] has analyzed the structural and electronic properties of PbSe NCs with different radius. Fig.6 shows the optimized structures of PbSe NCs in $Pb_{44}Se_{44}$, $Pb_{140}Se_{140}$ and $Pb_{240}Se_{240}$ form and both Pb and Se atoms have rocksalt structure. The Pb-Se bond lengths and bond angles are appreciably distorted contrast to that of its bulk counterpart as represented in Fig.7. In fact, Fig.7 clearly shows the deviation of bond length of the Pb-Se with respect to the distance of the bonds from the center of the NC and the bond lengths of Pb-Se NCs are in close match with its bulk crystal bond length (3.103 Å). The average Pb-Se bond length of the NC in the ground state is smaller than the bulk crystal bond length and decreases with the size. The formula for calculating formation energy of NC (PbSe) can be given as

(1)

where the calculated total energy for NC is E_{NC} with calculated bulk crystal total energy E_{bulk} and total number of Pb and Se atoms are represented by N for NC.

Srivastava et al. have also investigated the structural properties of various bulk as well as NCs such as AlAs [57,58], AlSb [59] and AlN [60]. The structural properties of bulk AlAs in zincblende (B3), wurtzite (B4), NiAs (B8), CsCl (B2) and NaCl (B1) type phases have been analyzed through first principle density functional theory (DFT) approach. Using local density approximation (LDA) with Perdew-Zunger (PZ) type parameterization the equilibrium lattice constant *(a)* of original B3 type phase of bulk AlAs is calculated as 5.64 Å and also from GGA-PBE (5.72 Å) and GGArevPBE (5.77 Å). The stability analysis of bulk AlAs in B8 type phase, corresponds to the *c/a* ratio of 1.59 Å, where the equilibrium lattice constants *a* and *c* are 3.72 Å and 5.91 Å, respectively [57]. Similar approach has been used for analyzing the structural stability of AlAs NCs in B3, B4, B2 and B1 type phases at ~0.9 nm, where the lattice parameters for

all the four stable phases of AlAs NCs [58] such as B3, B4, B1 and B2 are 5.761, 4.283, 5.350 and 3.207 Å, respectively. The total energy values for the B3, B4, B1 and B2 type phases of AlAs NCs are −468.52, −468.46, −467.84 and −467.77 eV, respectively.Fig.8 shows the variation of energy with respect to volume for B3 and B1 type phase of AlAs NCs. The calculated lowest total energy confirms that the B3 phase as the most stable one at lower pressure and the B1 type phase can be considered as the high pressure phase. On comparison with bulk AlAs, stability trend of the low pressure phases are same in bulk and nano-dimension but the high pressure phases are different.

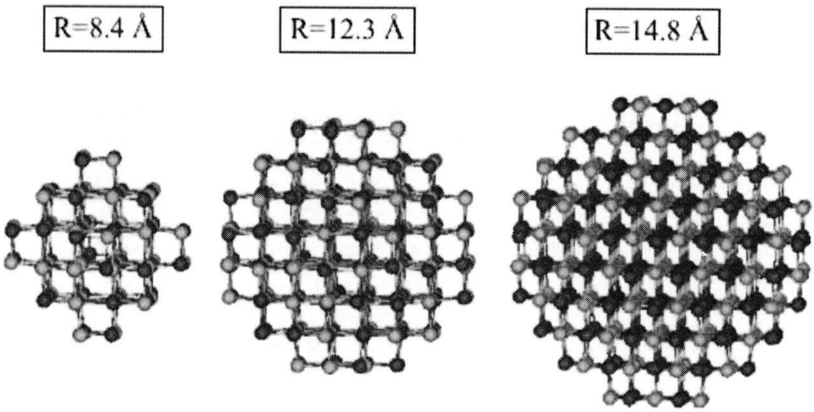

Figure 6. Three different sizes optimized geometries of PbSe NCs. The dark circles indicate Pb atoms and the light circles for Se atoms [**56**].

The stability of AlSb nanocrystal of ~1 nm size has also been analyzed in B3, B1 and B2 type phases theoretically. The article reports that computed lattice constant (6.235 Å) for bulk AlSb in B3 type phase is in close match with its experimental as well as other theoretical counterparts. However the computed lattice constants of B3, B1 and B2 type phase of AlSb NCs are 5.33 Å, 5.92 Å and 3.57 Å respectively. The findings conclude that B3 type phase is most stable with lowest total energy and highest binding energy [59]. In an another report by Tyagi et al. ab-initio pseudopotential approach has been used for calculating the lattice constants of 1-D AlN NC in B3, B1 and B2 type phase as 4.58 Å, 4.48 Å and 2.98 Å respectively [60]. Besides these theoretical works few experimental work has also been performed as zinc-blende GaP NCs have been synthesized by three different methods and it has been found that NCs transferred irreversibly into four new phases of GaP NCs with different lattice

constants (5.8872, 5.5493 and 7.0339 Å) and the largest one is about 1.3 times of the reported data, which has been characterized by X-ray diffraction and transmission electron microscopy [61].

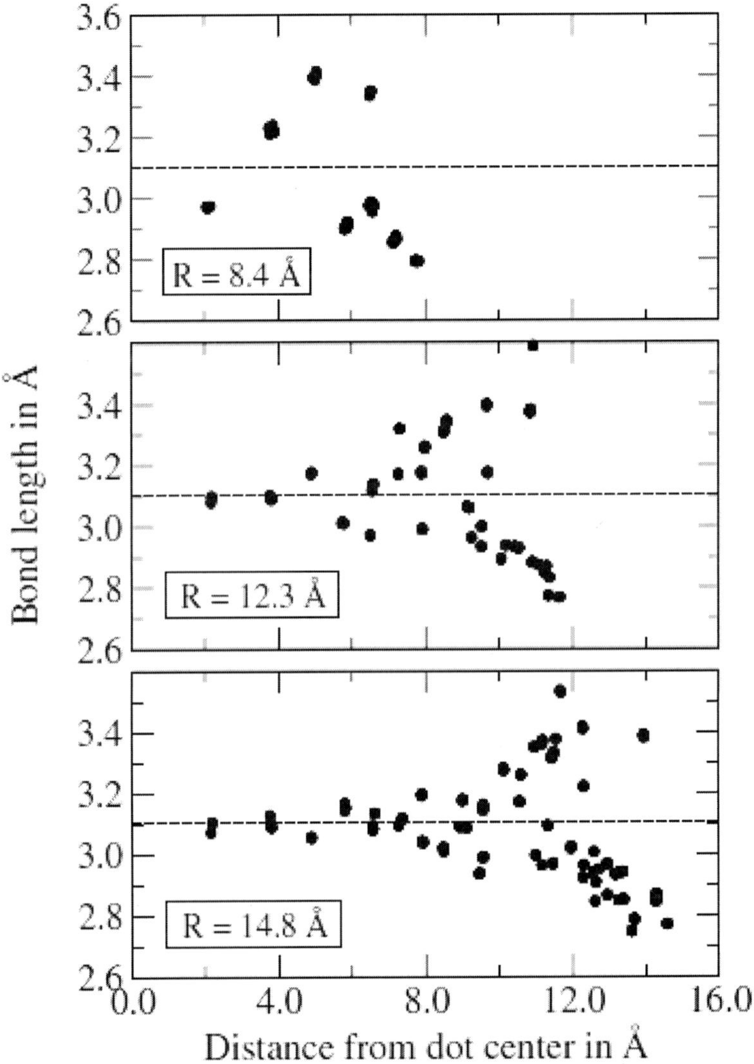

Figure 7. The variation of all Pb-Se bonds as a function of distance of the bond from the geometric centre of the NC. The dash lines shows the computed bulk Pb-Se bond length Ref.[56].

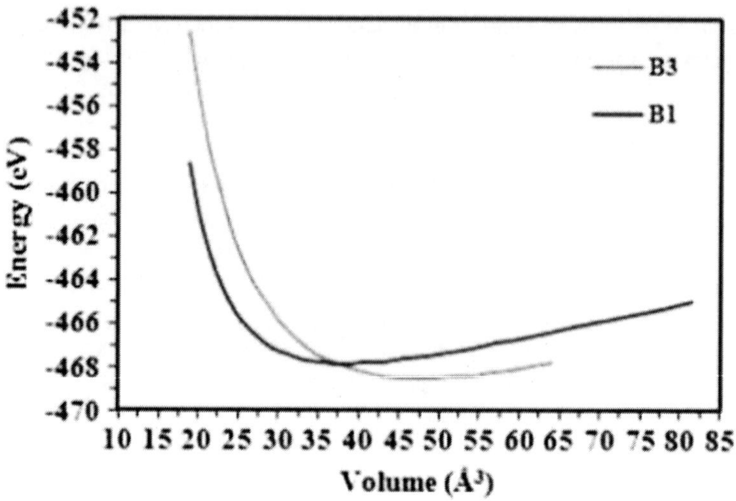

Figure 8. Energy as a function of volume for AlAs nanocrystals Ref.[**58**].

4.2. Electronic Properties

Electronic structures in nano regimes are extremely enviable for the future potential electronic devices. On the other hand, nano structures are still a challenge, where different experimental as well as theoretical approaches have been employed to understand the variation of electronic structures of NCs as a function of its size. The study of quantum confinement phenomena in semiconductor NCs is the subject of intense research which in fact determines their electronic behavior. A theoretical approach effective mass approximation (EMA) [62] is employed firstly to understand quantum confinement effects on the electronic band gap as a function of size of NCs. The confining potential for the electron and hole are assumed infinite in most of the EMA calculations. In EMA approach, the wave function of both electron as well as hole get vanishes at and beyond the surface of the NCs without any tunneling. To date numerous theoretical methods have been developed such as first-principles [63], semiempirical pseudopotential [64] and tight binding (TB) method [65] to see the effect of size on the electronic band structure of various NCs. The TB method has received great attention because of its realistic description of structural and dielectric properties in terms of chemical bonds and ease of handling large systems. For the description of hole wave function the effective bond orbital model and EMA calculations are used for of electron wave function by Einevoll [66], and Nair and his coworkers [67]. To see the effect of NC size on the electronic band structures various pseudopotential and TB calculations have been performed. UV-visible absorption spectroscopy tool has

been utilized to study the variation of band gap as a function of NC size. There are reports on the individual shifting of highest occupied molecular orbital of the valence band and the lowest unoccupied molecular orbital of the conduction band region with NC size, using various methods of energy spectroscopy e.g. X-ray absorption and photoemission [68]. Another semi empirical pseudopotential method has been employed to calculate the electronic structures of Si, CdSe [69] and InP [70] nanocrystals. An appreciable progress has been observed in the accuracy and the results of the electronic band structure calculations of NCs while using TB and pseudopotential approaches. In comparison to EMA, the reason behind the popularity of TB might be low computational effort, basic picture of atomic orbitals and hopping interactions for a predefined range. The ab-initio electronic band structure computation may lead to perceptive TB scheme and show accuracy because it use physical and real basis set. In 1989 Lippens and Lannoo [71] have used semi-empirical TB approach with sp^3s^* orbital basis to compute the variation in the energy gap of semiconductor NCs as a function of size and the results so obtained are in good agreement with the EMA potential, however the sp^3s^* TB model have a tendency to underestimate the energy gap. This model cannot reproduced the lowest lying conduction band, however the electron-hole interaction improves the nearest neighbor sp^3s^* model by the inclusion of spin-orbit coupling. In a study on InP [72] NCs where the sp^3d^5 orbital basis is used for the anion and the sp^3 basis for the cation with the next nearest neighbor interactions, gives good agreement with the experimental values. In sp^3d^5 orbital basis addition of the next nearest neighbor interactions for the anion gives correct explanation of the electronic band structure of semiconductors in its bulk as well as NCs [73]. Diaz et al.[74] have studied the electronic structure and optical spectra of GaAs NCs for a wide range of sizes by using both sp^3s^* and $sp^3s^*d^5$ nearest-neighbor TB models and found that the enclosure of *d* orbitals into a minimal basis set is essential for a good explanation of the lowest lying states of electron, particularly in the well confined system. Franceschetti [56] has analyzed the electronic properties of PbSe NCs within the radius range of 8.4-14.8 Å. Wannier function approach has been employed to study the band gap dependence on the different shapes (spherical, cubical, and tetrahedral) of CdS NCs and the report suggests that energy gap depends on the number of atoms present in the NC and not on their arrangements. Fig.9 shows the variation of energy gap variation as a function of the number of unit cells in the NC of different shapes, indicates that the energy gap slightly depends on the arrangements of the atoms in the NC [75].

Using first-principle DFT approach Srivastava et al. have analyzed the electronic properties of various bulk/NCs such as AlN, AlSb, AlAs [57-60] etc. in their different structural phases. The detailed discussion of the electronic structures of 1-D AlN NC in its various possible phases like zinc blende (B3), NaCl (B1) and

CsCl (B2) is as follows. The band structure of B3 type phase of AlN NC as shown in Fig.10(a) has a band gap of about 1.6eV at Γ point, defends the semiconducting nature.Fig.10(b) shows the band structure plot for B1 type phase of AlN NC, clearly depicts a band gap of around 1.5eV at Γ point however the B2 type phase is purely metallic in nature. The band gaps of all the AlN 1-D NCs are quite smaller as compared to bulk crystal, which contradicts to the quantum size effect where usually enhancement in the band gaps has been observed. This abnormal AlN band gaps is due to the surface effects. The surface states related with the tri coordinated N and Al atoms at lateral facets in NC creates new energy levels at valence band edges as well as conduction band edges that directly narrows the band gaps relative to their bulk counterparts.

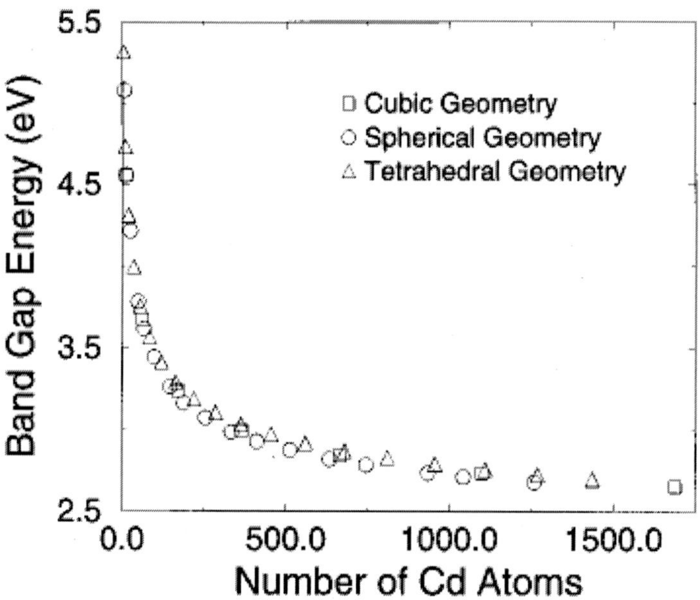

Figure 9. Dependence of band gaps on the number of Cd atoms for various shapes NC. Ref.[**75**].

The density of states for the most stable B3 type phase of AlN NC has also been analyzed in the energy range of -5.0 to 5.0 eV shown in Fig. 11, with three visible distinct peaks, where a highest peak appears in the valence band region near -2.3eV and satisfies the band structure plot by showing the unavailability of peaks at the Fermi level.

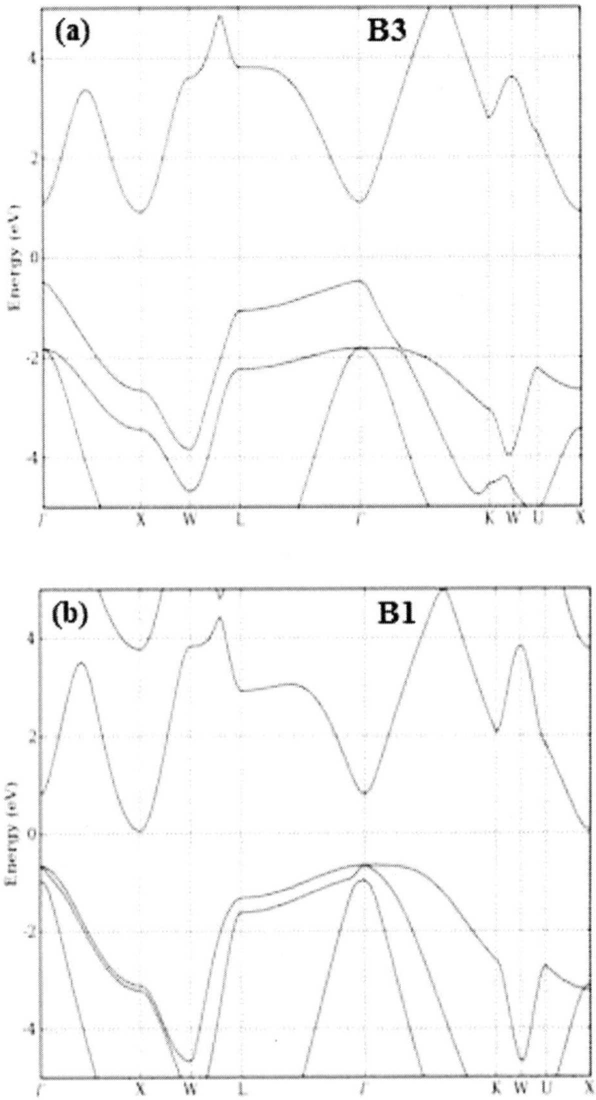

Figure 10. a-b) Electronic Band structure of AlN NC in B3 and B1 type phase Ref.[**60**].

4.3. Optical Properties

It is well known that the band gaps of different semiconductors such as III-nitrides, II-VI group, IV group elements can be tuned by varying the

composition in their alloys which make these systems prominent for various the applications like in optoelectronic devices, to use them for a spectral range starting from deep infrared to distant ultraviolet region. Quantum confinement has a great impact on the optical properties of semiconductor NCs. The phenomena of absorption and emission in the semiconductors are governed by their fundamental direct band gap. The increment in the energy band gap and strengthening of the oscillations with decrease in size of NCs has been revealed by spectroscopic studies. Due to the advancements in the synthesis, characterization and size selections of direct band structures of III-V and II-VI semiconductors, the spectroscopic studies of quantum confinement effects in these systems become much easier. The potential candidature of semiconductor NCs as light absorbers in solar-cell devices of third generation, has attracted number of researchers. For example PbSe, PbS, CdSe, PbTe, InAs and Si NCs shows much efficient multiple-exciton generation i.e. the chance of producing multiple electron-hole pairs from a single, high-energy photon, which thus increased the solar cell power [76]. For the solar energy conversion the semiconductor NCs requires a good match between the NC absorption spectrum and the solar spectrum. Keeping in mind that the quantum confinement effect tends to open up the energy band gap, the narrow band gap semiconductor NCs such as Ge, InAs, group-III antimonides and lead chalcogenides [77] are the potential candidates for the solar cell applications. Due to indirect band gap bulk Si is a poor light emitting material. In Si the excited electron-hole pairs have very wide radiative lifetime (ms), hence opposing non-radiative recombination rule and cause most of the excited electron-hole pairs to recombine non-radiatively. Si NCs formed by aerosol technique or by thermal precipitation of Si atoms rooted in SiO_2 layers, give strong visible and near infrared photoluminescence (PL), similar characteristics as in porous Si [78]. Optical properties of isolated Si NCs of ~4 nm size have been analysed with spectroscopic ellipsometry in the photon energy range of 1.1-5.0 eV. Si NCs reveals a major reduction in the dielectric functions and optical constants and whereas shows a large blue shift (~0.6 eV) in the absorption spectrum. The dielectric function and optical constants of the Si NC are shown in Fig. 12 and 13, where overall spectral features of Si NC are similar to bulk Si. Nevertheless, the Si NC shows a remarkable drop in the optical constants and dielectric function in contrast to bulk Si. It is well recognized that decrease in the static dielectric constant becomes significant as the size of the structure approaches to the nano regime [79].

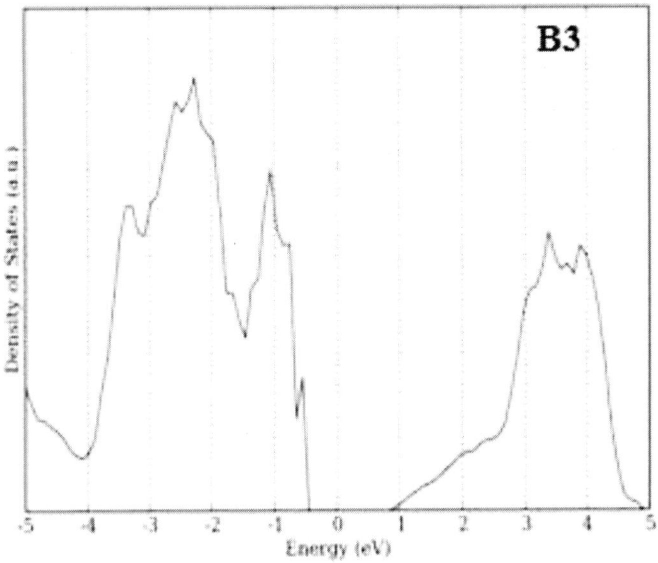

Figure 11. Density of states of AlN NC in different structural B3 phase Ref.[**60**].

Photoluminescence (PL) properties of GaAs NCs in SiO_2 matrices formed by sequential ion implantation and thermal annealing, have been studied and observe broad PL due to GaAs NCs appears in the red spectral region, where the spectral shape of the red PL band depends on the hydrogen concentration in the sample. These GaAs NCs in SiO_2 film with low defects density will be used as the material for the future excitonic devices [80]. The blinking (or ON-OFF) behavior has been observed in single CdSe NC PL, which gets attributed to an intermittent photo ionization and subsequent neutralization of the NC [81]. The optical properties are related to the edge transition of the electronic band gaps in a semiconductor. Hence it is necessary to study the size dependence of the electronic band gap and the related exciton energy in semiconductor NCs. Few earlier experiments found that the lack of a large overlap between absorption and emission spectra in CdSe NC can improve the efficiency of light-emitting diodes (LEDs) due to the reduction in re-absorption [82]. Photovoltaic response has been investigated in PbSe NCs of various sizes and conjugated polymers. The devices composed with these NCs shows good diode characteristics and sizable photovoltaic response in a spectral range from the ultraviolet to infrared [83]. A very interesting observation about the NCs is that their band gap can be tuned over a large energy range simply via synthetic control over the size of the NCs [84]. NC having absorption onset in the near to mid infrared (IR) will also strongly absorb solar photons of higher energy. Generation of multiexcitons in PbSe NCs from a single photon absorption event is observed to

take place in picosecond and occurs with up to 100% efficiency depending upon the excess energy of the absorbed photon. Thermodynamic conversion efficiency in solar cells is found to be 43.9% in concentrated solar lighting. The computed percentage depends upon the hypothesis that the absorption of an individual photon with energy above a semiconductor band gap results in formation of a single exciton and the photon energy which is more than the energy band gap get lost by electron-phonon interactions. Numerous techniques have existed to raise the power exchange efficiency of solar cells together with the improvement of multi-junction cells, intermediate band devices, hot electron extraction and carrier multiplication [85,86]. In Si_{35} and Si_{66} cores passivated with oxide shows lowering of energy band gap by 2.4 eV and 1.5 eV. The oxide and hydrogen passivated NCs are optically forbidden indirect band gap and optically allowed direct band gap transitions respectively. However the highest occupied molecular orbitals and lowest unoccupied molecular orbitals are delocalized in both cases. Theoretical prediction get confirmed through the experimental observation that hydrogen passivated Si NCs emit blue light whereas oxide passivated Si NCs emit yellow-red light [87].

4.4. Mechanical Properties

Mechanical properties of nano-materials are mainly focused on their constitutive response as well as on the fundamental physical mechanisms. Numbers of characteristics of mechanical behavior are presented to-date, like young modulus, shear modulus, bulk modulus, elastic constants and compressive strength etc. However literature shows that particularly in semiconducting NCs, these properties are yet not very much exploited; a very few reports are available on the bulk modulus of NCs which actually tells about the hardness of the material and inversely shows the compressibility of the same structure. The first-principles calculations using the LDA as exchange correlation functional, the variation in the bulk modulus of silicon nanocrystal have been investigated. On decreasing the size of the NC an enhancement in the bulk modulus has been observed [88]. In case of CdSe NCs structural transformation has been studied using high pressure X-ray diffraction and also calculates the bulk modulus (B_0), defined as the reciprocal of the volume compressibility, as ±74 GPa for rock salt phase of NCs and in close match with others report data for CdSe bulk and NCs. However due to the overlap of the numerous wurtzite diffraction peaks the exact value of B_0 has not achieved [89]. Another example is GaN where its bulk and NCs have been studied using X-ray diffraction and calculated bulk moduli are reported as 187 and 319 GPa for the wurtzite phase of bulk and NCs respectively, although the respective NaCl phases are found to have very similar bulk moduli 208 and 206 GPa [90]. A

in-situ high-pressure study observes semiconductor to metal phase transitions in ZnS materials with average grain sizes of 10 µm and 11 nm. Bulk modulus and its pressure derivative (B_0'), of B3 phase has been estimated from the Birch-Murnaghan equation of state in a pressure range from 0 to 9 GPa as B_0 = 72±7 GPa and B_0' = 9±3 for 11 nm ZnS and B_0 = 68±3 GPa and B_0' = 7±1 for 10 µm ZnS NCs [91]. Another example is X-ray diffraction study of pressure-induced phase transformations in ZnO nanorods shows that the reduction of the particle size for ZnO crystallites leads to a significant increase of the bulk modulus and the pressure range of the coexistence of the wurtzite and rocksalt phases suggest that hardness of ZnO nanorods is higher than its bulk counterpart [92]. Raman study of nanocrystalline ZnO with different average crystalline sizes, predicts that the average elastic modulus of nanocrystalline ZnO shows a non-monotonic variation with crystalline size, suggests that the non-monotonic behavior is due to interplay between the elastic properties of the individual grains and those of the intergranular region [93]. A study on AlN NCs /nanowires reveals that the wurtzite structure of AlN NCs and nanowires have the B_0 ~215.8 GPa and ~208.8 GPa, respectively, however the B_0 of rocksalt phase of AlN NCs and nanowires are ~312.6 GPa, and 324.9 GPa, respectively. This shows that the decrease of particle size can appreciably lead to an increase of the B_0 [94].

4.5. Magnetic Properties

Scientists divides the information technology into two parts one to exploits the degree of freedom of electronic charge in semiconductors to advances the information and other in magnetic materials where the spin degree of freedom is used to preserve the information. Spin of the charge carriers plays a slight role on semiconductor devices since most of the well known semiconductor devices are based on non-magnetic Si and GaAs which shows very less effect of spin. In contrast the superior spin related features have been acknowledged in magnetic semiconductors and diluted magnetic semiconductors (DMS) because of the coexistence of both magnetic and semiconductor properties. Nano magneto electronics is a newly developed field, where the two degrees of freedom, the charge and spin of the carriers, are utilized simultaneously to create new spintronics devices. Mn^{2+} or Co^{2+} the transition metal ions are the important dopants as they act as optically active impurities and localized spins in semiconductors and responsible for the optical and magnetic functionalities. Magnetic circular dichroism (MCD) spectroscopy of Mn doped and ZnS coated CdS NCs gives the electronic structures of Mn impurities and calculate the amount of Mn composition in NCs. Fig.14 shows the temperature dependence of the Zeeman splitting (ΔE_z) of 0.2, 2, and 10 mol % Mn doped CdS NCs, where the MCD peak signal is allotted to Zeeman splitting of the lowest exciton. The

study reveals that MCD signal directly depends on the Zeeman splitting and the first derivative of the absorption spectrum, though the Zeeman splitting is smaller than the bandwidth of the lowest exciton absorption [95].

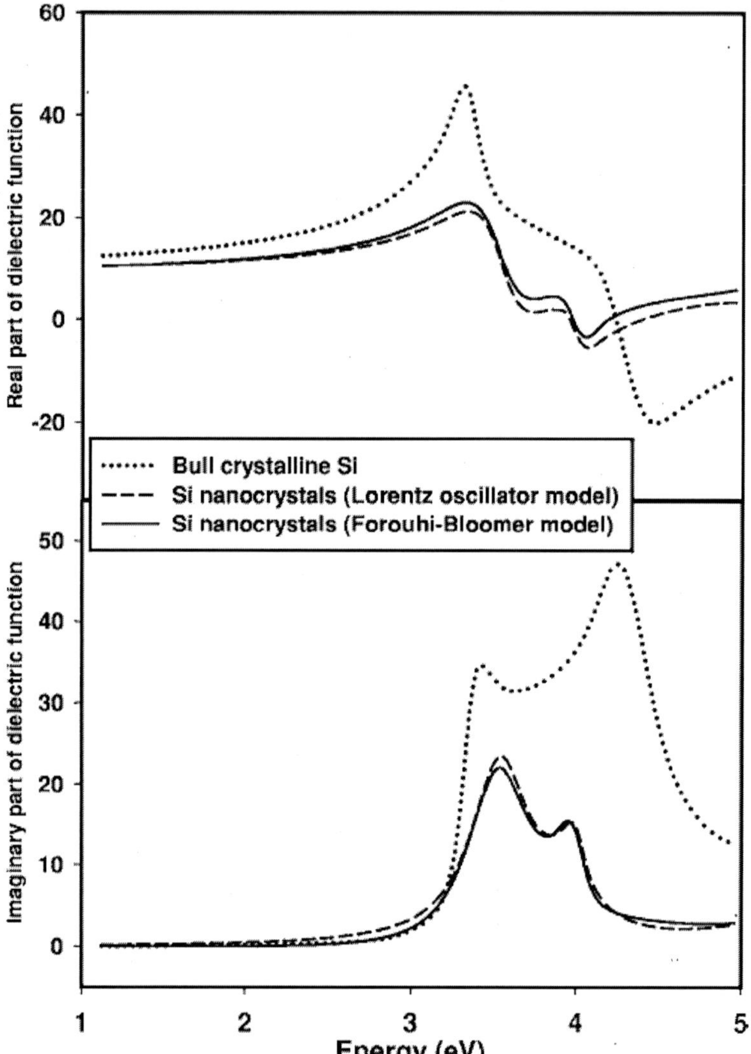

Figure 12. Real (ε1) and imaginary (ε2) parts of the complex dielectric function of the Si NC obtained from the spectral fittings based on the Lorentz oscillator model and the FB model. The dielectric function of bulk Si is also included for comparison Ref.[**79**].

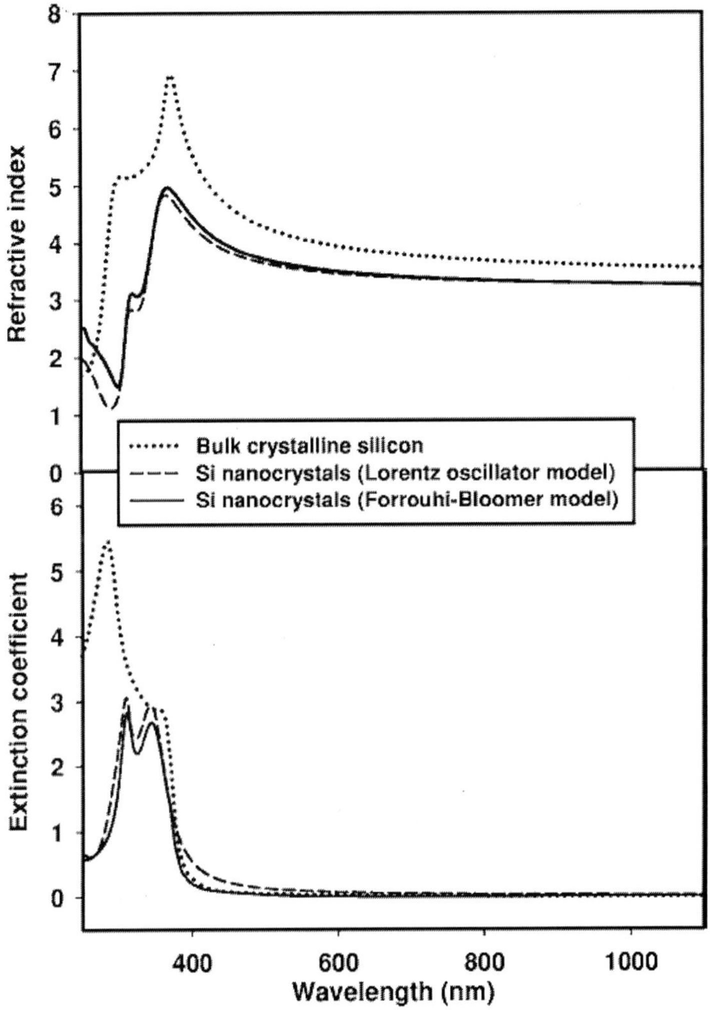

Figure 13. Refractive index (n) and extinction coefficient (k) of the Si NC and bulk Si as function of wavelength Ref.[**79**].

Magnetic ion doped semiconductors are called diluted magnetic semiconductors (DMSs) [96] and these DMSs are expected to be a key material for spintronics devices due to their spin-dependent effects [97]. According to the literature not much report are available on the magnetic properties of semiconductor NCs. However, Magnetic properties of ZnS and ZnO NCs doped with various concentrations of a transition metals have been analyzed. Over a wide range of dopant concentrations without changing the size (1.6 and 4.7

nm) of the NC, the study clearly reveals the magnetic properties of such doped systems, remaining paramagnetic down to the lowest temperature, can give precise information concerning the dopant level in such samples [98,99]. ZnO NC with Co doping and without capping agent shows ferromagnetic and paramagnetic behavior, respectively. Co intra d-d transition raises in ferromagnetic NCs which may be because of the profound bound states at the NCs surface.Fig. 15 shows the variation of magnetization as a function of magnetic field for Co doped ZnO NCs and in case of washed NCs (also called as-prepared) the magnetization is directly proportional to the applied magnetic field lacking hysteresis, which shows paramagnetic behavior. However the ferromagnetic nature with hysteresis has been seen in the O capped NCs sample. As the high magnetic field increases the paramagnetic behavior is also predicted, which indicates that paramagnetic and ferromagnetic NCs coexist in these samples [100].

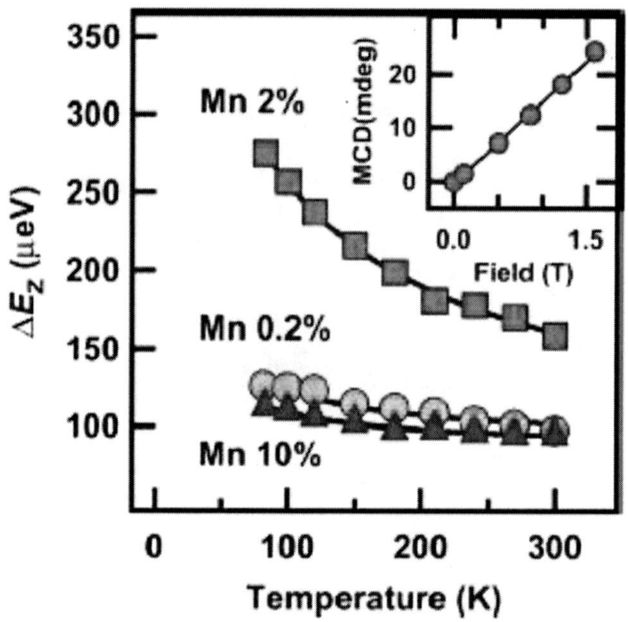

Figure 14. a) Dependence of the Zeeman splitting energy on Temperature of CdS NCs doped with 0.2, 2, and 10 mol % Mn ions. The inset shows the magnetic field dependence of the MCD signal intensity in 2mol % Mn doped CdS NCs Ref.[95].

Mn doped GaN NCs with the average diameters ranging from 4-18 nm has been prepared at low temperature and reveals that the NCs exhibit ferromagnetism with magnetization and Curie temperature values increasing with the concentration of Mn and particle size [101]. The magnetic properties of hydrogenated Si NCs doped with pairs of Mn atoms were investigated using spin density functional theory, where the two sites occupied by Mn, shows strong dependence of obtained formation energies and total magnetic moments. Pairs at sites with the same character tend to ferromagnetic spin arrangements, significantly influenced by their noncollinearity [102]. As a theoretical understanding one Ab-initio pseudopotentials approach has been employed to investigate the electronic and magnetic properties of Mn doped Ge, GaAs, and ZnSe NCs. To predict that the ferromagnetic and half-metallicity trends found in the bulk are preserved in the NCs. Due to localization of Mn states, they are less affected by quantum confinement. As a result in NCs, the Mn-related impurity states become much deeper in the gap with decreasing size [103].

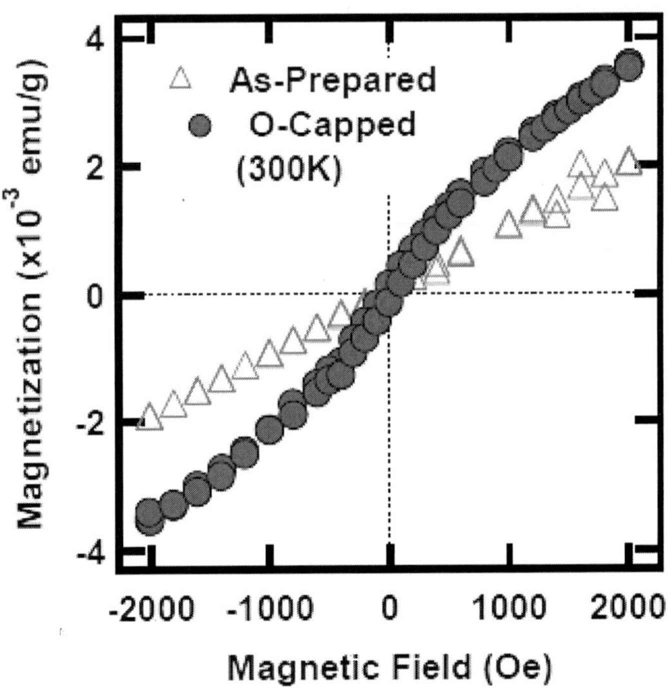

Figure 15. Variation of magnetization with respect to magnetic field at 300K of Co doped ZnO NCs Ref.[**100**].

4.6. Structural Phase Transition

The effects of crystal size on the stability of NC is of considerable interest because for utilizing a material in several applications, information about crystal structure at that particular temperature and pressure is must. In nano-materials the reduction in the size of the crystal changes its entire properties; this may be because of the change in the lattice parameters for a given phase and also due to pressure effects. If we assume a spherical NC, then pressure applied by the surface on the core of the NC will lead to the relaxation of the first atomic planes and therefore to an increase or a decrease in the lattice parameter. This effect is well seen in semiconductors, due to the adsorption state of the surface and surface itself reconstructs themselves to minimize the number of dangling bonds, which produces more stresses in the upper layers. Observation shows that in NC, these stresses will also contribute to the surface pressure and to the reduction or extension of atomic bonds. One obvious question arises that how will the relative stability of different possible solid structures change for NCs with respect to bulk materials? One way to answer this question is to use the pressure to force nanostructured materials to get convert it from one solid structure to another. It has been known classically that when pressure is applied the structure of the solid changes and hence its density as well as volume, that leads to the overlapping of the electron shells and structural phase transformation may take place. In III-V compound semiconductors the zinc blende to rocksalt phase transition are generally expected and intermediate phase NiAs are also seen sometimes and further the structure stabilized in a high pressure CsCl phase. The efforts on the high pressure behavior of NC starts from the year 1994 by S. H. Tolbert and A. P. Alivisatos [104] as both are the pioneer workers in the field of high pressure structural transformation in NCs. They discussed the stability of CdSe NCs in wurtzite and rocksalt phases and observed the phase transition in the range of 3.6 to 4.9 GPa for crystallites radius ranging from 21 to 10 Å. There are also some reports on the high pressure behaviour of Si and CdS NCs, shows that the smaller the crystallite higher the transition pressure, which has been explained by them in terms of surface energy differences between the phases involved [105,106]. Qadri et al. [107] reported that the effect of reduced grain size in PbS NCs elevates the transition pressure, while the compressibility increases with decreasing grain size. An enhancement of transition pressure has been observed in ZnO NCs [108], ZnS [5] and PbS [6] in comparison to its bulk counterpart. In an another report the pressure induced I–II transition has been studied for nanocrystalline Ge of size 13, 49 and 100 nm using synchrotron x-ray diffraction, where the bulk modulus and the transition pressure increases with decrease in particle size for both Ge-I and Ge-II, but the percentage volume collapse at the transition remains constant [109]. The hexagonal AlN nanocrystals have a particle size of ~10 nm (Fig. 16) shows wurtzite to rocksalt phase transition at around 14.5 GPa, which is significantly lower than the

transition pressure of 22.9 GPa observed for the bulk AlN by using the same technique.Fig.17 clearly illustrates that at the phase transition pressure of 14.5 GPa the high-pressure rocksalt phase is about 20.5% denser than the hexagonal wurtzite phase [110].

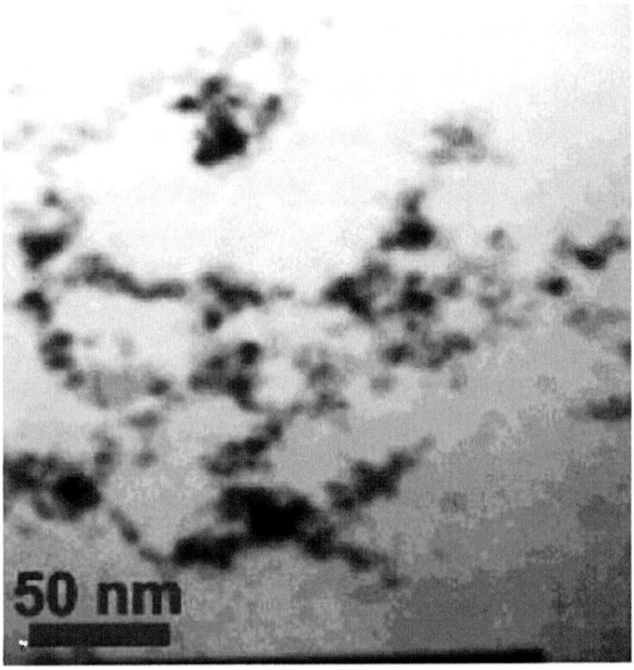

Figure 16. TEM image of AlN NCs Ref.[**110**].

Wen et al. [111] have studied the pressure induced phase transition in CdSe and ZnO NCs and found that the transition pressure increases as the nanocrystals size reduces. A pressure induced first-order structural phase transition from wurtzite to rocksalt type structure has been observed in GaN NCs at around 48.8 GPa using x-ray diffraction technique [112]. Theoretically [58] the structural stability and high pressure behavior of AlAs NCs has been analysed in various phases such as wurtzite (B4), zincblende (B3), CsCl (B2) and NaCl (B1), and observes the structural transformations from B3→B1 at around 8.9 GPa, B3→B2 at 7.12 GPa and from B3→B4 at 3.88 GPa, which is probably being the first report. The calculated values of transition pressures for AlAs NCs are lower in comparison to its bulk counterpart [57, 113,114]. The amount of volume collapse at the transition pressure has also been analyzed and the computed volume collapse for B3→B4, B3→B2 and B3→B1 transitions are found to be 5.8%, 3.5% and 1.5% respectively as illustrated in Fig.18.

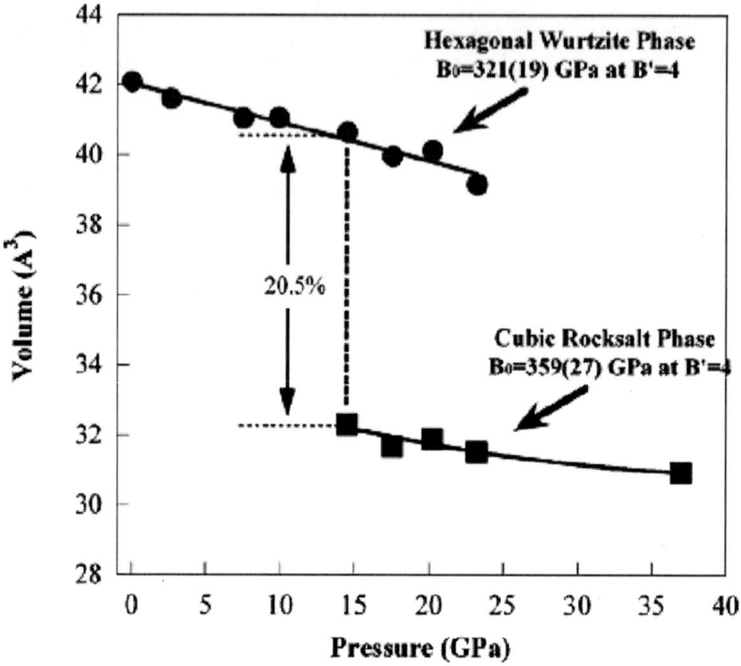

Figure 17. Equation of state plot and the relative volumetric variation of AlN upon the phase transition at 14.5 GPa.

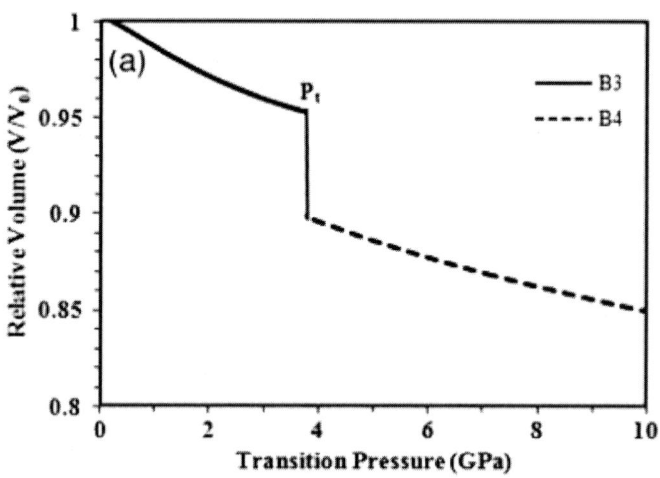

Figure 18. Relative volume as a function transition pressure for AlAs NCs Ref.[**58**].

Similar approach has been employed to investigate the structural stability of ~1nm sized AlSb NC [58] in its B3, B1 and B2 type phases under high compression and confirms that the B3 type phase is the most stable amongst the other considered phases, further under compression, the original B3 type phase of AlSb NC transforms to B1 type phase at a pressure of about 8.9 GPa, which is comparatively larger than that of bulk crystal. This discussion concludes that the transition pressure strictly depends on the size of the NC and may vary accordingly.

5. CONCLUSION

In this chapter we have described four topics; first, we have explained the fundamentals behind the peculiar properties of semiconductor NCs, the factors that makes them different from their bulk counterparts. Second, recent applications of NCs in electronic, optoelectronics, photonics as well as in medicines have been discussed. Third, we referred the present synthesis and characterization techniques for these tiny crystals, discussed various recent methods of controlled synthesis on growth and size of NCs. Fourth, we have remarked on the various properties of semiconductor NCs such as structural, electronic, optical, mechanical, magnetic properties of NCs with a detailed description of the structural transformations of semiconductor NCs under the application of pressure. We believe that the discussion on semiconductor NCs with recent results and ideas will be certainly be helpful, especially to variety of scientific community like physicists, chemists, and biologists, nano researchers/scientists for the advancements of science and technology, and its exploitation in variety of applications.

ACKNOWLEDGEMENT

Authors are grateful to ABV-Indian Institute of Information Technology and Management, Gwalior for providing the infrastructural support. One of us Neha Tyagi is thankful to ABV-IIITM for the award of Ph.D scholarship.

REFERENCES

1. M. Achermann, M. A. Petruska, S. Kos, D. L. Smith, D. D. Koleske, V. I. Ekimov, Energy-transfer pumping of semiconductor nanocrystals using an epitaxial quantumwell. Nature 2004429642646

2. W. U. Huynh, J. J. Dittmer, A. P. Alivisatos, Hybrid Nanorod-Polymer Solar Cells. Science 2002295556424252427

3. D. L. Klein, R. Roth, A. K. L. Lim, A. P. Alivisatos, P. L. A. Mc Euen, transistor. single-electron, from. a. made, selenide. cadmium, nanocrystal, Nature 19973896652699701

4. L. B. Boinovich, A. M. Emelyanenko, Hydrophobic materials and coatings: principles of design, properties and applications. Russian Chemical Reviews 2008777583600

5. J. Z. Jiang, L. Gerward, D. Frost, R. Secco, J. Peyronneacu, J. S. Olsen, Grain-size effect on pressure-induced semiconductor-to-metal transition in ZnS. Journal of Applied Physics 1999861166086610

6. J. Z. Jiang, L. Gerward, D. Frost, R. Secco, J. Peyronneacu, J. S. Olsen, J. Truckenbrodt, Phase transformation and conductivity in nanocrystal PbS under pressure. Journal of Applied Physics 200087526582660

7. J. J. Shiang, A. V. Kadavanich, R. K. Grubbs, A. P. Alivisatos, Symmetry of Annealed Wurtzite CdSe Nanocrystals: Assignment to the C-3v Point Group. Journal of Physical Chemistry 199599481741717422

8. V. L. Colvin, M. C. Schlamp, A. P. Alivisatos, Light-emitting diodes made from cadmium selenide nanocrystals and a semiconducting polymer. Nature 19943706488354357

9. X. Peng, L. Manna, W. Yang, J. Wickham, E. Scher, A. Kadavanich, A. P. Alivisatos, Shape control of CdSe nanocrystals. Nature 200040467735961

10. M. Achermann, M. A. Petruska, D. D. Koleske, M. H. Crawford, V. I. Klimov, Nanocrystal-based light-emitting diodes utilizing high-efficiency nonradiative energy transfer for color conversion. Nano Letters 20066713961400

11. N. Tessler, V. Medvedev, M. Kazes, S. H. Kan, U. Banin, Efficient Near-Infrared Polymer Nanocrystal Light-Emitting Diodes. Science 2002295555915061508

12. H. Mueller, M. A. Petruska, M. Achermann, D. J. Werder, E. A. Akhadov, D. D. Koleske, M. A. Hoffbauer, V. I. Klimov, Multicolor light-emitting diodes based on semiconductor nanocrystals encapsulated in GaN charge injection layers. Nano Letters 20055610391044

13. M. Peralvarez, J. Barreto, J. Carreras, A. Morales, D. N. Urrios, Y. Lebour, C. Dominguez, B. Garrido, Si-nanocrystal-based LEDs fabricated by ion implantation and plasma-enhanced chemical vapour deposition. Nanotechnology 200920405201110

14. M. S. Gudiksen, K. N. Maher, L. Ouyang, Hongkun. Park, Electroluminescence from a Single-Nanocrystal Transistor, Nano Letters 200551122572261

15. D. Kahng, S. M. A. Sze, floating-gate, application. its, memory. to, The. devices, System. Bell, Journal. Technical, 196746412881295

16. S. Tiwari, F. Rana, H. Hanafi, A. Hartstein, E. F. Crabbe, K. A. Chan, nanocrystals. silicon, memory. based, Applied Physics Letters 1996681013771379

17. T. C. Chang, F. Y. Jian, S. C. Chen, Y. T. Tsai, Development in nanocrystal memory. Materials today 20111412608615

18. G. Ammendola, V. Ancarani, V. Triolo, M. Bileci, D. Corso, I. Crupi, L. Perniola, C. Gerardi, S. Lombardo, B. De Salvo, Nanocrystals memories for FLASH device applications. Solid-State Electronics 200448914831488

19. J. K. Kim, H. J. Cheong, Y. Kim, J. Y. Yi, H. J. Bark, S. H. Bang, J. H. Cho, Rapid-thermal-annealing effect on lateral charge loss in metal-oxide-semiconductor capacitors with Ge nanocrystals. Applied Physics Letters 2003821525272529

20. X. B. Lu, P. F. Lee, J. Y. Dai, Synthesis and memory effect study of Ge nanocrystals embedded in LaAlO3 high-k dielectrics. Applied Physics Letters 20058620311113

21. H. Du, C. Chen, R. Krishnan, T. D. Krauss, J. M. Harbold, F. W. Wise, M. G. Thomas, J. Silcox, Optical Properties of Colloidal PbSe Nanocrystals. Nano Letters 200221113211324

22. K. R. Choudhury, Y. Sahoo, T. Y. Ohulchanskyy, P. N. Prasad, Efficient photoconductive devices at infrared wavelengths using quantum dot-polymer nanocomposites. Applied Physics Letters 200573110 EOF

23. D. Qi, M. Fischbein, M. Drndic, S. Selmic, Efficient polymer-nanocrystal quantum-dot photodetectors. Applied Physics Letters 200593103 EOF

24. X. , B. Yuan, Z. Yan, The photodetector of Ge nanocrystals/Si for 1.55 μm operation deposited by pulsed laser deposition, Optics Communications 20062601337339

25. D. Guimard, R. Morihara, D. Bordel, K. Tanabe, Y. Wakayama, M. Nishioka, Y. Arakawa, Fabrication of InAs/GaAs quantum dot solar cells with enhanced photocurrent and without degradation of open circuit voltage. Applied Physics Letters 2010203507 EOF

26. Gur, N. A. Fromer, M. L. Geier, A. P. Alivisatos, Air, Air-Stable All-Inorganic Nanocrystal Solar Cells Processed from Solution. Science 20053105747462465

27. N. Y. Gross, M. S. Harari, M. Zimin, S. Kababya, A. Schmidt, N. Tessler, Molecular control of quantum-dot internal electric field and its application to CdSe-based solar cells. Nature Materials 20111012974979

28. V. Svrcek, D. Mariotti, T. Nagai, Y. Shibata, I. Turkevych, M. Kondo, Photovoltaic Applications of Silicon Nanocrystal Based Nanostructures Induced by Nanosecond Laser Fragmentation in Liquid Media. Journal of Physical Chemistry C 20111151250845093

29. Suman. V. S. R. Katteboinaa, P. Chandrasekhar, S. Balaji, Nanocrystals. a. Drug, Formulation. Novel, for. Approach, Soluble. Poorly, Drugs, International Journal of PharmTech Research 2009

30. E. M. M. Liversidge, G. G. Liversidge, Nanoparticles. Drug, Poorly. Formulating, Compounds. Water-Soluble, Toxicologic Pathology 20083614348

31. P. T. Tran, E. R. Goldman, G. P. Anderson, J. M. Mauro, H. Mattoussi, Use of luminescent CdSe-ZnS nanocrystal bioconjugates in quantum dot-based nanosensors. Physica Status Solidi B 2002427 EOF432 EOF

32. B. Dubertret, P. Skourides, D. J. Norris, V. Noireaux, A. H. Brivanlou, A. Libchaber, In vivo imagining of quantum dots encapsulated in phospholipid micelles. Science 20022985599175917 62

33. W. J. Parak, R. Boudreau, M. L. Gros, D. Gerion, D. Zanchet, C. M. Micheel, S. C. Williams, A. P. Alivisatos, C. Larabell, Cell motility and metastatic potential studies based on quantum dot imaging of phagokinetic tracks. Advanced Materials 20021412882885

34. R. Taylor, M. M. Fang, S. M. Nie, Probing specific sequences on single DNA molecules with bioconjugated fluorescent nanoparticales. Analytical Chemistry 200072919791986

35. W. J. Parak, T. Pellegrino, C. Plank, Labelling of cells with quantum dots. Nanotechnology 2005R9R25.

36. Brus, Electronic wave functions in semiconductor clusters: experiment and theory. Journal of Physical Chemistry 1986901225552560

37. L. Steigerwald, A. P. Alivisatos, J. M. Gibson, T. D. Harris, R. Kortan, A. J. Muller, A. M. Thayer, T. M. Duncan, D. C. Douglass, L. E. Brus, Surface derivatization and isolation of semiconductor cluster molecules. Journal of American Chemical Society 19881101030463050

38. K. Yong, T. , Y. Sahoo, M. T. Swihart, P. N. Prasad, Shape Control of CdS Nanocrystals in One-Pot Synthesis. Journal of Physical Chemistry C 2007111624472458

39. W. Wang, B. Poudel, J. Y. Huang, D. Z. Wang, S. Kunwar, Z. F. Ren, Synthesis of gram-scale germanium nanocrystals by a low-temperature inverse micelle solvothermal route. Nanotechnology 200516811261129

40. Chen, J. D. Holmes, S. R. Garcia, M. A. Morris, Facile Synthesis of Monodisperse ZnO Nanocrystals by Direct Liquid Phase Precipitation. Journal of Nanomaterials 2011201119

41. Yu, X. Hao, D. Cui, Q. Wang, X. Xu, M. Jiang, Synthesis of aluminium nitride nanocrystals and their catalytic effect on the polymerization of benzene. Nanotechnology 200329 EOF32 EOF

42. H. Yang, X. Yao, D. Huang, Sol gel synthesis and photoluminescence of AlP nanocrystals embedded in silica glasses. Optical Materials 2007297747752

43. H. Q. Yang, X. Yao, D. M. Huang, X. J. Wang, H. Z. Shi, B. L. Zhang, S. X. Liu, Y. Fang, Sol-gel synthesis and photoluminescence of III-V semiconductor InAs nanocrystals embedded in silica glasses. Journal of Nanoscience and Nanotechnology 2005557869

44. H. Yang, D. Huang, X. Wang, X. Gu, Fujian. Wang, S. Xie, X. Yao, Sol, Sol-Gel Synthesis of Luminesent InP Nanocrystals Embedded in Silica Glasses. Journal of Nanoscience and Nanotechnology 200551017371740

45. J. Nayak, R. Mythili, M. Vijayalakshmi, S. N. Sahu, Size quantization effect in GaAs nanocrystals. Physica E: Low-dimensional Systems and Nanostructures 2004227 EOF233 EOF

46. J. M. Yuk, K. Kim, Z. Lee, M. Watanabe, A. Zettl, T. W. Kim, Y. S. No, W. K. Choi, J. Y. Lee, Direct fabrication of zero- and one-dimensional metal nanocrystals by thermally assisted electromigration. ACS Nano 20104629993004

47. C. W. Chiu, T. W. Liao, K. Y. Tsai, F. M. Wang, Y. W. Suen, C. H. Kuan, Fabrication method of high-quality Ge nanocrystals on patterned Si substrates by local melting point control. Nanotechnology 2011275604 EOF

48. H. Namatsu, K. Kurihara, M. Nagase, T. Makino, Fabrication of 2-nm-wide silicon quantum wires through a combination of a partially-shifted resist pattern and orientation-dependent etching. Applied Physics Letters 1997705619621

49. C. B. Murray, D. J. Norris, M. G. Bawendi, Synthesis and characterization of nearly monodisperse CdE (E = sulfur, selenium, tellurium) semiconductor nanocrystallites. Journal of American Chemical Society 19931151987068715

50. J. O. Mc Caldin, Current approaches to p-n junction in wider band gap II-VI semiconductors. Journal of Vaccum Science and Technology A 1990

51. E. Alphandery, R. J. Nicholas, N. J. Mason, S. G. Lyapin, P. C. Klipstein, Photoluminescence of self-assembled InSb quantum dots grown on GaSb as a function of excitation power, temperature, and magnetic field. Physical Review B 2002115322 EOF

52. Y. Xie, Y. T. Qian, W. Z. Wang, S. Y. Zhang, Y. H. A. Zhang, Synthetic. Benzene-Thermal, to. Route, Ga. N. Nanocrystalline, Science 1996272527019261927

53. M. D. Dvorak, B. L. Justus, D. K. Gaskill, D. G. Hendershot, Nonlinear absorption and refraction of quantum confined InP nanocrystals grown in porous glass. Applied Physics Letters 1995667804806

54. S. S. Kim, K. I. Bang, J. Kwak, K. S. Lim, Growth of Silicon Nanocrystals by Low-Temperature Photo Chemical Vapor Deposition. Japanese Journal of Applied Physics 2006L46L49.

55. Costales, M. A. Blanco, E. Francisco, C. J. F. Solano, A. M. Pendas, Theoretical Simulation of AlN Nanocrystals. Journal of Physical Chemistry C 20081121766676676

56. Franceschetti, Structural and electronic properties of PbSe nanocrystals from first principles. Physical Review B 2008075418 EOF

57. Srivastava, N. Tyagi, R. K. Singh, Pressure induced phase transformation and electronic properties of AlAs. Materials Chemistry Physics 201166 EOF71 EOF

58. Srivastava, N. Tyagi, High pressure behavior of AlAs nanocrystals: the first-principle study. High Pressure Research iFirst 2012201215

59. Tyagi, A. Srivastava, Structural Phase Transition and Electronic Properties of AlSb Nanocrystal. Proc. Intl. Conf. on Nanosci. Eng. Tech. (ICONSET 2011) 978-1-4673-0073-5/11/$26.00 @2011 IEEE IEEEXplore 20112011421423

60. N. Tyagi, A. Srivastava, Electronic properties of AlN nanocrystal: A first principle study. XVI National Seminar on Ferroelectrics and Dielectrics and (NSFD-XVI), American Institute of Physics Conference Proc. 20111372252262

61. Z. , G. , Y. Bai, J. , D. Cui, L. , X. Hao, P. , L. Wang, M. , Q. Wang, L. , X. Xu, G. , Increase of lattice constant in GaP nanocrystals. Journal of Crystal Growth 2002486 EOF490 EOF

62. Al. L. Efros, A. L. Efros, Interband absorption of light in a semiconductor sphere. Soviet Physics Semiconductors 1982167772775

63. F. Buda, J. Kohanoff, M. Parrinello, Optical properties of porous silicon: A first-principle study. Physical Review Letters 199269812721275

64. M. V. R. Krishna, R. A. Friesner, Exciton spectra of semiconductor clusters. Physical Review Letters 1991675629632

65. J. M. Jancu, R. Scholz, F. Beltram, F. Bassani, spds. Empirical, calculations. tight-binding, cubic. for, General. semiconductors, method, parameters. materials, Physical Review B 1998571164936507

66. Einevoll G.T.Confinement of excitons in quantum dots. Physical Review B 199245734103417

67. S. V. Nair, L. M. Ramaniah, K. C. Rustagi, Electron states in a quantum dot in an effective-bond-orbital model. Physical Review B 1992451159695979

68. T. V. Buuren, L. N. Dinh, L. L. Chase, W. J. Siekhaus, L. J. Terminello, Changes in the Electronic Properties of Si Nanocrystals as a Function of Particle Size. Physical Review Letters 1998801738033806

69. L. W. Wang, A. Zunger, semiempirical. Local-density-derived, Physical. pseudopotentials, B. Review, 199551241739817416

70. H. Fu, A. Zunger, Local-density-derived semiempirical nonlocal pseudopotentials for InP with applications to large quantum dots. Physical Review B 199755316421653

71. E. Lippens, M. Lannoo, Calculation of the band gap for small CdS and ZnS crystallites. Physical Review B 198939151093510942

72. S. Sapra, R. Viswanatha, D. D. Sarma, An accurate description of quantum size effects in InP nanocrystallites over a wide range of sizes. Journal of Physics D: Applied Physics 2003361315959

73. S. Sapra, N. Shanthi, D. D. Sarma, Realistic tight-binding model for the electronic structure of II-VI semiconductors. Physical Review B 2002205202 EOF

74. J. G. Diaz, G. W. Bryant, Electronic and optical fine structure of GaAs nanocrystals: The role of d orbitals in a tight-binding approach. Physical Review B 2006075329 EOF

75. Ari. Mizel, M. L. Cohen, Electronic energy levels in semiconductor nanocrystals: A Wannier function approach. Physical Review B 1997561167376741

76. M. C. Beard, K. P. Knutsen, P. Yu, J. M. Luther, Q. Song, W. K. Metzger, R. J. Ellingson, A. J. Nozik, Multiple exciton generation in colloidal silicon nanocrystals. Nano Letters 20077825062512

77. D. Schaller, J. M. Pietryga, S. V. Goupalov, M. A. Petruska, S. A. Ivanov, V. I. Klimov, Breaking the Phonon Bottleneck in Semiconductor Nanocrystals via Multiphonon Emission Induced by Intrinsic Nonadiabatic Interactions. Physical Review Letters 2005196401 EOF

78. G. Cullis, L. T. Canham, P. D. J. Calcott, The structural and luminescence properties of porous silicon. Journal of Applied Physics 1997823909965

79. L. Ding, T. P. Chen, Y. Liu, C. Y. Ng, S. Fung, Optical properties of silicon nanocrystals embedded in a SiO2 matrix. Physical Review B 2005

80. Y. Kanemitsu, H. Tanaka, T. Kushida, K. S. Min, H. A. Atwater, Luminescence properties of GaAs nanocrystals fabricated by sequential ion implantation. Journal of Luminesence 2000432 EOF434 EOF

81. M. Nirmal, B. O. Dabbousi, M. G. Bawendi, J. J. Macklin, J. K. Trautman, T. D. Harris, L. E. Brus, Fluorescence intermittency in single cadmium selenide nanocrystals. Nature 19963836603802804

82. M. C. Schlamp, X. Peng, A. P. Alivisatos, Improved efficiencies in light emitting diodes made with CdSe(CdS) core/shell type nanocrystals and a semiconducting polymer. Journal of Applied Physics 1997821158375842

83. X. Jiang, R. D. Schaller, S. B. Lee, J. M. Pietryga, V. I. Klimov, A. A. Zakhidov, Pb, PbSe nanocrystal/conducting polymer solar cells with an infrared response to 2 micron, Journal Materials Research 200722822042210

84. D. Schaller, V. I. Klimov, High Efficiency Carrier Multiplication in PbSe Nanocrystals: Implications for Solar Energy Conversion. Physical Review Letters 2004186601 EOF

85. M. A. Green, generation. Third, solar. photovoltaics, for. cells, 20, beyond, Physica E 200265 EOF70 EOF

86. J. Nozik, Spectroscopy and hot electron relaxation dynamics in semiconductor quantumwells and quantum dots. Annual Review of Physical Chemistry 200152193231

87. Z. Zhou, L. Brus, R. Friesner, Electronic Structure and Luminescence of 1.1- and 1.4-nm Silicon Nanocrystals: Oxide Shell versus Hydrogen Passivation. Nano Letters 200332163167

88. Cherian, C. Gerard, P. Mahadevan, N. T. Cuong, R. Maezono, Size dependence of the bulk modulus of semiconductor nanocrystals from first-principles calculations. Physical Review B 2010235321 EOFhttp%3A%2F%2Fprb.aps.org%2Fabstract%2FPRB%2Fv82%2Fi23%2Fe235321%3Bhttp%3A%2F%2Feprints.iisc.ernet.in%2F35877%2F EOF

89. H. Tolbert, A. P. Alivisatos, The wurtzite to rock salt structural transformation in CdSe nanocrystals under high pressure. Journal of Chemical Physics 19951021146424656

90. J. Jorgensen, E. , J. M. Jakobsen, J. Z. Jiang, L. Gerward, J. S. Olsen, High-pressure X-ray diffraction study of bulk- and nanocrystalline GaN. Journal of Applied Crystallography 2003362920925

91. J. Z. Jiang, Phase transformations in nanocrystals. Journal of Materials Science 20045103 EOF5110 EOF

92. X. Wu, Z. Wu, L. Guoc, C. Liu, J. Liu, X. Li, H. Xu, Pressure-induced phase transformation in controlled shape ZnO nanorods. Solid State Communications 2005780 EOF784 EOF

93. Panchal, S. Ghosh, S. Gohil, N. Kulkarni, P. Ayyub, Non-monotonic size dependence of the elastic modulus of nanocrystalline ZnO embedded in a nanocrystalline silver matrix. Journal of Physics: Condensed Matter 2008345224 EOF

94. W. Lei, D. Liu, J. Zhang, Q. L. Cui, G. T. Zou, Comparative studies of structural transition between AIN nanocrystals and nanowires. Journal of Physics: Conference Series 200812116200618

95. S. Taguchi, A. Ishizumi, T. Tayagaki, Y. Kanemitsu, Mn, Mn-Mn couplings in Mn-doped CdS nanocrystals studied by magnetic circular dichroism spectroscopy. Applied Physics Letters 20099417173101173103

96. K. Ando, K. Takahashi, T. Okuda, M. Umehara, Magnetic circular dichroism of zinc-blende-phase MnTe. Physical Review B 199246191228912297

97. J. K. Furdyna, magnetis. Diluted, semiconductor, Journal of Applied Physics 1988R29R64.

98. S. A. Wolf, A. A. Awschalom, R. A. Buhrman, J. M. Daughton, S. Molnar, M. L. Roukes, A. Y. Chtchelkanova, D. M. Treger, A. Spintronics, Electronics. Spin-Based, for. Vision, Future. the, Science 2001294554614881495

99. D. D. Sarma, R. Viswanatha, S. Sapra, A. Prakash, M. G. Hernandez, Magnetic Properties of Doped II-VI Semiconductor Nanocrystals. Journal of Nanoscience and Nanotechnology 20055915031508

100. S. Taguchi, T. Tayagaki, Y. Kanemitsu, Luminescence and magnetic properties of Co doped ZnO nanocrystals. IOP Conf. Series: Materials Science and Engineering 2009601202914

101. K. Biswas, K. Sardar, C. N. R. Rao, Ferromagnetism in Mn-doped GaN nanocrystals prepared solvothermally at low temperatures. Applied Physics Letters 20068913132503132505

102. Panse, R. Leitsmann, F. Bechstedt, Magnetic interaction in pairwise Mn-doped Si nanocrystals. Physical Review B 2010125205 EOF

103. Huang, A. Makmal, J. R. Chelikowsky, L. Kronik, Size dependent spintronics properties of diluted magnetic semiconductors nanocrystals. Physical Review Letters 2005

104. S. H. Tolbert, A. P. Alivisatos, Size Dependence of a First Order Solid-Solid Phase Transition: The Wurtzite to Rock Salt Transformation in CdSe Nanocrystals. Science 19942655170373376

105. S. H. Tolbert, A. P. Alivisatos, The Wurtzite to Rock-Salt Structural Transformation in Cdse Nanocrystals under High-Pressure. Journal of Chemical Physics 19951021146424656

106. S. H. Tolbert, A. B. Herhold, L. E. Brus, A. P. Alivisatos, Pressure-induced structural transformations in Si nanocrystals: Surface and shape effects. Physical Review Letters 1996762343844387

107. S. B. Qadri, J. Yang, B. R. Ratna, E. F. Skelton, J. Z. Hu, Pressure induced structural transitions in nanometer size particles of PbS. Applied Physics Letters 1996691522052207

108. J. Z. Jiang, J. S. Olsen, L. Gerward, D. Frost, D. Rubie, J. Peyronneau, Structural stability in nanocrystalline ZnO. Europhysics Letters 20005014853

109. H. Wang, J. F. Liu, Y. He, Y. Wang, W. Chen, J. Z. Jiang, J. S. Olsen, L. Gerward, High-pressure structural behaviour of nanocrystalline Ge. Journal of Physics: Condensed Matter 2007156217 EOF

110. Z. Wang, K. Tait, Y. Zhao, D. Schiferl, C. Zha, H. Uchida, R. T. Downs, Size, Size-Induced Reduction of Transition Pressure and Enhancement of Bulk Modulus of AlN Nanocrystals. Journal of Physical Chemistry B 20041083111506l1508

111. S. Li, Z. Wen, Q. Jiang, Pressure-induced phase transition of CdSe and ZnO nanocrystals. Scripta Materialia 2008595526529

112. Q. Cui, Y. Pan, W. Zhang, X. Wang, J. Zhang, T. Cui, Y. Xie, J. Liu, G. Zou, Pressure-induced phase transition in GaN nanocrystals. Journal of Physics: Condensed Matter 200214441104111044

113. Mujica, R. J. Needs, A. Munoz, First-principles pseudopotential study of the phase stability of the III-V semiconductors GaAs and AlAs. Physical Review B 1995521288818892

114. J. Cai, N. Chen, Theoretical study of pressure-induced phase transition in AlAs: From zinc-blende to NiAs structure. Physical Review B 2007174116 EOF

Chapter 3

Rare Earth Ion-Doped Upconversion Nanocrystals: Synthesis and Surface Modification

Hongjin Chang [1], Juan Xie [2], Baozhou Zhao [1], Botong Liu [1], Shuilin Xu [1], Na Ren [1], Xiaoji Xie [1], Ling Huang [1,*] and Wei Huang [1,2,*]

[1]Key Laboratory of Flexible Electronics (KLOFE) and Institute of Advanced Materials (IAM), Jiangsu National Synergistic Innovation Center for Advanced Materials (SICAM), Nanjing Tech University (NanjingTech), Nanjing 211816, China

[2]Key Laboratory for Organic Electronics and Information Displays and Institute of Advanced Materials (IAM), National Synergistic Innovation Center for Advanced Materials (SICAM), Nanjing University of Posts and Telecommunications, Nanjing 210023, China

ABSTRACT

The unique luminescent properties exhibited by rare earth ion-doped upconversion nanocrystals (UCNPs), such as long lifetime, narrow emission line, high color purity, and high resistance to photobleaching, have made them widely used in many areas, including but not limited to high-resolution displays, new-generation information technology, optical communication, bioimaging, and therapy. However, the inherent upconversion luminescent properties of UCNPs are influenced by various parameters, including the size, shape, crystal structure, and chemical composition of the UCNPs, and even the chosen synthesis process and the surfactant molecules used. This review will provide a complete summary on the synthesis methods and the surface modification

strategies of UCNPs reported so far. Firstly, we summarize the synthesis methodologies developed in the past decades, such as thermal decomposition, thermal coprecipitation, hydro/solvothermal, sol-gel, combustion, and microwave synthesis. In the second part, five main streams of surface modification strategies for converting hydrophobic UCNPs into hydrophilic ones are elaborated. Finally, we consider the likely directions of the future development and challenges of the synthesis and surface modification, such as the large-scale production and actual applications, stability, and so on, of the UCNPs.

KEYWORDS

rare earth; nanocrystal; upconversion; synthesis; surface modification

1. INTRODUCTION

On 29 December 1959, Richard Feynman predicted at the annual American Physical Society meeting that: *"if we use a method to control the arrangement of things at a small scale, so that we can get a lot of features beyond imagination, also can see the performance of the material to produce rich change."* The tiny material is what we call nanomaterial today, which is usually referred to as that in the three dimensional space, where at least one dimension (1D) is at the nanoscale range (1–100 nm) or materials are composed of them as a basic unit. As the size of the material decreases to the nanometer scale, the material will show a lot of new features that the bulk counterpart does not possess, such as the small size effect, surface effect of nanomaterials, quantum confinement effect, and macroscopic quantum tunneling effect [1,2,3]. In the past decades, nanomaterials have gained wide attention from all over the world, and have now been widely applied in many fields.

Rare earth (RE) elements possess unique electronic configuration where the 4f electrons are effectively shielded by the closely lied 5s and 5p subshells. Typically, the electron transitions of RE ions are mainly derived from the inner *4f-4f* or *4f-5d*transitions, and thus the spectroscopic properties of RE ions are barely perturbed by the local chemical microenvironment, which imparts RE compounds including the complexes' and nanomaterials' unique spectroscopic characteristics, such as rich energy levels, long luminescence decay time, narrow emission line, and high color purity in contrast to those of quantum dots and organic dyes [4,5,6]. Due to the preserved electron

transitions, the emission wavelengths of each lanthanide ion largely depend on its own electronic configuration, and the combination (at certain ratio) of different lanthanide ions is also widely adopted to realize various luminescent materials with adjustable luminescence for different purposes.

If these RE ions were doped into proper nanocrystals, a series of new luminescent characteristics related with the original features of according RE ions may be observed. In addition, upconversion nanocrystals (UCNPs) have shown high chemical stability, biological compatibility, long luminescence lifetime, and tunable emission wavelength [7,8,9], making them tremendously exploited in biolabeling [10], bio-detection [11,12], bioimaging [13,14], FRET-based sensing [15], drug delivery [16], and volumetric 3D display [17].

Despite the substantially shielded transitions of RE ions, the luminescent properties of UCNPs are also affected by their size, shape, crystal structure, and chemical composition of the materials. For example, reduced particle size will cause increased surface area, which would introduce more defects on the nanoparticle surface and consequently hamper the luminescent efficiencies, though the small nanoparticles are more advantageous to the biological applications. To facile the applications of UCNPs, it is of primary necessity to develop according viable and robust methodologies to synthesize target nanocrystals with desired size, shape, crystal structure, chemical composition, and most importantly, the proper surface functional groups for anticipated applications.

In recent years, various attempts have been reported for synthesizing UCNPs in a controlled manner, including thermal decomposition, thermal coprecipitation, hydro/solvothermal, combustion, microwave, and so on. In the meantime, the nanocrystals, after surface modification, such as SiO_2 encapsulation, polymer encapsulation, ligand oxidation, and ligand exchange, can be easily coupled with DNA, protein, and other functional molecules, and facilitate expected applications. However, there are still some challenges in the synthesis of desired nanocrystals, and the surface modification usually involves extra experimental steps and lowers the luminescence efficiency at certain extent depending on the chosen method. Thus, a proper synthesis method and a suitable strategy for designed surface modification are highly desired. Herein, we attempt to provide a comprehensive overview of the state-of-the-art synthetic methods and the surface modification strategies for UCNPs reported in the past decades.

2. SYNTHETIC APPROACHES

2.1. Thermal Decomposition

The thermal decomposition process, comprising of dissolution of organic and/or inorganic precursors in organic solvent with high-boiling point, is a traditional method for preparation of inorganic nanocrystals. The typical experimental procedure of thermal decomposition method is composed of: (1) a given amount of $RE(CF_3COO)_3$ precursors is added into a mixture of oleic acid (OA), 1-octadecene (OD), and sometimes oleylamine (OM) at room temperature; (2) the solution is heated to 165 °C for 30 min with vigorous magnetic stirring to remove water and oxygen under argon protection; (3) the solution is heated to high temperature (usually >300 °C) for a certain period of time under argon protection and the nanocrytals are then collected from the reaction mixture after cooling down to room temperature. Yan [18,19] first prepared high-monodisperse LaF_3 triangular nanoplates (Figure 1a) and hexagonal SmF_3 nanoparticles (Figures 1b,c) via the thermal decomposition process using $La(CF_3COO)_3$ and $Sm(CF_3COO)_3$ as precursors. Using this method, Capobianco [20] and Nann [21] synthesized $NaYF_4$ nanocrystals with narrow size-distribution. Later, this approach was extended as a common process to synthesize high-quality UCNPs including but not limited to $NaLaF_4$ [22], $NaGdF_4$ [23], $LiYF_4$ [24], KY_3F_{10} [25], $BaYF_5$[26], REOF [19,27,28], and REOCl [29]. The reaction temperature, time, and the molar ratio of OA, OD, and sometimes OM in the reaction mixture have been demonstrated to exert different effects on the final nanocrystals. It should be noted that sometimes OM shall be introduced as a necessary component to adjust the reaction environment so that the final product with different morphologies and dimensions or a brand new product can be obtained [27,28,29].

The most prominent advantages of this method are that the products are of high quality, with pure crystal phase, and strong upconversion emission. However, this method also suffers from disadvantages including: (1) presynthesis of $RE(CF_3COO)_3$ precursors is typically required; (2) the decomposition of trifluoroacetates simultaneously produces toxic fluorinated and oxyfluorinated carbon species and thus careful handling of the reactions in fully ventilated chemical hood is required; (3) the anaerobic and water-free reaction environment further increases the operation difficulty. In the themolysis method, one of the key factors for achieving size-tunable and monodispersed UCNPs requires a proper selection of the coordinating ligands. Yan and co-worker reported that oleylamine ligand is a delicate buffer for F^- ions, the lighter the rare earth, the more OM it requires, owing to the fact that the basicity of the RE oxide gradually decreases along with the increasing of the atomic number of the RE series [19].

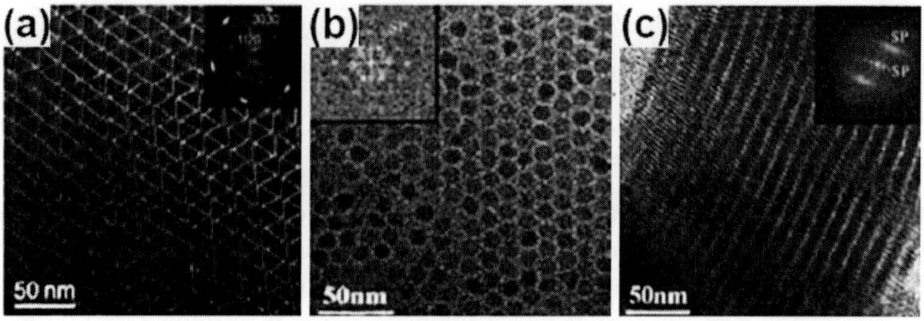

Figure 1. (**a**) TEM image of edge-to-edge super lattices of LaF_3 triangular nanoplates. Reproduced from [18]. Copyright 2005, American Chemical Society. (**b,c**) TEM images of edge-to-edge and face-to-face super lattices of SmF_3 hexagonal nanoplates, respectively. Reproduced from [19]. Copyright 2006, John Wiley and Sons.

2.2. Thermal Coprecipitation

Due to the limitations of the thermal decomposition method, thermal coprecipitation approach is developed and has now been used as one of the most convenient methods for UCNPs synthesis. The experimental procedure is generally composed of: (1) RE salts were mixed with a solution of OA, OD, and sometimes OM, at certain ratio, which was heated to 165 °C for 30 min and then cooled down to room temperature; (2) a methanol solution of NH_4F and AOH (A = Li, Na, K) was added to the mixture and stirred for 30 min; (3) after removal of the methanol and residual water by evaporation, the reaction mixture was heated to high temperature (usually >300 °C) under argon protection, which produces desired nanocrystals. The benefits of the coprecipitation method include operational simplicity, the lack of toxic by-products, and the wide application across various materials.

With continuous efforts, many kinds of UCNPs have been directly synthesized by this method. For example, in 2002, van Veggel and co-workers [30] have synthesized the down conversion LaF_3 nanocrystals doped with RE^{3+} (RE = Eu, Er, Nd, and Ho) ions. This approach was later expanded by Yi and Chow [31], who prepared upconversion LaF_3 nanophosphors with smaller particle size (5 nm). In addition to LaF_3, $LuPO_4$:Yb/Tm, $YbPO_4$:Er, $NaYF_4$:Yb/Er(Tm), $NaGdF_4$:Yb/Er, and $Y_3Al_{15}O_{12}$(YAG):Yb/Tm nanophosphors were synthesized using this method [5,32,33,34]. Recently, our group has synthesized the first Scandium-based fluoride nanocrystals Na_xScF_{3+x}:Ln using the coprecipitation method [35] (Figure 2). Interestingly, the crystal structure evolution from pure monoclinic Na_3ScF_6 to pure hexagonal $NaScF_4$ phase was observed by tuning the ratio of OA and OD (Figure 2a). In addition, the hexagonal $NaScF_4$:Yb/Er

crystals emit strong red upconversion (665 nm) under 980 nm laser excitation, different from those of the traditional hexagonal $NaYF_4$:Yb/Er nanocrystals, which usually emit strong green (547 nm) upconversion luminescence (Figure 2b).

Despite the general usefulness and obvious improvement of the thermal coprecipitation method for UCNPs synthesis, it suffers from the long-time and continuous operation of the experimental process, which usually takes more than 5 h, including the removal of methanol solvent, water generated during the synthesis, and the controlled crystal growth at a certain high-temperature. Moreover, large scale synthesis of UCNPs using this method is still a great challenge.

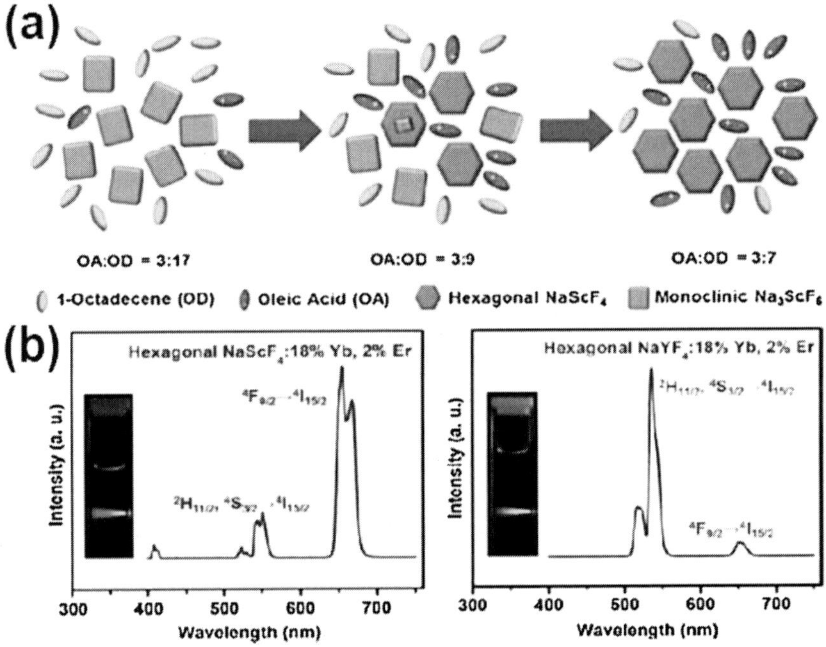

Figure 2. (**a**) Schematic illustration of the crystal structure evolution at varying polarities of the reaction medium. (**b**) Upconversion luminescence spectra of hexagonal-phase $NaScF_4$:Yb/Er and $NaYF_4$:Yb/Er nanocrystals. Reproduced from [35]. Copyright 2012, American Chemical Society.

2.3. Hydro/Solvothermal Synthesis

Hydro/solvothermal is a process of chemical reaction between negative ions and positive ions that usually precipitate from the solvent under high temperature and high pressure, generating nanoscale materials in the solvent after proper processing. This method has now become a widely-employed synthetic approach for UCNPs since it is easy to operate, does not require stringent operation of the experimental process, and moreover, the reaction temperature of hydro/solvothermal method is usually lower than those used for thermal decomposition and thermal co-precipitation synthesis for UCNPs.

The commonly used surfactants for UCNPs preparation include ethylenediamine tetraacetic acid (EDTA) [36,37], cetyltrimethylammonuim bromide (CTAB) [38], trisodium citrate (Na_3Cit) [39], linoleate acid [40], oleic acid [41,42,43,44,45,46]. As early as 2002, Li and coworkers [47,48,49] used RE nitrate and KOH to synthesize $Ln(OH)_3$ nanowires, nanotubes, and nanoparticles (Figure 3a–c). Long and uniform nanobelts were obtained when NaOH/KOH and RE acetates were used (Figures 3d,e). Later on, EDTA was used as complexing agent and CTAB as structure-directing agent to synthesize β-$NaYF_4$:Yb/Er nanotubes, nanorods, and nanospheres [34] (Figure 4a–d). Different from Li's work, Wang and co-workers [50] synthesized upconversion $NaYF_4$ nanocrystals using oleic acid-mediated hydrothermal synthesis (Figure 4e–h).

Recently, via a hybrid thermal decomposition/solvothermal method, LaF_3:Yb/Er/Tm/Ho nanoplates with multicolor upconversion luminescence were synthesized [51,52]. The feasibility of the hydro/solvothermal methods was demonstrated by several groups. For example, Zhang has synthesized $BaYF_4$ [53], $CePO_4$ [54], YPO_4 [55], Gd_2O_3 [56], $GdVO_4$ [57], and Lin's group reported the uniform microstructured YPO_4 [58], YVO_4 [59], LnF_3 [60], and $NaYF_4$ [45] via the sodium citrate assisted hydrothermal route. In recent years, Liu and co-workers reported the synthesis of $KMnF_3$ nanocrystals with only single-band UC emission [61]. Zhao and Hao's group [62,63] represented a strategy for the rationale manipulation of green and red upconversion emission, and the pure red emission of $NaYF_4$:Yb/Er nanocrystals has been achieved by Mn^{2+} doping.

However, the disadvantages of this method are substantially difficult to overcome. Firstly of all, there are too many parameters including the reaction temperature, surfactant type and concentration, reactant concentration, solvent and the composition of it, and reaction time, to consider in order to filter out the most optimized experimental conditions for one specific reaction. Furthermore, the obtained nanocrystals usually have large size distribution, and sometimes the by-product residues stay on the surface of the nanocrystals,

which are difficult to remove. Therefore, it remains challenging to develop a general and facile hydro/solvothermal method for the synthesis of high-quality UCNPs.

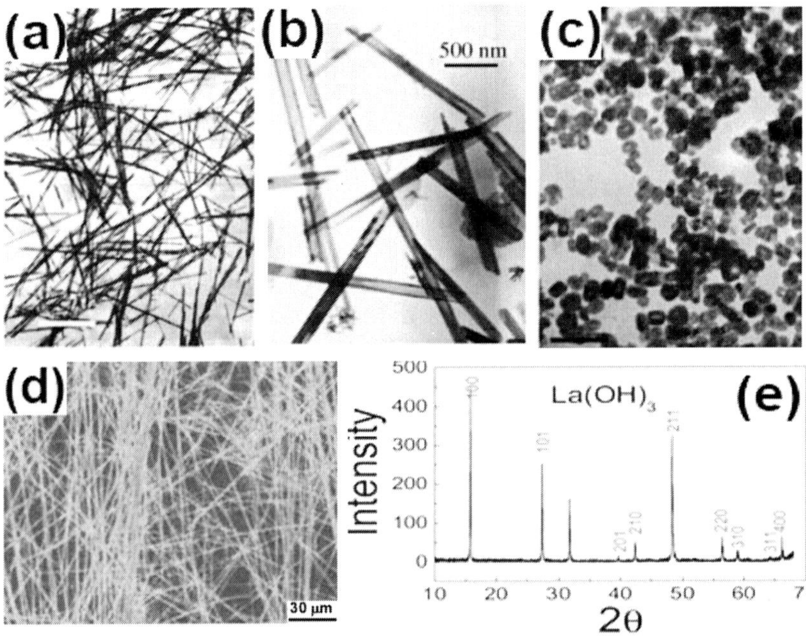

Figure 3. TEM images of (**a**) $La(OH)_3$ nanowires. Reproduced from [47]. Copyright 2002, John Wiley and Sons; (**b**) Y_2O_3 nanotubes. Reproduced from [48]. Copyright 2003, John Wiley and Sons; (**c**) LaF_3 nanoparticles. Reproduced from [49]. Copyright 2003, John Wiley and Sons; (**d**) $La(OH)_3$ nanobelt; (**e**) a typical XRD pattern of the as-synthesized $La(OH)_3$. Reproduced from [50]. Copyright 2007, John Wiley and Sons.

2.4. Sol-Gel Method

Sol-gel process is a typical wet-chemical technique for synthesizing UCNPs, which can be generally divided into three types: (1) sol-gel route based on the hydrolysis and condensation of molecular precursors; (2) gelation route based on condensation of the aqueous solutions containing metal-chelates; and (3) polymerizable complex route [64]. In sol-gel process, a RE nitrate salt or metal alkoxide is generally used as the starting reactants. The reaction is started by mixing the reactants in liquid phase through hydrolysis and condensation

reactions, followed by annealing at high temperature for a certain period of time.

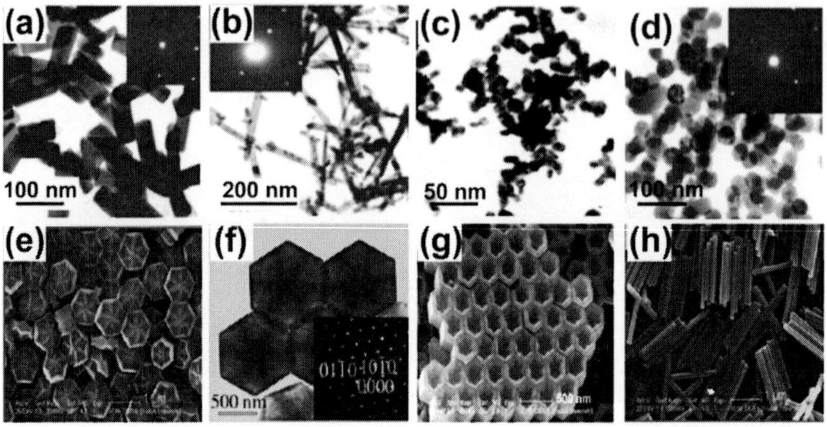

Figure 4. TEM and SEM images of $NaYF_4$:Yb/Er nanocrystals prepared under different hydro/solvothermal conditions. (**a,b**) TEM images of $NaYF_4$:Yb/Er nanocrystals synthesized in acetic acid and ethanol in the presence of CTAB, respectively; (**c,d**) TEM images of $NaYF_4$:Yb/Er nanocrystals using EDTA in acetic acid and ethanol, respectively. Reproduced from [34]. Copyright 2005, John Wiley and Sons; SEM images of (**e,f**) flower-patterned hexagonal disks; (**g**) hexagonal nanotubes; and (**h**) nanorods of β-$NaYF_4$. Reproduced from [51]. Copyright 2007, John Wiley and Sons.

In 2002, Prasad *et al.* [65] developed a sol emulsion-gel method that produces Er^{3+} doped ZrO_2 nanophosphors. In order to reduce the segregation of particles and ensure compositional homogeneity, Lin and co-workers fabricated an inorganic YVO_4:Eu thin film phosphor by combining the pechini-type sol-gel process with inkjet printing. The mixed solution of metal salt precursors, citric acid, and poly(ethylene glycol) was directly used as ink to deposit patterns on ITO-coated glass substrate. After calcination at 600 °C in air, the YVO_4:Eu patterns at the micrometer-scale were formed on the substrate [66]. In another report, Song and co-works successfully fabricated many kinds of inverse opal photonic crystals (PCs) by the sol-gel method with a PMMA latex sphere template, including YVO_4:Dy [67], TiO_2:Sm [68], YBO_3:Eu [69], and $LaPO_4$:Ce,Tb [70]. In addition, the sol-gel process was also developed for synthesizing various UCNPs using metal oxides as host materials such as TiO_2:Er, $BaTiO_3$:Er (Figure 5), $Lu_3Ga_5O_{12}$:Er, and YVO_4:Yb/Er [71,72,73].

For the sol-gel method, the annealing procedure (temperature and time) is the key step in the preparation process, which can seriously determine the quality of the samples. It should be noted that, although the sol-gel method can be used for large-scale production and the products usually offer high luminescence intensity due to the high crystallinity formed at high annealing temperature, the sol-gel derived nanocrystals generally have broad particle size distribution, irregular morphology, and are insoluble in water, which compose the shortcomings of this method.

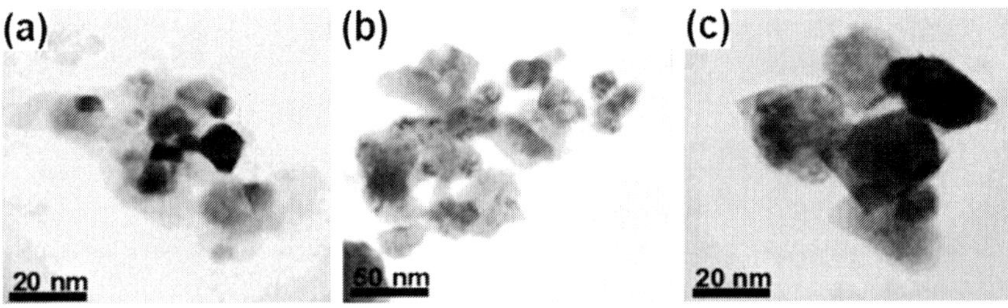

Figure 5. TEM images of the 1.0 mol% Er^{3+}-doped $BaTiO_3$ nanoparticles obtained after heating to three different temperatures: (**a**) 700 °C; (**b**) 850 °C; and (**c**) 1000 °C. Reproduced from [71]. Copyright 2003, American Chemical Society.

2.5. Microemulsion

To synthesize nanomaterials via the microemulsion method, it usually requires surfactant, co-surfactant, organic solvent, water, and the initial reagents. In a typical microemulsion solution, amphiphilic surfactants form a monolayer at the oil-water interface, with the hydrophobic tails of the surfactant molecules dissolved in the oil phase and the hydrophilic head groups in the aqueous phase. Preparation of RE fluoride nanomaterials usually needs two separate microemulsion systems and the two microemulsions containing RE ions and fluorine ions, respectively, get mixed to initiate the reaction. Over the past few years, Lemyre and coworkers [74] have reported that upconversion YF_3 nanoparticles can be prepared in a reverse water-in-cyclohexane microemulsion system stabilized by polyoxyethylene isooctylphenyl ether (IGEPAL CO-520 or NP-5was supplied by Aldrich) (Figure 6a). Qin and coworkers [75,76] prepared YF_3 upconversion nanophosphors, using CTAB and 1-pentanol instead of NP-5 (Figure 6b).

Figure 6. TEM images of (**a**) YF_3 nanophosphors prepared in the NP-5 stabilized microemulsion system. Reproduced from [74]. Copyright 2005, American Chemical Society. (**b**) YF_3 nanobundles synthesized in the CTAB and NP-5 stabilized microemulsion system. Reproduced from [75]. Copyright 2008, American Chemical Society.

The use of microemulsion has many advantages, such as the low-cost for equipment, easy operation, the small size of the UCNPs, and the controlled morphology of products by adjusting the dosage of the surfactant, solvent, as well as the aging time. However, this technique has many challenges including the small amount of products generated, difficulty of sample separation and narrow scope synthesis. More importantly, it is difficult to achieve massive production to meet industry requirements.

2.6. Combustion Synthesis

Compared to sol-gel and hydro/solvothermal synthesis, combustion method for UCNPs synthesis can be finished in a short period of time. The process of combustion synthesis is an oxidation-reduction reaction in essence. Metal nitrates are usually selected as oxidizer and the source for metal ions while the organic compounds as reducing agent and fuel. There are two requirements for organic fuel selection: (1) the reaction occurred between the fuel and nitrate must be relatively mild, producing nontoxic gases; (2) it is better to select the fuels that can complex with the metal ions, enhance the solubility of metal ions, and prevent the separation by crystallization of the metal salts in the precursor solution.

During the combustion synthesis process, combustion wave spreads the reaction materials in a self-sufficient situation without requiring extra heat in the process of the whole reaction. This time- and energy-saving method was

used to synthesize oxide and oxysulfide nanomaterials. For example, Capobianco, Luo, and Zhang's groups have respectively synthesized a variety of oxide and oxysulfide nanophosphors (Y_2O_3, $Gd_3Ca_5O_{12}$, La_2O_2S, and Gd_2O_3) [77,78,79,80] via this method.

2.7. Flaming Synthesis

Flaming synthesis is another powerful method for producing RE oxide nanomaterials, which can be divided into four stages: (1) precursor reaction; (2) nucleation; (3) growth and polymerization; and (4) ion deposition. Notably, the flaming synthesis differs from the typical combustion synthesis that all reactions take place in the gas-phase and form fine powders. The core advantages of the technique are time-saving and low-cost. In 2007, Ju [81] reported the synthesis of Y_2O_3:Yb/Er(Tm, Ho) nanophosphors using this method. However, since the oxidation reaction happens along with the flaming process, this method is mainly limited to oxide nanomaterial synthesis, and it is almost impossible to synthesize other types of UCNPs, such as fluoride, phosphate, vanadate, and so on. It should also be noted that the flaming synthesis offers an opportunity for large-scale synthesis of RE doped oxides.

2.8. Electrospinning

During the electrospinning process, the precursor solution is spun through four stages (cocoons, stretching, refinement, and curing) under the effect of high voltage electrostatic field. Simple operation, good repeatability, and wide application scope are the advantages of this approach. For example, Song [82] have reported $NaYF_4$/PVP composite nanofibers, with diameter in the range of 300–800 nm, prepared through electrospinning. Nevertheless, this method is still in its infantile stage. New recipes for the synthesis of uniform and small-sized (<20 nm) nanoparticles with controllable morphology are still required.

2.9. Microwave Synthesis

Microwave synthesis contains solid microwave and liquid microwave method. The former mixes up RE oxides with ammonium bifluoride and ammonium fluoride, and the nanoparticles were directly synthesized via microwave. The latter dissolves the RE salts and the fluoride source into solvent. Then, these raw materials react with each other when the solvent is heated by microwave.

The method was extended to produce PrF_3 hemispheres with diameter of about 31 nm [83].

The advantage of microwave synthesis is generally composed of: (1) it can selectively heat the samples to high temperature while the rest of the microwave device remains at room temperature; (2) microwave can heat the reaction system uniformly, cause less side reaction and the product is relatively simple; (3) fast heat and low energy consumption; (4) improve the structure and properties of synthetic material by adding proper surfactant.

Besides the above synthetic strategies elaborated, combinatorial approaches have also been employed to produce micro- or nano-scale RE fluoride crystals, which after proper lanthanide doping, can upconvert near-infrared light to visible frequencies, enabling the applications of such materials to biological imaging, telecommunication, and solar energy conversion [84]. For example, Liu [85] have synthesized $NaYF_4$:Yb,Er upconversion nanocrystals with characteristic upconversion luminescence spectra in a continuous capillary. Later, the research group improved the condition of the reaction prevented the growth of β-$NaYF_4$ [86]. On the other side, Zhu *et al.* synthesized LaF_3 [87] and $LaPO_4$ [88] nanoparticles doped with Ce^{3+} and Tb^{3+} using microcapillary flow reactors heated with microwaves. However, they also suffer from the wide quality variation of the UCNPs synthesized from different vessels and the large-scale production.

3. SURFACE MODIFICATION OF UCNPS

Because of the influence of impurities and lattice defects on the synthesized UCNPs, the quantum yield of UCNPs is lower than the corresponding bulk materials. In addition, the UCNPs are mostly insoluble in water since they are prepared from the organic environment and usually surrounded by hydrophobic surfactant molecules. Thus, it is important to develop appropriate strategies to make them hydrophilic and in the meantime maintain their upconversion efficiency to satisfy various purposes. For example, the ideal luminescent nanocrystals used for biocompatible purposes should meet several requirements, including: (1) high luminescence efficiency and low background noise; (2) good solubility and stability in biological environment; (3) good biological compatibility; and (4) proper size (below 100 nm).

However, a notorious weak point of UCNPs is the inherent low upconversion efficiency. Currently, the urgent task is to improve the upconversion efficiency, and there have been several groups working on this topic via the core-shell strategy.

The introduction of an inert crystalline shell of an undoped material around each doped nanocrystal provides an effective option to improve the luminescence efficiency of UCNPs. The shell usually has the same composition as the core host crystal, which can effectively reduce the surface quenching effect. In such structures, all dopant ions are confined in the interior core of the nanocrystals, effectively suppressing the non-radiative energy transfer from RE ions to the surface quenching sites, which results in improved upconversion luminescence efficiency. A significant demonstration was made by Yi and Chow [89] who reported a luminescence enhancement of nearly 30 times on 8 nm $NaYF_4$:Yb/Tm nanocrystals coated with a 1.5 nm thick $NaYF_4$ shell. Later on, Chen and co-works exploited a strategy to achieve dual-mode luminescence from identical Eu^{3+} ions in monodisperse hexagonal-phase $NaGdF_4$ nanocrystals that consist of the $NaGdF_4$:Yb/Tm core and the $NaGdF_4$:Eu shell. Typical red downconversion luminescence of Eu^{3+} has been detected via the sensitization of Gd^{3+} ions. By using Yb^{3+} and Tm^{3+} embedded in the cores as double sensitizers, intense upconversion luminescence of Eu^{3+} in the shells can be achieved in $NaGdF_4$:Yb/Tm@$NaGdF_4$:Eu core-shell nanocrystals upon excitation at 976 nm. The upconversion intensity of Eu^{3+} in core-shell nanocrystals is found about one order of magnitude higher than the counterparts of the triply-doped core only, due to the inhibition of the deleterious cross-relaxations between Tm^{3+} and Eu^{3+} ions in core-shell nanocrystals that are reasonably separated in space [90].

Recently, Capobianco [91] proposed a strategy to significantly enhance the intensity of the upconversion by employing a novel $NaGdF_4$:Yb/Er@$NaGdF_4$:Yb active-core@active-shell architecture (Figure 7). The active-shell serves two purposes: (i) minimize the quenching centers and channels; and (ii) transfer absorbed near infrared energy to the luminescence centers (emitters).

Utilizing the core-shell strategy, Zhang and co-workers [92] have broken through the well accepted upper limit of the concentration quenching threshold, e.g., from ~2 mol% to 5 mol% for Er^{3+}, by a designed multi-layer core-shell structure, which contains four parts: the illuminating core ($NaYF_4$:Yb/Er), the first separating shell ($NaYF_4$:Yb), the second illuminating shell ($NaYF_4$:Yb/Er) and the final inactive shell ($NaYF_4$:Yb). The separating layer effectively inhibits the energy transfer process between the Er^{3+} ions in the inside and outside layers and largely reduces the possibility of excitation energy trapping by defects, resulting in an effective upconversion luminescence enhancement [92]. Other core-shell examples include CeF_3:Tb@LaF_3 [93], $NaYF_4$:Yb/Er@$NaYF_4$ [89,94], $NaGdF_4$:Yb/Er@$NaGdF_4$ [95], and KYF_4@KYF_4: Yb/Er [96]. Liu [97] introduced a rationale core-shell strategy that provides precise control over the concentration of dopant in the core and shell layers of nanoparticles. A small amount of Nd^{3+} ions is doped into the core, while a high concentration of Nd^{3+} ions (~20 mol%) is selectively doped in the shell layers for

effective harvesting of light at 800 nm. Chen and co-workers developed a facile strategy based on successive layer-by-layer (LBL) injection of shell precursors for the synthesis of $LiLuF_4:Ln^{3+}$ core-shell UCNPs [98]. It is reported that the size of nanocrystals can be tuned by adjusting the amount of NH_4F (the more the amount of NH_4F solution used, the bigger the size of crystals) and the shell thickness of core-shell nanocrystals can also be controlled by the volume of the core [99]. Besides, Yan has proved in the structure of $NaYF_4$:Yb,Er@CaF_2 that the thickness of the CaF_2 shell were controlled by adjusting the [Ca]/[RE] ratio [100]. Although this approach provides tunable emission intensity in nanocrystals, the luminescence quantum yields of the nanocrystals are limited due to the weak ligand fields and high energy oscillations. Further improvement of the luminescence efficiency of UCNPs can be expected through the controlled growth of a rationally designed crystalline inner shell. Generally speaking, such core-shell structure improves the optical properties of the nanocrystals, but usually does not change the chemical functional group binding on the surface. So, proper surface modification is needed in order to make practical applications of UCNPs, especially for bio-related purposes.

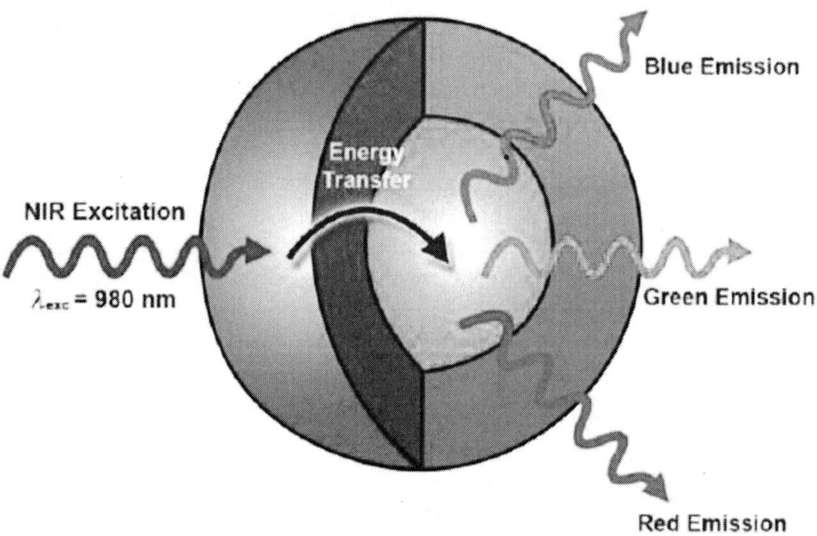

Figure 7. Schematic illustration of the active core-active shell nanoparticle architecture showing the absorption of NIR near infrared light by the Yb^{3+}-rich shell (red) and subsequent energy transfer to the Er^{3+},Yb^{3+}-doped core (green), which leads to upconverted blue, green, and red emissions. Reproduced from [91]. Copyright 2009, John Wiley and Sons.

3.1. SiO_2 Encapsulation

SiO_2 encapsulation involves the growth of an amorphous silica shell on the UCNPs core. Due to the rich –OH groups on the SiO_2 coated UCNPs, it becomes possible to further functionalize the UCNPs with expected chemical groups such as $–NH_2$, –COOH, polymers, and more importantly a wide variety of biomolecules. Nann and Capobianco [101] obtained surface-functionalized YF_3 nanocrystals using this technique. One year later, Zhang and coworkers [102] succeeded in growing a silica shell with adjustable thickness in the range of 2–10 nm on the surface of PVP stabilized cubic $NaYF_4$:Yb/Er/Tm nanocrystals. Later on, they used the microemulsion method to coat $NaYF_4$ nanocrystals with SiO_2, which resulted in monodispersed SiO_2-coated UCNPs (Figure 8) [103].

Alternatively, Song and co-workers [104] showed the conversion of hydrophobic $NaYF_4$:Yb/Er upconversion nanophosphors into hydrophilic ones by amphiphilic silane modification with ultrathin thickness (~1 nm) at room temperature. In this strategy, the coating layers can also provide the possibility for loading $Eu(TTA)_3(TPPO)_2$ complex with downconversion luminescence where they realized the dual mode temperature sensing and dual mode cell imaging within the physiological environment [104]. There are other successful examples to modify the surface of the UCNPs [105,106,107,108] via the SiO_2 encapsulation method, such as LaF_3:Ln@SiO_2, $NaYF_4$:Yb/Er/Fe_3O_4@SiO_2, $NaYF_4$:Er,Tm,Ho@SiO_2, NdF_3@SiO_2, and $NaYF_4$@SiO_2@ quantum dots [103].

The SiO_2 coating provides a versatile platform for multi-functionalization of UCNPs. Nevertheless, the method is time consuming and difficult for large scale synthesis. On the other hand, the coated SiO_2 layer may also affect the luminescence intensity of the UCNPs through light scattering, which prevents it from becoming the ideal surface modification recipe because of the already low upconversion efficiency of the UNCPs. Thus, there is still much work to do to improve the general applicability of this technique.

3.2. Polymer Encapsulation

Polymer encapsulation typically involves the absorption of an additional amphiphilic polymer onto the nanophosphor surface through the hydrophobic-hydrophobic attraction between the original surfactant molecules and the hydrocarbon chains of the polymer. Using this strategy, Chow and co-workers successfully coated a polyacrylicacid (PAA) layer on $NaYF_4$:Yb/Er@$NaYF_4$ core-shell nanoparticles [89]. The hydrophobic core-shell nanoparticles were rendered hydrophilic by amphiphilic PAA coating.

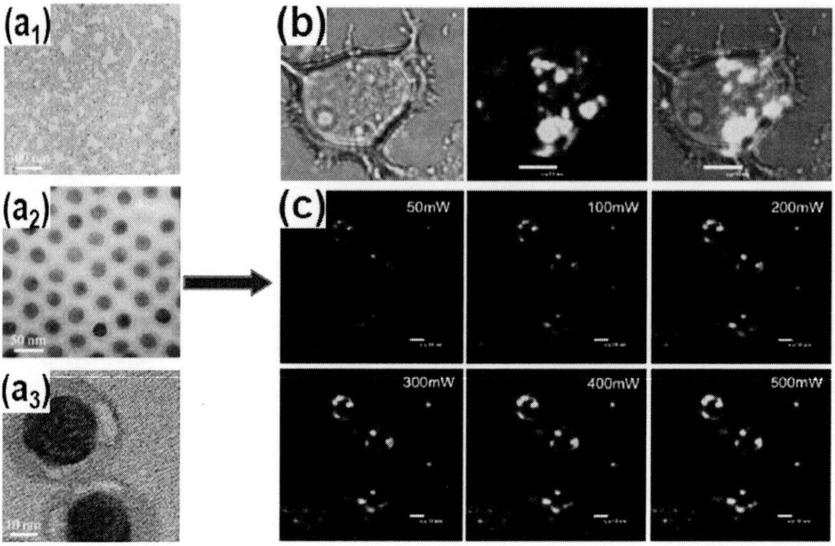

Figure 8. Silica-coated $NaYF_4$:Yb/Er nanocrystals and their application for cell imaging. (**a**1–C**a**3) TEM images of silica-coated $NaYF_4$:Yb/Er UCNPs upconversion nanoparticles at different magnifications; (**b**) Confocal fluorescence image of MCF-7 cells using silica-coated $NaYF_4$:Yb/Er nanospheres (Left: bright-field, middle: upconversion image under 980 nm excitation, and Right: superimposed images of MCF-7 cells incubated with the nanoparticles for 24 h); (**c**) Confocal fluorescence images of MCF-7 cells with the nanospheres, excited by a 980 nm laser with different power intensities. Reproduced from [103]. Copyright 2008, John Wiley and Sons.

Through amphiphilic coating, the multi-functionalization of upconversion nanoparticles could also be realized. For instance, Feng and co-workers fabricated upconversion detection nanocomposites, which were formed by coating the amphiphilic polymer (C18PMH-PEG) on the $NaYF_4$:Yb/Er,Tm nanophosphors based on the hydrophobic–hydrophobic interaction. The polymer modified nanoparticles were then assembled for the selective luminescence detection of mercury ions in water. Using the ratiometric upconversion luminescence emission as a detection signal, the detection limit of $Hg^{2+}$for this nanoprobe in aqueous solution is 8.2 ppb, which is much lower than that (329 ppb) determined by the UV/Vis technology [109]. Liu modified UCNPs with a polyethylene glycol (PEG) grafted amphiphilic polymer, via a hydrophobic interaction-based supramolecular chemistry strategy, for targeted intracellular drug delivery and UCL imaging. This work reveals the great potential of UCNPs for multifunctional drug delivery and biomedical imaging applications (Figure 9) [110]. In contrast to the conventional methods, K.

Prud'homme group [111] reported the successful preparation of colloidal UCNPs stable in buffers and serum media (Leibovitz L-15 media with added fetal bovine serum) using FNP and PEG surface coatings. These polymer-modified UCNPs provide promising new materials for applications in bioimaging and photodynamic therapy.

The polymer-coating method is easier when compared to SiO_2 encapsulation. The *in-situ* coating of polymers on the surface of UCNPs sheds light on the development of multi-functional upconversion nanoparticles. However, the coating through hydrophobic interaction is not stable, and better strategies for a robust polymer coating are needed to satisfy various application requirements.

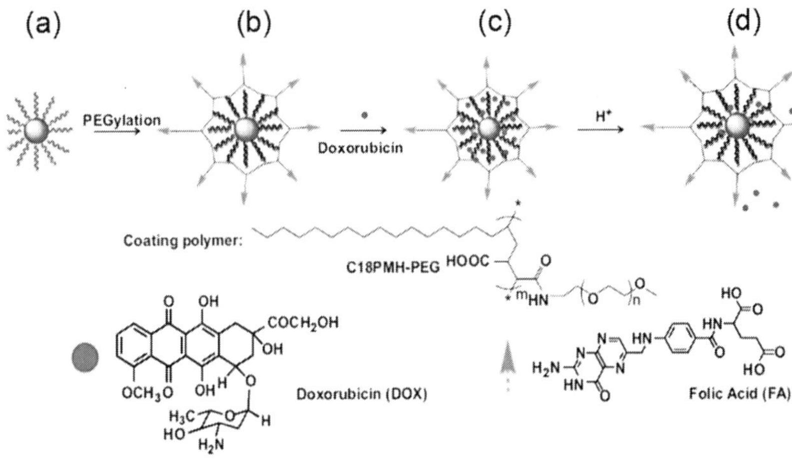

Figure 9. Schematic illustration of the UCNP-based drug delivery system. (**a**) As-synthesized oleic acid capped UCNPs; (**b**) C18PMH-PEG-FA functionalized UCNPs; (**c**) DOX loading on UCNPs. DOX molecules are physically adsorbed into the oleic acid layer on the nanoparticle surface by hydrophobic interactions; (**d**) Release of DOX from UCNPs triggered by decreasing pH. Reproduced from [110]. Copyright 2011, Elsevier Ltd.

3.3. Ligand Oxidation

The ligand oxidation technique involves oxidation of the native ligands that contain unsaturated carbon-carbon bonds to generate a pendant hydrophilic functional group. It is reported that a large number of carboxylic acid groups can be generated on the UCNPs surface after oxidation, which not only renders

the nanocrystals good solubility in water, but also provides tailorability to biological molecules via direct chemical coupling (Figure 10). It is also reported that the oxidation process has no obvious negative effect on the shape, crystal structure, chemical composition, and luminescence properties of the upconversion nanomaterials [112]. However, this method is limited only to those ligands that contain unsaturated carbon-carbon bonds. The oxidation process may also cause ligand removal due to the harsh experimental conditions. It would be ideal if the modulation of ligand could be achieved under mild experimental conditions, which cause no or less damage to the surfaces of the UCNPs. Biomedical research faces a major challenge in that the lengthy period of oxidation may lead to the excessive formation of the brown MnO_2 side product, which is not easy to separate and further weakens the upconversion fluorescence [23]. So, the surface functionalization of UCNPs to render these nanoparticles dispersible in aqueous media is very important. Capobianco and co-workers show that the Ln-OA (Ln = lanthanide; OA = oleate) surface of the Ln-UCNPs is replaced by Ln-OH whose state of charge can be tuned by pH and the efficiency of upconversion luminescence can be enhanced by replacing OH with OD. Furthermore, they have studied the effect of different acids, *i.e.*, HCl, H_3PO_4, and HF on the surface properties of the oleate-free Ln-UCNPs [113].

3.4. Ligand Exchange

Ligand exchange involves the displacement of original hydrophobic ligands that have weak coordination with the RE ions on the surface of UCNPs, by ligands that have stronger coordination capability (with the RE ions) and hydrophilic functional groups. As an early example, Chow and co-workers [114] have demonstrated the preparation of water-soluble $NaYF_4$:Yb/Er nanoparticles with oleylamine ligands via the ligand exchange method (Figure 11). A large variety of ligands have been reported, including poly(acrylicacid) (PAA) [115], poly(ethyleneglycol) (PEG)-phosphate [116], mercaptopropionic acid (MPA) [117], citrate [118]. Li's research group [119] demonstrated a new generation of ^{18}F-labeled lanthanide nanoparticles of $NaYF_4$ co-doped with $Gd^{3+}/Yb^{3+}/Er^{3+}$ as multimodality nanoprobes for PET, MR and UCL imaging. The presence of Yb^{3+} and Er^{3+} co-doped in the $NaYF_4$ nanoparticles gives rise to intense UCL emission in the visible region for luminescent imaging, and 60% Gd^{3+} doping provides the paramagnetic relaxation for MRI. Successful labeling of the lanthanide nanoparticles with ^{18}F gave a product suitable for PET imaging [119]. It is noted that after the ligand exchange, most of these commonly used ligand molecules on the UCNP surface carry additional functional groups to facilitate further biofunctionalization and bioconjugation. However, the exchange efficiency of the technique is difficult to evaluate since the surfactant

molecules cannot be completely replaced by the oleylamine molecules, and the exchange process is also affected by the pH value of the solution, the concentrations of both the surfactant and the oleylamine molecules, and even the ionic strength.

Besides the above surface modification strategies, Prabhas V. Moghe and coworkers [120] have developed a new approach to rendering $NaYF_4$:Yb,Er nanoparticles stable in coacervated HSA nanoshells functionalized with cyclic arginine-glycine-aspartic acid (cRGD) tripeptide. They observed that the composite particles were highly biocompatible *in vitro*, capable of selectively targeting cancerous cell lines exhibiting higher expression of cancer-specific integrin markers, and amenable to fluorescence imaging with high fidelity.

4. CONCLUSIONS

In summary, this review has summarized the recent development of UCNPs with emphasis on the synthetic methods and the surface modification strategies. With careful control of the reaction conditions, we can now obtain high quality, monodispersed UCNPs with various chemical components. However, each individual synthetic strategy has their unique advantages associated with substantial shortcomings. Our subsequent goal is to develop a general synthetic approach which could meet the requirements of large scale synthesis and multifunctionalities for UCNPs.

In parallel, the controlled surface modification method now can produce UCNPs with high colloidal stability, biocompatibility, and tailorable chemical functionalities. The above discussion has also demonstrated that the rapid development of surface modification method facilitates the applications of UCNPs in detection, bioimaging, therapy, and solar cells. Despite the rapid development, these exiting methods still can only be used in a narrow scope. Strategies with universal comparability for various UCNPs may be the next goal of this field. Alternatively, methods that provide easy surface modulation capability for certain types of nanocrystals could be another option for multifunctionalization.

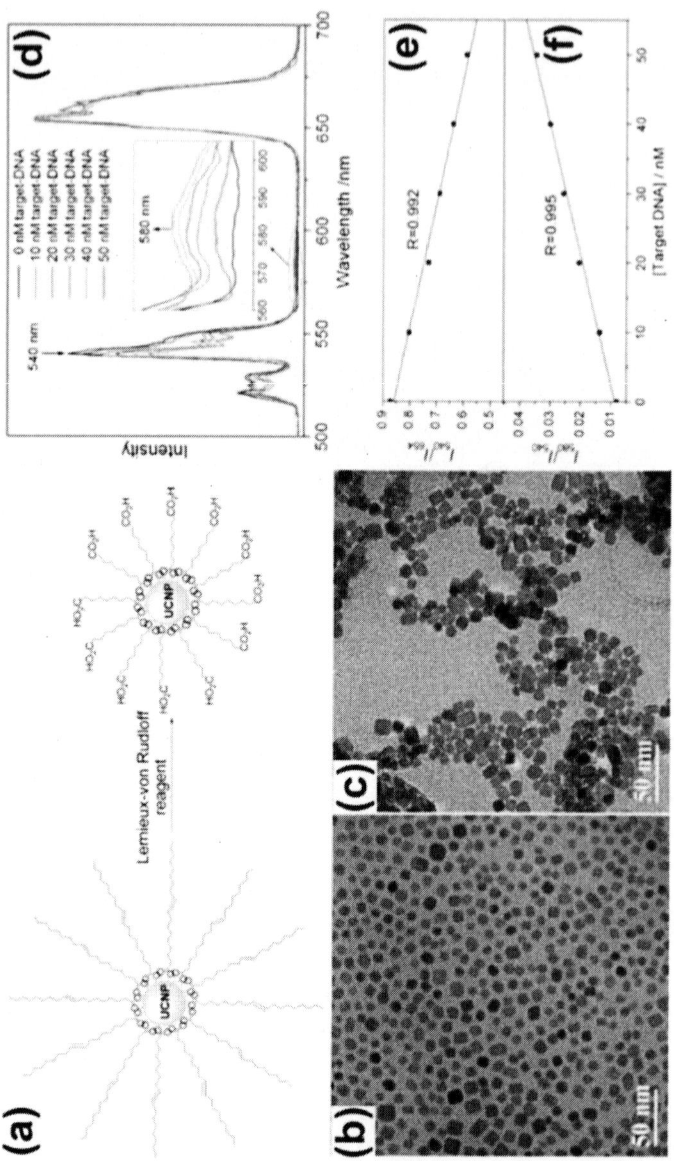

Figure 10. (**a**) Schematic illustration of the ligand oxidization process; (**b,c**) TEM images of the $NaYF_4$:Yb/Er nanoparticles before and after ligand oxidization, respectively; (**d**) Luminescence spectra of a mixture of streptavidin-functionalized $NaYF_4$:Yb/Er nanoparticles, capture-DNA, and reporter-DNA in the presence of different concentrations of target-DNA under continuous-wave excitation at 980 nm. The linear relationships between target-DNA concentration and the intensity ratios of (**e**) I540/I654 and (**f**) I580/I540. Reproduced from [112]. Copyright 2008, American Chemical Society.

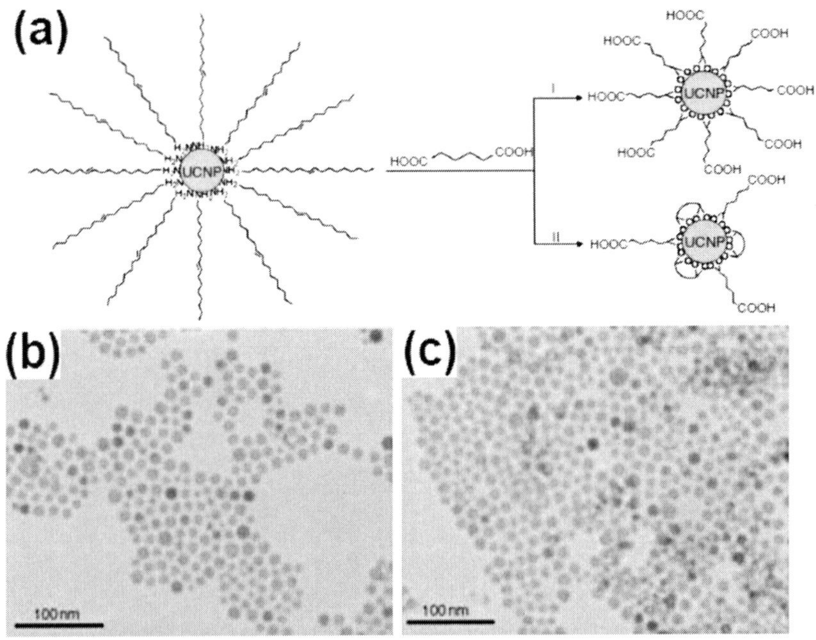

Figure 11. (**a**) Schematic illustration showing the ligand-exchange reactions on OM-stabilized upconversion nanophosphors. (**b,c**) TEM images of $NaYF_4$:Yb/Er nanophosphors prior to and after ligand exchange reactions, respectively. Reproduced from [114]. Copyright 2006, John Wiley and Sons.

ACKNOWLEDGEMENT

We thank the financial support from the National Natural Science Foundation of China (No.: 21301090) and the Natural Science Foundation of Jiangsu Province (No.: BK20130923).

AUTHOR CONTRIBUTIONS

Hongjin Chang, Ling Huang, and Wei Huang proposed the structure of the manuscript and wrote the paper. Juan Xie, Baozhou Zhao, Botong Liu, Shuilin Xu, and Na Ren helped to search for the reference and copyright collection. Xiaoji Xie performed the manuscript editing and language coloring.

CONFLICTS OF INTEREST

The authors declare no conflict of interest.

REFERENCES

1. Law, M.; Goldberger, J.; Yang, P.D. Semiconductor nanowires and nanotubes. Annu. Rev. Mater. Sci. 2004, 34, 83–122.

2. Yoffe, A.D. Low-dimensional systems: Quantum size effects and electronic properties of semiconductor microcrystallites (zero-dimensional systems) and some quasi-two-dimensional systems. Adv. Phys. 2002, 42, 173–266.

3. Su, V.C.; Chen, P.H.; Lin, R.M.; Lee, M.L.; You, Y.H.; Ho, C.I.; Chen, Y.C.; Chen, W.F.; Kuan, C.H. Suppressed quantum-confined stark effect in InGaN-based LEDs with nano-sized patterned sapphire substrates. Opt. Exp. 2013, 21, 30066–30071.

4. Wang, F.; Liu, X.G. Recent advances in the chemistry of lanthanide-doped upconversion nanocrystals. Chem. Soc. Rev.2009, 38, 976–989.

5. Yi, G.S.; Lu, H.C.; Zhao, S.Y.; Yue, G.; Yang, W.J.; Chen, D.P. Synthesis, characterization, and biological application of size-controlled nanocrystalline NaYF4:Yb,Er infrared-to-visible up-conversion phosphors. Nano Lett. 2004, 4, 2191–2196.

6. Carlos, L.D.; Ferreira, R.A.; Ribeiro, S.J. Lanthanide-containing light-emitting organic-inorganic hybrids: A bet on the future. Adv. Mater. 2009, 21, 509–534.

7. Zhou, J.; Liu, Z.; Li, F.Y. Upconversion nanophosphors for small-animal imaging. Chem. Soc. Rev. 2012, 41, 1323–1349.

8. Wang, F.; Han, C.M.; Li, F.Y. Upconversion-nanophosphor-based functional nanocomposites. Adv. Mater. 2013, 25, 5287–5303.

9. Chen, J.; Guo, C.R.; Wang, M.; Wang, L.P.; Mao, C.B.; Xu, S.K. Controllable synthesis of NaYF4:Yb,Er upconversion nanophosphors and their application to in vivo imaging of Caenorhabditis elegans. J. Mater. Chem. 2011, 21, 2632–2638.

10. Wang, F.; Han, Y.; Liu, X.G. Simultaneous phase and size control of upconversion nanocrystals through lanthanide doping. Nature 2010, 463, 1061–1065.

11. Hampl, J.; Hall, M.; Mufti, N.A.; Yao, Y.M.; Mac Queen, D.B.; Wright, W.H.; Cooper, D.E. Upconverting phosphor reporters in immunochromatographic assays. Anal. Biochem. 2001, 288, 176–187.

12. VandeRijke, F.; Zijlmans, H.; Li, S.; Vail, T.; Raap, A.K.; Niedala, R.S.; Tanke, H.J. Up-converting phosphor reporters for nucleic acid microarrays. Nat. Biotechnol. 2001, 19, 273–276.

13. Lim, S.F.; Riehn, R.; Ryu, W.S.; Khanarian, N.; Tung, C.-K.; Tank, D.; Austin, R.H. In vivo and scanning electron microscopy imaging of upconverting nanophosphors in Caenorhabditis elegans. Nano Lett. 2006, 6, 169–174.

14. Chatterjee, D.K.; Rufaihah, A.J.; Zhang, Y. Upconversion fluorescence imaging of cells and small animals using lanthanide doped nanocrystals. Biomaterials 2008, 29, 937–943.

15. Zhang, P.; Rogelj, S.; Nguyen, K.; Wheeler, D. Design of a highly sensitive and specific nucleotide sensor based on photon upconverting particles. J. Am. Chem. Soc. 2006, 128, 12410–12411.

16. Gai, S.L.; Yang, P.P.; Li, C.X.; Wang, W.X.; Dai, Y.L.; Niu, N.; Lin, J. Nanomorphology and charge generation in bulk heterojunctions based on low-bandgap dithiophene polymers with different bridging atoms. Adv. Funct. Mater. 2010,20, 1166–1172.

17. Hinklin, T.R.; Rand, S.C.; Laine, R.M. Transparent, polycrystalline upconverting nanoceramics: Towards 3-D displays.Adv. Mater. 2008, 20, 1270–1273.

18. Zhang, Y.W.; Sun, X.; Si, R.; You, L.P.; Yan, C.H. Single-crystalline and monodisperse LaF3 triangular nanoplates from a single-source precursor. J. Am. Chem. Soc. 2005, 127, 3260–3261.

19. Sun, X.; Zhang, Y.W.; Du, Y.P.; Yan, Z.G.; Si, R.; You, L.P.; Yan, C.H. From trifluoroacetate complex precursors to monodisperse rare-earth fluoride and oxyfluoride nanocrystals with diverse shapes through controlled fluorination in solution phase. Chem. Eur. J. 2007, 13, 2320–2332.

20. Boyer, J.C.; Cuccia, L.A.; Capobianco, J.A. Synthesis of colloidal upconverting NaYF4:Er3+/Yb3+ and Tm3+/Yb3+monodisperse nanocrystals. Nano Lett. 2007, 7, 847–852.

21. Wang, H.Q.; Thomas, N. Monodisperse upconverting nanocrystals by microwave-assisted synthesis. ACS Nano 2009,3, 3804–3808.

22. Yi, G.S.; Lee, W.B.; Chow, G.M. Synthesis of LiYF4, BaYF5, and NaLaF4 optical nanocrystals. J. Nanosci. Nanotechnol.2007, 7, 2790–2794.

23. Naccache, R.; Vetrone, F.; Mahalingam, V.; Capobianco, J.A. Controlled synthesis and water dispersibility of hexagonal phase NaGdF4:Ho/Yb nanoparticles. Chem. Mater. 2009, 21, 717–723.

24. Mahalingam, V.; Vetrone, F.; Naccache, R.; Speghini, A.; Capobianco, J.A. Colloidal Tm3+/Yb3+-doped LiYF4nanocrystals: Multiple luminescence spanning the UV to NIR regions via low-energy excitation. Adv. Mater. 2009, 21, 4025–4028.

25. Mahalingam, V.; Vetrone, F.; Naccache, R.; Speghini, A.; Capobianco, J.A. Structural and optical investigation of colloidal Ln3+/Yb3+ co-doped KY3F10 nanocrystals. J. Mater. Chem. 2009, 19, 3149–3152.

26. Vetrone, F.; Mahalingam, V.; Capobianco, J.A. Near-infrared-to-blue upconversion in colloidal BaYF5:Tm3+,Yb3+nanocrystals. Chem. Mater. 2009, 21, 1847–1851.

27. Zhang, P.D.; Zhang, Y.W.; Zheng, Z.G.; Sun, L.D.; Yan, C.H. Highly luminescent self-organized sub-2-nm EuOF nanowires. J. Am. Chem. Soc. 2009, 131, 16364–16365.

28. Du, Y.P.; Zhang, Y.W.; Sun, L.D.; Yan, C.H. Luminescent monodisperse nanocrystals of lanthanide oxyfluorides synthesized from trifluoroacetate precursors in high-boiling solvents. J. Phys. Chem. C 2008, 112, 405–415.

29. Du, Y.P.; Zhang, Y.W.; Sun, L.D.; Yan, C.H. Atomically efficient synthesis of self-assembled monodisperse and ultrathin lanthanide oxychloride nanoplates. J. Am. Chem. Soc. 2009, 131, 3162–3163.

30. Stouwdam, J.W.; van Veggel, F.C.J.M. Near-infrared emission of redispersible Er3+, Nd3+, and Ho3+ doped LaF3nanoparticles. Nano Lett. 2002, 2, 733–737.

31. Yi, G.S.; Chow, G.M. Colloidal LaF3:Yb,Er, LaF3:Yb,Ho and LaF3:Yb,Tm nanocrystals with multicolor upconversion fluorescence. J. Mater. Chem. 2005, 15, 4460–4464.

32. Heer, S.; Kompe, K.; Gudel, H.U.; Haase, M. Highly efficient multicolour upconversion emission in transparent colloids of lanthanide-doped NaYF4 nanocrystals. Adv. Mater. 2004, 16, 2102–2105.

33. Heer, S.; Lehmann, O.; Haase, M.; Gudel, H.U. Blue, green, and red upconversion emission from lanthanide-doped LuPO4 and YbPO4 nanocrystals in a transparent colloidal solution. Angew. Chem. Int. Ed. 2003, 42, 3179–3182.

34. Zeng, J.H.; Su, J.; Li, Z.H.; Yan, R.X.; Li, Y.D. Synthesis and upconversion luminescence of hexagonal-phase NaYF4:Yb3+,Er3+ phosphors of controlled size and morphology. Adv. Mater. 2005, 17, 2119–2123.

35. Teng, X.; Zhu, Y.H.; Wei, W.; Wang, S.C.; Huang, J.F.; Naccache, R.; Hu, W.B.; Han, Y.; Zhang, Q.C.; Fan, Q.L.; et al. Lanthanide-doped NaxScF3+x nanocrystals: Crystal structure evolution and multicolor tuning. J. Am. Chem. Soc. 2012,134, 8340–8343.

36. Liang, L.F.; Xu, H.F.; Su, Q.; Konishi, H.; Jiang, Y.B.; Wu, M.M.; Wang, Y.F.; Xia, D.Y. Hydrothermal synthesis of prismatic NaHoF4 microtubes and NaSmF4 nanotubes. Inorg. Chem. 2004, 43, 1594–1596.

37. Zhang, M.F.; Fan, H.; Xi, B.J.; Wang, X.Y.; Dong, C.; Qian, Y.T. Synthesis, characterization, and luminescence properties of uniform Ln3+-doped YF3 nanospindles. J. Phys. Chem. C 2007, 111, 6652–6657.

38. Liang, X.; Wang, X.; Zhuang, J.; Peng, Q.; Li, Y.D. Branched NaYF4 nanocrystals with luminescent properties. Inorg. Chem. 2007, 46, 6050–6055.

39. Zhao, J.W.; Suan, Y.J.; Kong, X.G.; Tian, L.J.; Wang, Y.; Tu, L.P.; Zhao, J.L.; Zhang, H. Controlled synthesis, formation mechanism, and great enhancement of red upconversion luminescence of NaYF4:Yb3+,Er3+nanocrystals/submicroplates at low doping level. J. Phys. Chem. B 2008, 112, 15666–15672.

40. Wang, X.; Zhuang, J.; Peng, Q.; Li, Y.D. Hydrothermal synthesis of rare-earth fluoride nanocrystals. Inorg. Chem. 2006,45, 6661–6665.

41. Wang, X.; Li, Y.D. A general strategy for nanocrystal synthesis. Nature 2005, 437, 121–124.

42. Huo, Z.Y.; Chen, C.; Li, Y.D. Systematic synthesis of lanthanide phosphate nanocrystals. Chem. Eur. J. 2007, 13, 7708–7714.

43. Wang, L.Y.; Li, Y.D. Systematic synthesis of lanthanide phosphate nanocrystals. Na(Y1.5Na0.5)F6 single-crystal nanorods as multicolor luminescent materials. Nano Lett. 2006, 6, 1645–1649.

44. Wang, L.Y.; Li, Y.D. Controlled synthesis and luminescence of lanthanide doped NaYF4 nanocrystals. Chem. Mater.2007, 19, 727–734.

45. Wang, G.F.; Peng, Q.; Li, Y.D. Luminescence tuning of upconversion nanocrystals. Chem. Eur. J. 2010, 16, 4923–4931.

46. Wang, L.Y.; Li, P.; Li, Y.D. Down and upconversion luminescent nanorods. Adv. Mater. 2007, 19, 3304–3307.

47. Wang, X.; Li, Y.D. Synthesis and characterization of lanthanide hydroxide single-crystal nanowires. Angew. Chem. Int. Ed. 2002, 41, 4790–4793.

48. Wang, X.; Sun, X.M.; Yu, D.P.; Zou, B.S.; Li, Y.D. Rare earth compound nanotubes. Adv. Mater. 2003, 15, 1442–1445.

49. Wang, X.; Li, Y.D. Fullerene-like rare-earth nanoparticles. Angew. Chem. Int. Ed. 2003, 42, 3497–3500.

50. Hu, C.G.; Liu, H.; Wang, Z.L. La(OH)3 and La2O3 nanobelts-synthesis and physical properties. Adv. Mater. 2007, 19, 470–474.

51. Zhang, F.; Wan, Y.; Yu, T.; Zhang, F.; Shi, Y.; Xie, S.; Li, Y.; Xu, L.; Tu, B.; Zhao, D. Uniform nanostructured arrays of sodium rare-earth fluorides for highly efficient multicolor upconversion luminescence. Angew. Chem. Int. Ed. 2007, 46, 7976–7979.

52. Liu, C.; Chen, D. Controlled synthesis of hexagon shaped lanthanide-doped LaF3 nanoplates with multicolor upconversion fluorescence. J. Mater. Chem. 2007, 17, 3875–3880.

53. Pan, S.H.; Deng, R.P.; Feng, J.; Wang, S.Y.; Wang, S.; Zhu, M.; Zhang, H.J. Microwave-assisted synthesis and down and upconversion luminescent properties of BaYF5:Ln (Ln = Yb/Er, Ce/Tb) nanocrystals. CrystEngComm 2013, 15, 7640–7643.

54. Yang, M.; You, H.P.; Zheng, Y.H.; Liu, K.; Jia, G.; Song, Y.H.; Huang, Y.J.; Zhang, L.H.; Zhang, H. Hydrothermal synthesis and luminescent properties of novel ordered sphere CePO4 hierarchical architectures. J. Inorg. Chem. 2009,48, 11559–11565.

55. Yang, M.; You, H.P.; Huang, Y.H.; Zhang, H.J. Synthesis and luminescent properties of orderly YPO4:Eu3+ olivary architectures self-assembled by nanoflakes. CrystEngComm 2010, 12, 4141–4145.

56. Zheng, Y.H.; You, H.P.; Jia, K.; Liu, K.; Song, Y.H.; Yang, M.; Zhang, H.J. Highly uniform Gd2O3 hollow microspheres: Template-directed synthesis and luminescence properties. Langmuir 2010, 26, 5122–5128.

57. Jia, G.; You, H.P.; Liu, K.; Zheng, Y.H.; Guo, N.; Zhang, H.J. Facile hydrothermal synthesis and luminescent properties of large-scale GdVO4:Eu3+ nanowires. Cryst. Growth Des. 2009, 9, 5101–5107.

58. Li, C.X.; Hou, Z.Y.; Zhang, C.M.; Yang, P.P.; Li, G.G.; Xu, Z.H.; Fan, Y.; Lin, J. Controlled synthesis of Ln3+ (Ln = Tb, Eu, Dy) and V5+ ion-doped YPO4 nano-/microstructures with tunable luminescent colors. Chem. Mater. 2009, 21, 4598–4607.

59. Li, C.X.; Yang, J.; Yang, P.P.; Lian, H.Z.; Lin, J. Hydrothermal synthesis of lanthanide fluorides LnF3 (Ln = La to Lu) nano/microcrystals with multiform structures and morphologies. Chem. Mater. 2008, 20, 4317–4326.

60. Li, C.X.; Yang, J.; Quan, Z.W.; Yang, P.P.; Kong, D.Y.; Lin, J. Different microstructures of ss-NaYF4 fabricated by hydrothermal process: Effects of pH values and fluoride sources. Chem. Mater. 2007, 19, 4933–4942.

61. Wang, J.; Wang, F.; Wang, C.; Liu, Z.; Liu, X.G. Single-band upconversion emission in lanthanide-doped KMnF3nanocrystals. Angew. Chem. Int. Ed. 2011, 50, 10369–10372.

62. Tian, G.; Gu, Z.J.; Zhou, L.J.; Yin, W.Y.; Liu, X.X.; Yan, L.; Jin, S.; Ren, W.L.; G, M.; Li, S.G.; et al. Mn2+ dopant-controlled synthesis of NaYF4:Yb/Er upconversion nanoparticles for in vivo imaging and drug delivery. Adv. Mater.2012, 24, 1226–1231.

63. Zeng, S.J.; Yi, Z.G.; Lu, W.; Qian, C.; Wang, H.B.; Rao, L.; Zeng, T.M.; Liu, H.R.; Liu, H.G.; Fei, B.; et al. Simultaneous realization of phase/size manipulation, upconversion luminescence enhancement, and blood vessel imaging in multifunctional nanoprobes through transition metal Mn2+ doping. Adv. Funct. Mater. 2014, 24, 4051–4059.

64. Lin, J.; Yu, M.; Lin, C.K.; Liu, X.M. Multiform oxide optical materials via the versatile Pechini-type sol-gel process: Synthesis and characteristics. J. Phys. Chem. C 2007, 111, 5835–5845.

65. Patra, A.; Friend, C.S.; Kapoor, R.; Prasad, P.N. Upconversion in Er3+:ZrO2 nanocrystals. J. Phys. Chem. B 2002, 106, 1909–1912.

66. Cheng, Z.Y.; Xing, R.B.; Hou, Z.Y.; Huang, S.S.; Lin, J. Patterning of light-emitting YVO4:Eu3+ thin films via inkjet printing. J. Phys. Chem. C 2010, 114, 9883–9888.

67. Zhu, Y.S.; Xu, W.; Zhang, H.Z.; Wang, W.; Xu, S.; Song, H.W. Inhibited long-scale energy transfer in dysprosium doped yttrium vanadate inverse opal. J. Phys. Chem. C 2012, 116, 2297–2302.

68. Wang, W.; Song, H.W.; Liu, Q.; Bai, X.; Wang, Y.; Dong, B. Modified optical properties in a samarium doped titania inverse opal. Opt. Lett. 2010, 35, 1449–1451.

69. Liu, Q.; Song, H.W.; Wang, W.; Bai, X.; Wang, Y.; Dong, B.; Xu, L.; Han, W. Observation of lamb shift and modified spontaneous emission dynamics in the YBO3:Eu3+ inverse opal. Opt. Lett. 2010, 35, 2898–2900.

70. Zhu, Y.S.; Sun, Z.P.; Yin, Z.; Song, H.W.; Xu, W.; Wang, Y.F.; Zhang, L.G.; Zhang, H.Z. Self-assembly, highly modified spontaneous emission and energy transfer properties of LaPO4:Ce3+,Tb3+ inverse opals. Dalton Trans. 2013, 42, 8049–8057.

71. Patra, A.; Friend, C.S.; Kapoor, R.; Prasad, P.N. Fluorescence upconversion properties of Er3+-doped TiO2 and BaTiO3nanocrystallites. Chem. Mater. 2003, 15, 3650–3655.

72. Venkatramu, V.; Falcomer, D.; Speghini, A.; Bettinelli, M.; Jayasankar, C.K. Synthesis and luminescence properties of Er3+-doped Lu3Ga5O12 nanocrystals. J. Lumin. 2008, 128, 811–813.

73. Yang, K.S.; Zheng, F.; Wu, R.; Li, H.; Zhang, X. Upconversion luminescent properties of YVO4:Yb3+,Er3+ nano-powder by sol-gel method. J. Rare Earth 2006, 24, 162–166.

74. Lemyre, J.L.; Ritcey, A.M. Synthesis of lanthanide fluoride nanoparticles of varying shape and size. Chem. Mater. 2005,17, 3040–3043.

75. Wang, G.F.; Qin, W.P.; Zhang, J.S. Synthesis, growth mechanism, and tunable upconversion luminescence of Yb3+/Tm3+-codoped YF3 nanobundles. J. Phys. Chem C 2008, 112, 12161–12167.

76. Wang, G.F.; Qin, W.P.; Wei, G.D. Synthesis and upconversion luminescence properties of Yb3+/Tm3+-codoped BaSiF6nanorods. J. Fluorine Chem. 2009, 130, 158–161.

77. Vetrone, F.; Boyer, J.C.; Capobianco, J.A.; Speghini, A.; Bettinelli, M. Significance of Yb3+ concentration on the upconversion mechanisms in codoped Y2O3:Er3+,Yb3+ nanocrystals. J. Appl. Phys. 2004, 96, 661–667.

78. Pandozzi, F.; Vetrone, F.; Boyer, J.C.; Naccache, R.; Capobianco, J.A.; Speghini, A.; Bettinellin, M. A spectroscopic analysis of blue and ultraviolet upconverted emissions from Gd3Ga5O12:Tm3+,Yb3+ nanocrystals. J. Phys. Chem. B2005, 109, 17400–17405.

79. Luo, X.X.; Cao, W.H. Ethanol-assistant solution combustion method to prepare La2O3:Yb,Pr nanometer phosphor. J. Alloys. Compds. 2008, 46, 529–534.

80. Xu, L.L.; Yu, Y.N.; Li, X.G. Synthesis and upconversion properties of monoclinic Gd2O3:Er3+ nanocrystals. Opt. Mater.2008, 30, 1284–1288.

81. Qin, X.; Yokomori, T.; Ju, Y.G. Flame synthesis and characterization of rare-earth (Er3+, Ho3+, and Tm3+) doped upconversion nanocryphosphors. Appl. Phys. Lett. 2007, 90, 073104.

82. Dong, B.; Song, H.W.; Yu, H.Q.; Zhang, H.; Qin, R.F.; Bai, X.; Pan, G.H.; Lu, S.Z.; Wang, F.; Fan, L.B.; et al. Upconversion properties of Ln3+ doped NaYF4/polymer composite fibers prepared by electrospinning. J. Phys. Chem. C 2008, 112, 1435–1440.

83. Ma, L.; Chen, W.X.; Zheng, Y.Z.; Zhao, J.; Xu, Z.D. Microwave-assisted hydrothermal synthesis and characterizations of PrF3 hollow nanoparticles. Mater. Lett. 2007, 61, 2765–2768.

84. Chan, E.M. Combinatorial approaches for developing upconverting nanomaterials: High-throughput screening, modeling, and applications. Chem. Soc. Rev. 2014. [CrossRef]

85. Liu, H.; Jakobsson, O.; Xu, C.T.; Xie, H.; Laurell, T.; Andersson-Engels, S. Synthesis of NaYF4:Yb3+/Er3+ Upconverting Nanoparticles in a Capillary-Based Continuous Microfluidic Reaction System. In Colloidal Quantum Dots/Nanocrystals for Biomedical Applications; SPIE: Bellingham, WA, USA, 2011; Volume 7909, p. 790917.

86. Xu, C.T.; Zhan, Q.; Liu, H.; Somesfalean, G.; Qian, J.; He, S.; Andersson-Engels, S. Upconverting nanoparticles for pre-clinical diffuse optical

imaging, microscopy and sensing: Current trends and future challenges. Laser Photon. Rev.2013, 7, 663–697.

87. Zhu, X.; Zhang, Q.; Li, Y.; Wang, H. Redispersible and water-soluble LaF3:Ce,Tb nanocrystalsvia a microfluidic reactor with temperature steps. J. Mater. Chem. 2008, 18, 5060–5062.

88. Zhu, X.; Zhang, Q.; Li, Y.; Wang, H.J. Facile crystallization control of LaF3/LaPO4:Ce,Tb nanocrystals in a microfluidic reactor using microwave irradiation. J. Mater. Chem. 2010, 20, 1766–1771.

89. Yi, G.S.; Chow, G.M. Water-soluble NaYF4:Yb,Er(Tm)/NaYF4/polymer core/shell/shell nanoparticles with significant enhancement of upconversion fluorescence. Chem. Mater. 2007, 19, 341–343.

90. Liu, Y.S.; Tu, D.T.; Zhu, H.M.; Li, R.F.; Luo, W.Q.; Chen, X.Y. A strategy to achieve efficient dual-mode luminescence of Eu3+ in lanthanides doped multifunctional NaGdF4 nanocrystals. Adv. Mater. 2010, 22, 3266–3271.

91. Vetrone, F.; Naccache, R.; Mahalingam, V.; Morgan, C.G.; Capobianco, J.A. The active-core/active-shell approach: A strategy to enhance the upconversion luminescence in lanthanide-doped nanoparticles. Adv. Funct. Mater. 2009, 19, 2924–2929.

92. Liu, X.M.; Kong, X.G.; Zhang, Y.L.; Tu, L.P.; Wang, Y.; Zeng, Q.H.; Li, C.G.; Shi, Z.; Zhang, H. Breakthrough in concentration quenching threshold of upconversion luminescence via spatial separation of the emitter doping area for bio-applications. Chem. Commun. 2011, 47, 11957–11959.

93. Wang, Z.L.; Quan, Z.W.; Jia, P.Y.; Lin, C.K.; Luo, Y.; Chen, Y.; Fang, J.; Zhou, W.; O'Connor, C.J.; Lin, J. A facile synthesis and photoluminescent properties of redispersible CeF3,CeF3:Tb3+, and CeF3:Tb3+/LaF3 (core/shell) nanoparticles. Chem. Mater. 2006, 18, 2030–2037.

94. Mai, H.X.; Zhang, Y.W.; Yan, C.H. Highly efficient multicolor up-conversion emissions and their mechanisms of monodisperse NaYF4:Yb,Er core and core/shell-structured nanocrystals. J. Phys. Chem. C 2007, 111, 13721–13729.

95. Boyer, J.C.; Gagnon, J.; Capobianco, J.A. Synthesis, characterization, and spectroscopy of NaGdF4:Ce3+,Tb3+/NaYF4core/shell nanoparticles. Chem. Mater. 2007, 19, 3358–3360.

96. Helmut, S.; Pavel, P.; Otmanef, Z.; Markus, H. Synthesis and optical properties of KYF4/Yb,Er nanocrystals, and their surface modification with undoped KYF4. Adv. Funct. Mater. 2008, 18, 2913–2918.

97. Xie, X.J.; Gao, N.Y.; Deng, R.R.; Sun, Q.; Xu, Q.H.; Liu, X.G. Mechanistic investigation of photon upconversion in Nd3+-sensitized core-shell nanoparticles. J. Am. Chem. Soc. 2013, 135, 12608–12611.

98. Huang, P.; Zheng, W.; Zhou, S.Y.; Tu, D.T.; Chen, Z.; Zhu, H.M.; Li, R.F.; Ma, E.; Huang, M.D.; Chen, X.Y. Lanthanide-doped LiLuF4 upconversion nanoprobes for the detection of disease biomarkers. Angew. Chem. Int. Ed. 2014, 53, 1252–1257.

99. Wang, F.; Deng, R.R.; Liu, X.G. Preparation of core-shell NaGdF4 nanoparticles doped with luminescent lanthanide ions to be used as upconversion-based probes. Nat. Protoc. 2014, 9, 1634–1644.

100. Wang, Y.F.; Sun, L.D.; Xia, J.W.; Feng, W.; Zhou, J.C.; Shen, J.; Yan, C.H. Rare-earth nanoparticles with enhanced upconversion emission and suppressed rare-earth-ion leakage. Chem. Eur. J. 2012, 18, 5558–5564.

101. Darbandi, M.; Nann, T. One-pot synthesis of YF3@silica core/shell nanoparticles. Chem. Commun. 2006, 7, 776–778.

102. Li, Z.Q.; Zhang, Y. Monodisperse silica-coated polyvinylpyrrolidone/NaYF4 nanocrystals with multicolor upconversion fluorescence emission. Angew. Chem. Int. Ed. 2006, 45, 7732–7735.

103. Li, Z.Q.; Zhang, Y.; Jiang, S. Multicolor core/shell-structured upconvers fluorescent nanoparticles. Adv. Mater. 2008, 20, 4765–4769.

104. Chen, B.T.; Dong, B.; Wang, J.; Zhang, S.; Xu, L.; Yu, W.; Song, H.W. Amphiphilic silane modified NaYF4:Yb,Er loaded with Eu(TTA)3(TPPO)2 nanoparticles and their multi-functions: Dual mode temperature sensing and cell imaging.Nanoscale 2013, 5, 8541–8549.

105. Sivakumar, S.; Diamenteand, P.R.; van Veggel, F.C.J.M. Silica-coated Ln3+-doped LaF3 nanoparticles as robust down and upconverting biolabels. Chem. Eur. J. 2006, 12, 5878–5884.

106. Liu, Z.Y.; Yi, G.S.; Zhang, H.T.; Ding, J.; Zhang, Y.W.; Xue, J.M. Monodisperse silica nanoparticles encapsulating upconversion fluorescent and superparamagnetic nanocrystals. Chem. Commun. 2008, 6, 694–696.

107. Wang, M.; Mi, C.C.; Zhang, Y.Z.; Liu, J.L.; Li, F.; Mao, C.B.; Xu, S.K. NIR-responsive silica-coated NaYbF4:Er/Tm/Ho upconversion fluorescent nanoparticles with tunable emission colors and their applications in immunolabeling and fluorescent imaging of cancer cells. J. Phys. Chem. C 2009, 113, 19021–19027.

108. Yu, X.F.; Chen, L.D.; Li, M.; Xie, M.Y.; Zhou, L.; Li, Y.; Wang, Q.Q. Highly efficient fluorescence of NdF3/SiO2core/shell nanoparticles and the applications for in vivo NIR detection. Adv. Mater. 2008, 20, 4118–4123.

109. Li, X.H.; Wu, Y.Q.; Liu, Y.; Zou, X.M.; Yao, L.M.; Lia, F.Y.; Feng, W. Cyclometallated ruthenium complex-modified upconversion nanophosphors for selective detection of Hg2+ ions in water. Nanoscale 2014, 6, 1020–1028.

110. Wang, C.; Cheng, L.; Liu, Z. Drug delivery with upconversion nanoparticles for multi-functional targeted cancer cell imaging and therapy. Biomaterials 2011, 32, 1110–1120.

111. Budijono, S.J.; Shan, J.N.; Yao, N.; Miura, Y.; Hoye, T.; Austin, R.H.; Ju, Y.G.; Prud'homme, R.K. Synthesis of stable block-copolymer-protected NaYF4:Yb3+,Er3+ up-converting phosphor nanoparticles. Chem. Mater. 2010, 22, 311–318.

112. Chen, Z.G.; Chen, H.L.; Hu, H.; Yu, M.X.; Li, F.Y.; Zhang, Q.; Zhou, Z.G.; Yi, T.; Huang, C.H. Versatile synthesis strategy for carboxylic acid-functionalized upconverting nanophosphors as biological labels. J. Am. Chem. Soc. 2008,130, 3023–3029.

113. Bogdan, N.; Vetrone, F.; Ozin, G.A.; Capobianco, J.A. Synthesis of ligand-free colloidally stable water dispersible brightly luminescent lanthanide-doped upconverting nanoparticles. Nano Lett. 2011, 11, 835–840.

114. Yi, G.S.; Chow, G.M. Synthesis of hexagonal-phase NaYF4:Yb,Er and NaYF4:Yb,Tm nanocrystals with efficient up-conversion fluorescence. Adv. Funct. Mater. 2006, 16, 2324–2329.

115. Chen, G.Y.; Ohulchanskyy, T.Y.; Law, W.C.; Agren, H.; Prasad, P.N. Monodisperse NaYbF4: Tm3+/NaGdF4 core/shell nanocrystals with near-infrared to near-infrared upconversion photoluminescence and magnetic resonance properties. Nanoscale 2011, 3, 2003–2008.

116. Boyer, J.C.; Manseau, M.P.; Murray, J.I.; van Veggel, F.C.J.M. Surface modification of upconverting NaYF4nanoparticles with PEG-phosphate

ligands for NIR (800 nm) biolabeling within the biological window. Langmuir 2010,26, 1157–1164.

117. Nyk, M.; Kumar, R.; Ohulchanskyy, T.Y.; Bergey, E.J.; Prasad, P.N. High contrast in vitro and in vivophotoluminescence bioimaging using near Infrared to near infrared up-conversion in Tm3+ and Yb3+ doped fluoride nanophosphors. Nano Lett. 2008, 8, 3834–3838.

118. Cao, T.Y.; Yang, T.S.; Gao, Y.; Yang, Y.; Hu, H.; Li, F.Y. Water-soluble NaYF4:Yb/Er upconversion nanophosphors: Synthesis, characteristics and application in bioimaging. Inorg. Chem. Commun. 2010, 13, 392–394.

119. Zhou, J.; Yu, M.X.; Sun, Y.; Zhang, X.Z.; Zhu, X.J.; Wu, Z.H.; Wu, D.M.; Li, F.Y. Fluorine-18-labeled Gd3+/Yb3+/Er3+ co-doped NaYF4 nanophosphors formultimodality PET/MR/UCL imaging. Biomaterials 2011, 32, 1148–1156.

120. Naczynski, D.J.; Andelman, T.; Pal, D.; Chen, S.; Richard, E.; Riman, R.E.; Roth, C.M.; Moghe, P.V. Albumin nanoshell encapsulation of near-infrared-excitable rare-earth nanoparticles enhances biocompatibility and enables targeted cell imaging. Small 2010, 6, 1631–1640.

Chapter 4

Optical, Magnetic, and Structural Properties of Semiconductor and Semimagnetic Nanocrystals

Ricardo Souza da Silva[1], Ernesto Soares de Freitas Neto[2] and Noelio Oliveira Dantas[2]

[1] Instituto de Ciências Exatas e Naturais e Educação (ICENE), Departamento de Física, Universidade Federal do Triângulo Mineiro, Uberaba, Minas Gerais,, Brazil

[2] Laboratório de Novos Materiais Isolantes e Semicondutores (LNMIS), Instituto de Física, Universidade Federal de Uberlândia, Uberlândia, Minas Gerais,, Brazil

1. INTRODUCTION

Semiconductor and semimagnetic nanocrystals (NCs), grown in different host materials, have attracted considerable attention due to their unique properties which are caused by zero-dimensional quantum confinement effects. Several advances in controlled chemical synthesis of materials have provided ways to grow NCs and manipulate their size, shape, and composition using different methodologies [1-5]. The interesting properties of these nanoparticles can be explored in diverse technological applications, such as wavelength tunable lasers, light-emitting devices, solar cells, and spintronic devices among others [6-13].A detailed and comprehensive understanding on the properties of these

NCs should be achieved in order to target many of the possible technological applications.In this chapter, we will present our main results and discussions on the optical, magnetic, and structural properties of semiconductor and semimagnetic NCs that were successfully grown by the melting-nucleation approach or by the chemical precipitation method.

Optical processes in PbS NCs were investigated by employing the following experimental techniques: optical absorption (OA), photoluminescence (PL) and atomic force microscopy (AFM). The OA and PL peaks of these PbS NC samples showed a separation of about 0.05-0.20 eV, confirming thus the large Stokes shift. A comprehensive understanding on this large Stokes shift was achieved by investigating the radiative and nonradiative processes in these nanoparticles [14].

We will report evidences to the induced migration of Mn^{2+} ions in $Cd_{1-x}Mn_xS$ NCs by selecting a specific thermal treatment to each NC sample. The characterization of these magnetic dots was investigated by the electronic paramagnetic resonance (EPR) technique. The comparison of experimental and simulation of EPR spectra confirms the incorporation of Mn^{2+} ions both in the core and at the dot surface regions. The thermal treatment to a magnetic sample, via selected annealing temperature and/or time, affects the fine and hyperfine interaction constants which modifies the shape and the intensity of an EPR transition spectrum. The identification of these changes has allowed tracing the magnetic ion migration from core to surface regions of a dot as well as inferring on the local density of the magnetic impurity ions [15].

The properties of $Pb_{1-x}Mn_xS$ NCs embedded in a borosilicate glass matrix has been investigated by magnetic measurements. The data indicated that only a small fraction of the nominal Mn-doping was incorporated into the PbS NCs, in both 0.3% and 0.7% nominal doping ends. Moreover, low temperature magnetization and susceptibility data showed that most of the magnetic ions hosted by the of $Pb_{1-x}Mn_xS$ NCs are in a paramagnetic state[16, 25].

We also have employed the magnetic force microscopy (MFM) in order to study the magnetic moments of Mn-doped nanoparticles, namely: $Cd_{1-x}Mn_xS$ and $Pb_{1-x}Mn_xS$ NCs. In these measurements, the interaction between tip and NC magnetization induces the contrast observed in the MFM images. A dark area (light area) in this contrast is caused by attraction (repulsion) between tip and NC magnetization.Evidently, the magnetization in each NC is caused by the size-dependent *sp-d*exchange interactions, proving that Mn^{2+} ions are actually incorporated into the semimagnetic nanostructures. Therefore, all these results certainly demonstrate that MFM is a powerful technique that plays a very important role in order to investigate semimagnetic nanocrystals [4, 17].

$Zn_{1-x}Mn_xO$ NCs were successfully grown by the chemical precipitation method and their magnetic properties were effectively investigated by the EPR technique. Thus, we have confirmed the actual incorporation of Mn^{2+} ions into the hosting ZnO NCs, while the hexagonal wurtzite structure of these nanoparticles was preserved. The well known Mn^{2+} six hyperfine lines in the EPR spectra of the as-produced samples were clearly observed. In addition, as the Mn-concentration increases to a level of about 0.81% a broad EPR line is observed, thus confirming the onset of Mn-Mn exchange interaction [5].

The structural properties of these $Zn_{1-x}Mn_xO$ NCs were characterized by Raman spectroscopy and X-Ray Diffraction (XRD) measurements. The observed shift in the diffraction peaks toward lower angles, with increasing in the x-concentration, was attributed to incorporation of Mn^{2+} ions into the ZnO NCs. This analysis is strongly corroborated by results obtained by the Raman spectroscopy, where the data have also provided evidences of the replacement of zinc ions by manganese ions into the $Zn_{1-x}Mn_xO$ NCs. Besides the Raman features typical of the ZnO structure, the $Zn_{1-x}Mn_xO$ ($x > 0$) nanoparticles display an extra Raman peak at 659 cm^{-1}. This finding is a strong evidence of the replacement of zinc ions by manganese ions [5, 18].

The results of this chapter confirm the high quality of the semiconductor and semimagnetic NCs that were successfully grown by the melting-nucleation approach or by the chemical precipitation method. The comprehensive discussions that were presented on the properties of nanoparticles certainly demonstrate the great potential of these systems for various technological applications. We believe that this chapter can motivate further investigations and applications of other systems containing NCs.

2. SYNTHESIS OF NANOCRYSTALS

The development of nanocrystals (NCs) produced of controlled way for possible applications technologic, depends on the synthesis methodology adopted. We report the study of $Pb_{1-x}Mn_xS$ and $Cd_{1-x}Mn_xS$ NCs synthesized in borosilicate glass matrix template using the fusion method and $Zn_{1-x}Mn_xO$ NCs using the co-precipitation method.

2.1. Synthesis Of $Pb_{1-x}mn_xs$ Nanocrystals

$Pb_{1-x}Mn_xS$ NCs were produced by the fusion method in the glass matrix with the following nominal composition: $40SiO_2{\bullet}30Na_2CO_3{\bullet}1Al_2O_3{\bullet}25B_2O_3{\bullet}4PbO$

(%mol), herein quoted as SNABP glass matrix. The nominal composition of the nanocomposite was achieved by adding 2S (%wt) plus *x*Mn with respect the (1-*x*)Pb, with $x = 0$, 0.003 and 0.007. The samples were produced following two major preparation steps. In the first step the powder mixture was melted in an alumina crucible at 1200°C for 30 minutes, following a quick cooling of the crucible containing the melted mixture from 1200°C down to room-temperature. At the end of this step a first series of samples labeled SNABP:*x*Mn were produced for further characterization. In the second step thermal annealing of the previously-melted glass matrix (SNABP: *x*Mn samples) was carried out at 500°C for times different, with the purpose to enhance the diffusion of Pb^{2+}, Mn^{2+}, and S^{2-} species within the hosting matrix. Due to the thermal annealing procedure $Pb_{1-x}Mn_xS$ NCs were formed within the glass template [7].

2.2. Synthesis Of $Cd_{1-x}mn_xs$ Nanocrystals

$Cd_{1-x}Mn_xS$ NCs were synthesized in a glass matrix (SNAB) with a nominal composition of $40SiO_2.30Na_2CO_3.\ 1Al_2O_3.\ 29B_2O_3$ (%mol) + 2[CdO + S] (%wt), and Mn-doping concentration (x) varying with respect to Cd-content from 0 to 10%. The first step of sample preparation consisted of melting powder mixtures in an alumina crucible at 1200°C for 30 minutes. Then, the crucible containing the melted mixture underwent quick cooling to room temperature. In the second step, thermal annealing of the previously melted glass matrix was carried out at 560°C for 02 and 20 hours in order to enhance the diffusion of Cd^{2+}, Mn^{2+}, and S^{2-} species into the host matrix. As a result of the thermal annealing, CdS and $Cd_{1-x}Mn_xS$ (x>0) NCs were formed in the glass template, wich were denominated at two classes: i) SNAB: CdS NCs and ii) SNAB: $Cd_{1-x}Mn_xS$ NCs[7].

2.3. Synthesis Of $Zn_{1-x}mn_xo$ Nanocrystals

Preparation of the $Zn_{1-x}Mn_xO$ NC samples is based on the transformation of the aqueous-$[Zn(NH_3)_4]^{2+}$ metal-complex in the presence of aqueous-Mn^{2+}, sodium oleate and hydrazine sulfate at 80°C. The best chemical synthesis results were achieved by keeping the pH value of the reaction medium at 8.5 during the whole reaction process, which was adjusted by controlling the addition of 4 M sodium hydroxide aqueous solution. Briefly, a typical protocol used started by magnetically-stirring, at room-temperature and for ½-hour, 100 mL of 0.38 M zinc chlorite mixed with 100 mL of 1.6 M ammonia hydroxide in order to form the aqueous-$[Zn(NH3)_4]^{2+}$ complex. Then 1 mL of hydrazine sulfate and 0.08 g of sodium oleate were added into the previously-stirred solution. The obtained

reaction medium was then heated at 80°C using water-bath in order to transform the aqueous-$[Zn(NH3)_4]^{2+}$ complex, while keeping the pH value fixed at 8.5. The chemical synthesis was carried out for 2 h, while ammonia was observed to be release out from the reaction medium as the chemical process proceeded. The resulting precipitates ($Zn_{1-x}Mn_xO$; x≥0) were percolated and washed with distilled water and absolute ethanol for several times and further dried under at 500°C for 2h [5].

3. OPTICAL PROPERTIES OF NANOCRYSTALS

The optical properties of PbS, $Zn_{1-x}Mn_xO$ and $Cd_{1-x}Mn_xS$ NCs, synthesized by methodology describe in section 2, were investigated by Optical Absorption (OA), and/or Photoluminescence (PL) spectroscopy techniques. The obtained results will be presented and discussed as follows.

3.1. Optical Properties Of Pbs Nanocrystals

3.1.1. Optical Absorption and Photoluminescence

Because of the large exciton Bohr radius (18 nm), PbS QD-doped glasses exhibit strong three-dimensional quantum-confinement effects at moderate nanocrystal size. This combined with small band gap energy (0.41 eV at room temperature) of PbS with different thermal annealing, which result in different average sizes of PbS nanocrystals [2,14]. Room-temperature Photoluminescence and optical absorption spectra of PbS nanocrystals with different time annealing process are shown inFigure 1. The strong quantum confinement in these structures is clearly observed. The appearance of defined band peaks in both absorption and emission spectra demonstrates the high quality of our samples and relatively small size distribution of the PbS nanocrystals.

3.1.2. Size-Dependent Stokes Shift

The dependence of the Stokes shift is closely linked to the size of the nanocrystals for schemes of strong quantum confinement the discrete levels of transition electron become more evident such that the difference between the position of peak absorption and emission increases with decreasing the size of

the nanocrystals [14, 19]. With the absorption of a photon from of the valence band to the conduction band occurs the formation of electron-hole pairs (exciton). The exciton, once formed after absorption, cannot decay to the top of the valence band by a direct dipole transition and hence is denominated of dark exciton [20]. In the process of deexcitation eventually takes place with the help of phonons, thus giving rise to red shifted photons, known with Stokes shift.

The behavior of Stokes shift of PbS NCs of Figure 1 is represents in Figure 2. From the data presented is observed the decline in the Stokes shift with the increase of NCs size.

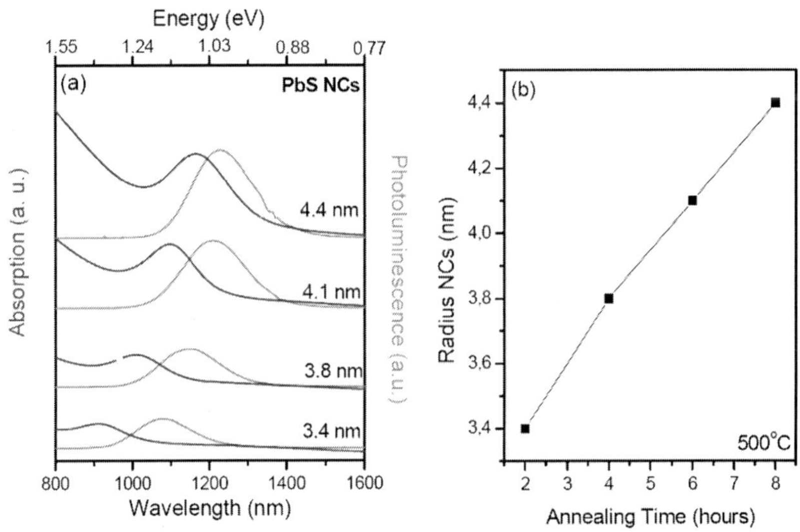

Figure 1. a). Room-temperature optical absorption and photoluminescence of PbS NCs synthesized in SNABP glass matrix. (b) Behavior of average size of PbS NCs with annealing time at 500°C.

The Stokes shift is originated by process of radiative decay from electron-hole recombination and a nonradiative decay via trapped states involves electron-phonons. As the size of the nanocrystal increase, the surface to volume ratio decreases, and there is a reduction in the overlap of the electron and hole wave functions. This is coincident with a decreased wave function overlap with the nanocrystal surface, which leads to less surface trapping and the decrease of Stokes shift [14]. A schematic of levels energy involved the process of

excitation and deexicitation is shows in Figure 3. The excitation occurs with the absorption of electrons of level $1S_h$ (fundamental state) to the excitation states $1S_e$ (OA_1) and $1P_e$ (OA_2) and the deexcitation process is characterized by a nonradiative recombination of levels $1S_e$ and $1P_e$ to the surface trapped states and a radiative recombination (PL) to the level $1S_h$, clear observed in Figure 1(a) to PbS NCs.

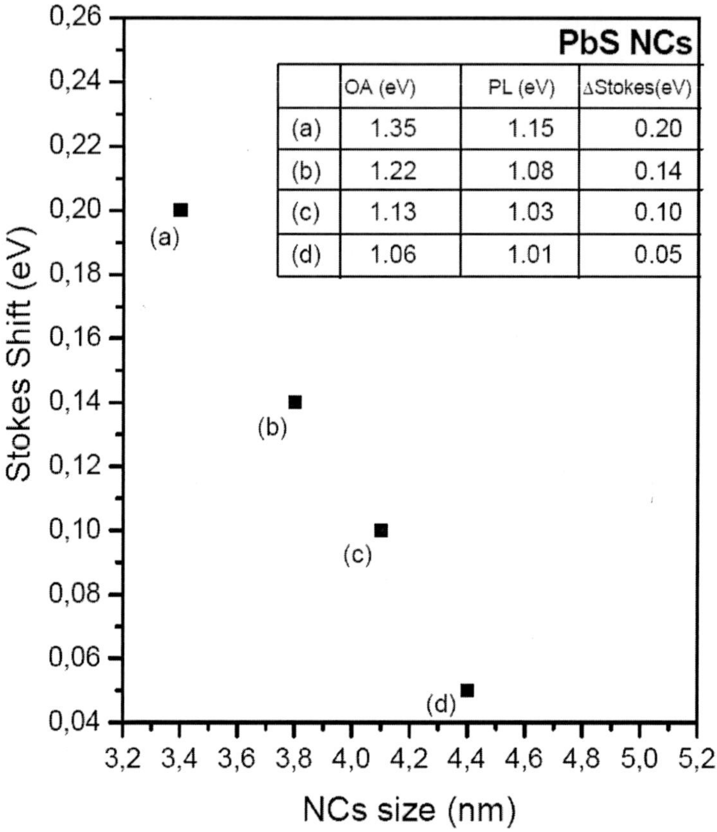

Figure 2. Behavior of Stokes shift to of PbS NCs synthesized in glass matrix.

3.2. Optical Properties Of Zn1-Xmnxo Nanocrystals

Optical absorption spectra provide strong evidences of the Mn^{2+} ions incorporation into the $Zn_{1-x}Mn_xO$ NCs different for the observed in samples

with ZnO NCs. With the introduction of impurities magnetic in semiconductor NCs the optical properties are completely modify due the exchange interactions (*sp-d*) between electronic subsystem of NCs and electrons originated in the partially filled of the Mn^{2+} ions. This exchange interactions causes the blue-shift of band gap observed in $Zn_{1-x}Mn_xO$ NCs in relationship to ZnO NCs that is proportional with the increase of x as show in Figure 4, for example, to x = 0 and 0.0081, being observed the blue-shift of band-gap of 3.33 eV (372 nm) to 3.41 eV (363 nm), respectively. This due the band gap of $Zn_{1-x}Mn_xO$ NCs semiconductor is between the ranges of 3.29 eV (gap ZnO bulk) at 4.2 eV (gap MnO bulk). The appearances of well-defined subband peaks in absorption spectra demonstrate the high quality of the synthesized samples and the relatively small size distribution of the NCs. Using this information with the energy of gap obtained by optical absorption spectra, the NCs size were determined by the effective mass model approximation ofequation 01 [21]:

$$E \cong E_{bulk} + \frac{\hbar^2 \pi^2}{2eR^2}\left(\frac{1}{m_e m_o} + \frac{1}{m_h m_o}\right) \tag{1}$$

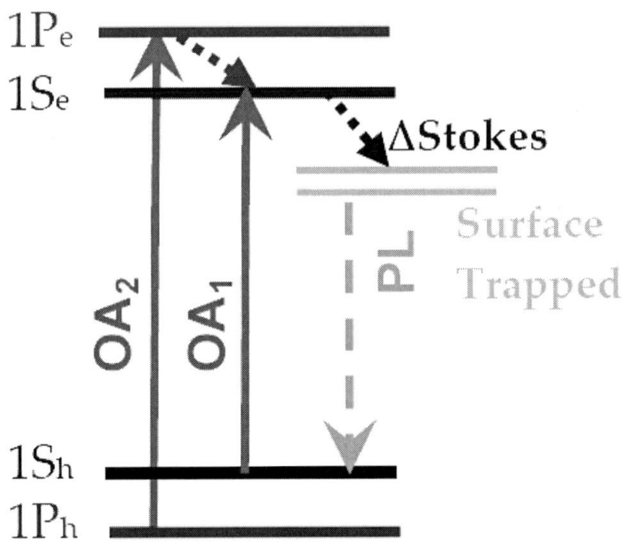

Figure 3. Model schematic used for explain the Stokes shift data difference between OA via radiative and nonradiative emission processes.

Where E is band gap of the nanocrystals, E_{bulk} is the band gap of the bulk material, R is the particle radius, m_e and m_h are effective mass of the electrons and holes, respectively, and m_o is the free electron mass. With the effective masses of electrons ($m_e = 0.28\ m_o$) and holes ($m_h = 0.59\ m_o$), we obtain the diameter of 4.1 nm for the as-prepared $Zn_{0.9919}Mn_{0.0081}$ NCs.

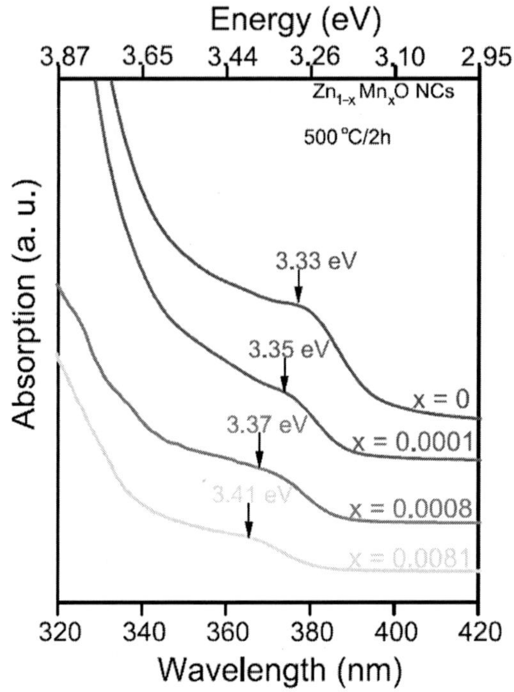

Figure 4. Room-temperature optical absorption spectra of $Zn_{1-x}Mn_xO$ NCs for x = 0, 0.001, 0.0008 and 0.0081.

3.3. Optical Properties Of Cd1-Xmnxs Nanocrystals

Figure 5 shows room-temperature OA spectra of the SNAB : $Cd_{1-x}Mn_xS$ set of samples with nominal concentrations x = 0, 0.001, 0.050, and 0.100. The quantum confinement can be clearly observed, and there are well-defined intersubband transition peaks in all OA spectra, which demonstrates the fairly good quality and narrow dot size distribution (~ 6%) in the samples. For a fixed doping concentration, an effective band gap reduction is observed, with the

quantum confinement regime decreasing for increasing nanocrystal size, R. Analyzing these OA spectra and using a quantum confinement model based on effective-mass approximation, it is possible to estimate the average radius R of these dot samples using the expression [28] $E_{conf}= E_g+ \hbar^2\pi^2/2\mu R^2- 1.8(e^2/\varepsilon R)$, where E_g is the energy band gap of the material (bulk), μ (ε) is the heavy-hole exciton reduced effective mass (dielectric constant),*e* is the elementary charge, and the last term is the electron–hole effective Coulomb interaction.

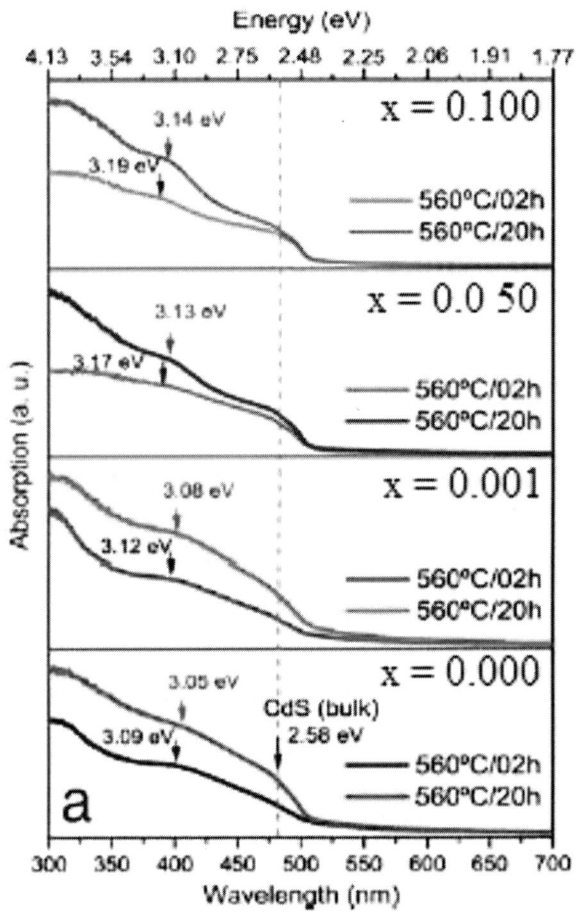

Figure 5. Panel (a): Room-temperature OA spectra of $Cd_{1-x}Mn_xS$ NCs embedded in SNAB glass matrix with concentration x = 0, 0.001, 0.050, and 0.100. The well-defined sub-band peaks demonstrate quantum confinement regimes and the relatively narrow dot size distributions. The energy shift of the OA peaks indicates the Mn incorporation in the dots.

The estimated average radii were R = 2.14 and R = 2.234 nm, for the samples annealed for 2 and 20 h, respectively. Under low magnetic impurity density and annealing temperature conditions, the dot size remained almost unchanged during Mn-ion incorporation induced by thermal treatment. In Fig. 3, we show the blue shift in the OA peaks, a quantity proportional to the concentration of magnetic ions. This shift changes from 3.09 eV in pure CdS (x = 0) dots to 3.19 eV in doped SNAB : $Cd_{0.900}Mn_{0.100}S$ samples treated at 560 °C for 2 h. For samples treated for 20 h, the values are 3.05 eV for x = 0 (undoped) and 3.14 eV for x = 0.100 (doped) dots.

4. MAGNETIC PROPERTIES OF DMS NANOCRYSTALS

The magnetic properties of DMS NCs are influenced by exchange interactions sp-d between the electronic subsystems of magnetic ions with the NCs, changing the configuration of confined electronic states. The $Zn_{1-x}Mn_xO$ and $Cd_{1-x}Mn_xS$ NCs were investigation by Electron Paramagnetic Resonance, $Cd_{1-x}Mn_xS$and $Pb_{1-x}Mn_xS$NCs investigation by Microscopy force Atomic (MFM) and $Pb_{1-x}Mn_xS$ NCs investigated by Magnetization measurements.

4.1. Evidencing The $Zn_{1-x}mn_xo$ Nanocrystal Growth By Electron Paramagnetic Resonance

The EPR spectra of the $Zn_{1-x}Mn_xO$ NC samples with x = 0, 0.0001, 0.0008 and 0.0081 are shown inFigure 6(a). We found ZnO NCs sample (x = 0) presenting a sharp EPR signal with *g* = 1.9568. However, the $Zn_{1-x}Mn_xO$ NC samples with low x-values exhibits a well-resolved EPR sextet in addition to some fine structure. They are stemmed from the hyperfine interaction between electron (*S*=5/2) and nuclear (*I* = 5/2) spins of the incorporated manganese ions, which are claimed to be located at different positions within the nanocrystal. For instance, the x = 0.0001 sample displays two sets of EPR sextets. Among them, the six well-defined EPR lines with hyperfine interaction splitting of 7.8 mT is assigned to isolated Mn^{2+}-ions substitutionaly incorporated into the $Zn_{1-x}Mn_xO$-core nanocrystal. While, the second set of EPR sextet structure, with hyperfine splitting of 8.0 mT, is due to Mn^{2+}-ions incorporated into the $Zn_{1-x}Mn_xO$-shell nanocrystal, at crystallographically-distorted sites near the NCs surface [27]. For this sample the smallest average diameter among the doped samples presents the strongest surface effect. Therefore, in comparison to the undistorted Mn-site (core-sites) the x = 0.0001 $Zn_{1-x}Mn_xO$ NCs sample holds enough distorted Mn-sites (shell-sites) to be probed by EPR in spite of low

doping concentration. In addition, the sensitivity of the EPR technique to prove both core-like as well as shell-like Mn-ions with concentration as low as x = 0.0001 is a strong indication of the monodispersity in size. As the Mn-ion concentration increases, the number of EPR lines also increases, and the identified hyperfine structure is now superimposed to a broad EPR background line, as shown in EPR spectrum of the x = 0.0008 NCs sample. The underlying physics can be understood in the following way. As the Mn-ion concentration increases, the amount of Mn-ions in shell-sites increases. Hence, the replacement of Zn-ion by Mn-ions in various crystal sites with different distortion of crystal field occurs, resulting in a multi-line hyperfine structure. Furthermore, the increased concentration of manganese in the shell of nanocrystals may lead to the formation of Mn ion cluster, which inducing a strong Mn-Mn interaction. Hence broad EPR background line emerges. As the manganese concentration goes over x = 0.0081, however, the multi-line hyperfine structure collapses and the EPR spectra are replaced by a symmetric, broad single line due to enhanced Mn-Mn interaction. In order to confirm this analysis we have performed EPR spectral simulation of Mn-doped ZnO NC samples using time dependent perturbation theory [27], in which the spin-Hamiltonian is described by $\hat{H} = \hat{H}_z + \hat{H}_0$. In the spin-Hamiltonian $\hat{H}_z = \mu_e \hat{S} \cdot g_e \cdot \vec{B}$ is the Zeeman term, μ_e, g_e, and $\vec{B}$

are the Bohr magneton, the Lande factor and the applied magnetic field, respectively. The second term of the spin-Hamiltonian is $\hat{H}_0 = D\left[S_z^2 - S(S+1)/3\right] + E\left(S_x^2 - S_y^2\right) + A\hat{S} \cdot \hat{I}$, where the first two terms describe the zero-magnetic field fine-structure splitting due to spin-spin interaction of electrons, which is nonzero only in environments with symmetries lower than cubic. The third term ($A\hat{S} \cdot \hat{I}$) is stemmed from the hyperfine interaction between electron and nuclear spins, leading to the observed six-line pattern. Since the interaction constants A, D and E strongly depend upon the local crystal field characteristics in which the Mn^{2+}-ion is located, the EPR spectrum varies when the local Mn^{2+}-ion crystal symmetry changes from $Zn_{1-x}Mn_xO$-core to $Zn_{1-x}Mn_xO$-shell. According to the pattern of the EPR spectra we found from our data, spectral simulation were performed by following a three step procedure, as illustrated in Figure 5 (a). Firstly, we computed the hyperfine structure. Second, a broad background resonance feature was simulated. Finally, we summed over these two spectra to end up with the EPR spectrum of the $Zn_{1-x}Mn_xO$ NCs sample with x = 0.0008. Figure 6 (b) displays the calculated spectra for x = 0.0001, 0.0008 and 0.0081. Excellent quantitative agreement between the simulated and the experimental spectra were achieved for instance in the case of the x = 0.0008 NCs (inset) sample using g = 2.0033, A = 7.8mT D = 6.1 mT, E= 0.5 mT, and 0.8 mT linewidth for the hyperfine structure simulation and g = 2.0033 and 60 mT linewidth for the broad background resonance calculation. Therefore, we could conclude that the EPR hyperfine six-

lines are due to $m_S = 1$ and $m_I = 0$ transitions, where m_S (m_I) stands for the projection of the spin S (I), the broad background is originated from the exchange narrowing due to the strong Mn-Mn interaction. Hence, the EPR simulations strongly support the picture that Mn-ions are incorporated into the hosting ZnO nanocrystals.

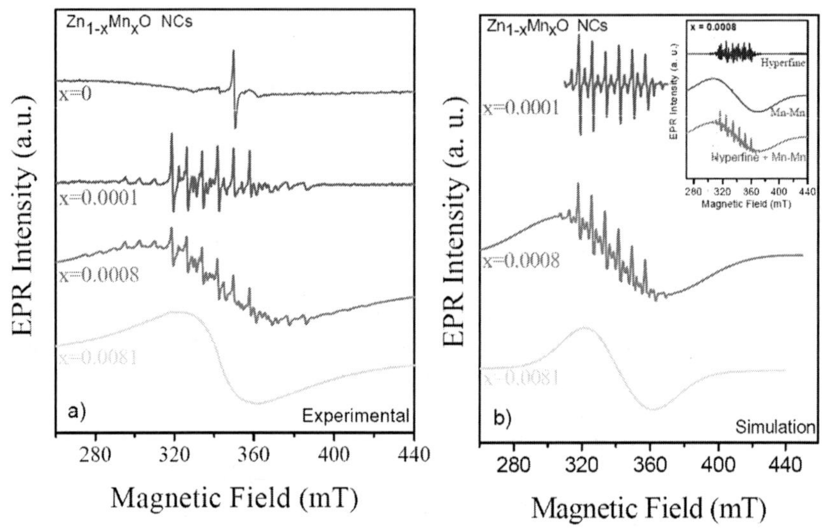

Figure 6. a). Room-temperature X-band EPR spectra of $Zn_{1-x}Mn_xO$ nanocrystals at different x values and (b) Simulated EPR spectra of $Zn_{1-x}Mn_xO$ nanocrystals for different x values. In inset illustration of processes used for EPR spectra simulation of the $Zn_{1-x}Mn_xO$ nanocrystal (x = 0.0008).

4.2. Confirming The Migration Process Of Mn^{2+} Ions In $Cd_{1-x}mn_xS$ Nanocrystals By Electron Paramagnetic Resonance

Theoretical model for explain the incorporation of magnetic impurities in nanocrystals are related in the literature [22-24], such as the "self-purification" mechanisms that are explained through energetic arguments. These mechanisms show that the formation energy of magnectic impurities increases when the NCs size decreases. Moreover, the binding energy of the impurities in

the crystalline faces is highly dependent on the semiconductor material, such as the crystal structure and NCs shape [23].

The energy required to replace a Cd^{2+} ion by an Mn^{2+} ion, called the formation energy, is greater for smaller $Cd_{1-x}Mn_xS$ NCs [22]. Thus, the present Mn^{2+} ions in smaller $Cd_{1-x}Mn_xS$ NCs are less stable, promoting the difusion of some of these impurities to regions closer to the dot surface, i.e., to the site S_{II}. This ability of the Mn^{2+} ion to diffuse through the nanocrystal is quite reasonable because the ionic radius of impurity (83 pm) is smaller than the ionic radius of the Cd^{2+} ion (95 pm). This mechanism, known as "self-purification" [24], is an intrinsic property of impurities (or defects) in semiconductor related to NCs size-dependent energetic arguments and explains the predominance of the signal S_{II} on the S_I in the EPR spectra shown in the Figure 7.

The EPR spectra shown in Figure 7(a) for selected SNAB : $Cd_{1-x}Mn_xS$ samples with x = 0.001, 0.050, and 0.100 display well-resolved transitions between a sixplet sublevel structure inserted into a broader horizontal S-like shaped EPR structure typical for free-like electron states with spin S = 1/2,•±1/2. These six sublevels are associated with the magnetic quantum numbers M_S= ±5/2,± 3/2, ± 1/2 that occur in fine exchange interaction induced transitions when Mn ions are present in a sample. Another much weaker interacting sixplet sublevel structure occurs because the hyperfine interaction coupling between the spin of localized electrons (S = 1/2) with nuclear spin (I = 5/2) of the incorporated Mn-ions in the doped samples. Among them, are six fairly welldefined EPR lines with fine interaction splitting described by the simulation signal SI, and assigned to dilute concentration of Mn^{2+}ions found in substitutional Cd places inside the CdS nanocrystal cell. However, the second EPR sixplet structure set described by the simulation signal SII is due to Mn ions located at crystallographically highly distorted sites near the dot surface.

In order to confirm this analysis we have performed EPR spectral simulation of Mn-doped CdS samples using time-dependent perturbation theory in which the spin Hamiltonian is given by $H = H_Z + H_0$, where $H_Z = \mu_e \vec{S} \cdot g_e \cdot \vec{B}$ describes the Zeeman interaction and μ_e, g_e, and $\vec{B}$ represent the Bohr magneton, the Landé g-factor, and the applied magnetic field, respectively. The second term, $H_0 = D\left[S_z^2 - S(S+1)/3\right] + E\left(S_x^2 - S_y^2\right) + A\hat{S}\cdot\hat{I}$, includes the zero magnetic field fine interaction (terms proportional to coupling constants D and E) between the electron spin and the crystal field.

This contribution only induces non-zero fine structure splitting in crystalline environments with symmetries lower than cubic. Finally, the term $A\hat{S}\cdot\hat{I}$ represents the hyperfine interaction between localized electrons and nuclear spins in the Mn ions and where each electron transition splits into six

additional levels characterized by the nuclear magnetic quantum numbers (M_I= ±5/2,±3/2,±1/2), producing, in principle, a total of 36 transitions. However, selection rules limit the number of allowed transitions, and the broader and stronger features observed in the EPR spectra are due to lines associated to the dipole-allowed ΔM_S= ±1 transitions with ΔM_I= 0. Since the interaction constants A, D, and E strongly depend on the local crystal field characteristics near the Mn^{2+}ion location, the EPR spectrum will be changed when substitutional Mn^{2+}ions move from the $Cd_{1-x}Mn_xS$ dot core (signal S_I) to the dot surface (signal S_{II}) region. On the other hand, any isolated Mn ion that is dispersed inside the glass matrix cannot be identified in the EPR spectra due to the absence of a well-defined crystal field of this amorphous material (glass). In addition, the Mn–Mn interactions would also be intensified if the formation of Mn clusters occurred in the glass environment. However, we consider that this later effect may be neglected, since the formation of Mn clusters is highly unlike due to the small amount of incorporated Mn ions in these samples.

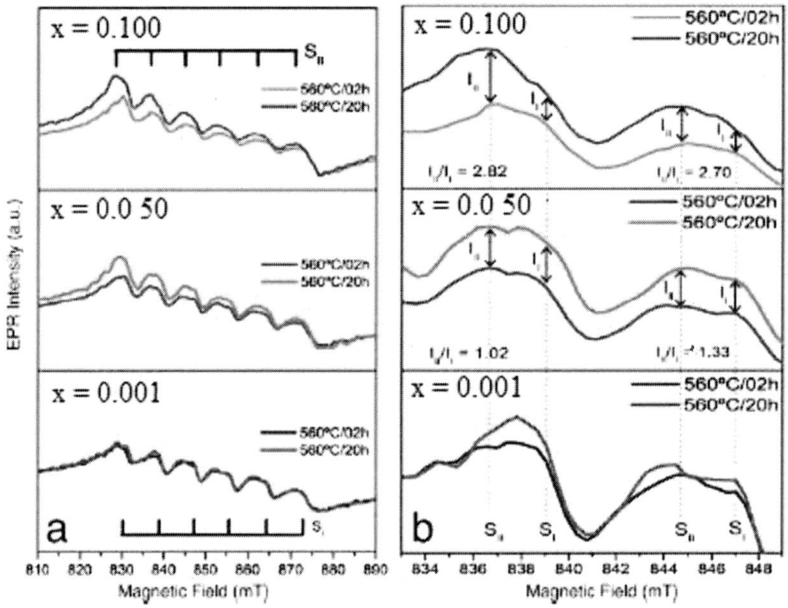

Figure 7. Panel (a): Room-temperature EPR spectra of selected $Cd_{1-x}Mn_xS$ NCs embedded in the SNAB glass matrix with Mn concentration x = 0.001, 0.050, 0.100. Panel (b) shows a zoom on two of the sublevel peaks, between 835 and 850 mT, in order to observe the intensity differences, marked as $I_{II}$and I_I. Noted the change of the ratio between EPR intensities when Mn ions are placed at the dot interface (I_{II}) or inside the core cell (I_I), in samples with concentrations x = 0.050 and x = 0.100. This ratio change indicates the interdiffusion of magnetic ions due to thermal treatment.

The other enhancement on the EPR signal intensity, as observed in Figure 7(a), can be associated to the increased concentration of Mn^{2+}ions. The larger the Mn density is, in samples subjected to thermal treatment for 20 h and also seen in the AFM images, the stronger is the EPR intensity. In addition, the EPR intensity can be strengthened due to migration of a fraction of magnetic ions from the glass matrix to the surface of the NCs, an effect which also increases the intensity of the broader background EPR peak due to Mn–Mn interaction.

The zoom in two of the sublevel exchange peaks, between 835 and 850 mT, displayed in Figure 7(b)shows the enhancement of all EPR lines for longer annealing times. However, the EPR intensity for surface Mn ions I_{II} increases whereas for the substitutional ions I_I decreases in both samples with nominal concentrations x = 0.050 and x = 0.100, when the thermal treatment increases from 2 to 20 h. The change in the ratio between intensities, I_{II} / I_I, confirms the migration of a fraction of Mn ions incorporated in the dot core to the surface region.

The energy required to replace $Cd^{2+}$by Mn^{2+}in the crystal, usually referred to as the "formation energy", is larger for small- size CdS NCs [9]. Therefore, Mn^{2+}ions in the less stable places inside $Cd_{1-x}Mn_xS$ dots are moved to interface region in a process of energy minimization that promotes the interdiffusion of a fraction of impurities from the core (signal S_I) to larger binding energy regions near [24] to the surface and, in this location, the magnetic ions generate the signal S_{II}. The ability of Mn^{2+}ions to migrate through the NC is quite reasonable because the ionicradius of this magnetic impurity (83 pm) is smaller than the ionic radius of the Cd^{2+}ion (95 pm). This mechanism seems to be an intrinsic general property of impurities (or defects) in semiconductor NCs [24], and the shape and the crystal structure of a NC determine which surface is more favorable for impurity binding; these facets are the (0001) crystalline planes in the CdS wurtzite structure [9]. Here, the larger ratio III/II between intensities of EPR peaks in samples with different annealing times, as shown in the Figure 7(b), corroborates this idea of controlled migration process.

The simulated signals for the EPR transitions in the $Cd_{0.900}Mn_{0.100}$ S sample, shown in Figure 8, confirm the presence of Mn^{2+} ions in two distinct sites: incorporation occurring in the core (signal S_I), and near the surface (signal S_{II}). The hyperfine interaction constants used to simulate these spectra were A_I= 8.1 mT and A_{II}= 8.4 mT for a magnetic system with spin S = 5/2, nuclear spin I = 5/2, and g-factor g_e= 2.005. The fine structure constants were D = 40 mT and E = 5 mT. We believe that these results have shown unambiguously that proper use of annealing temperature with different times may produce controlled diffusion of Mn^{2+}ions in magnetic dots. These findings are strongly supported by the fairly good agreement between simulated and experimental EPR resonant transitions.

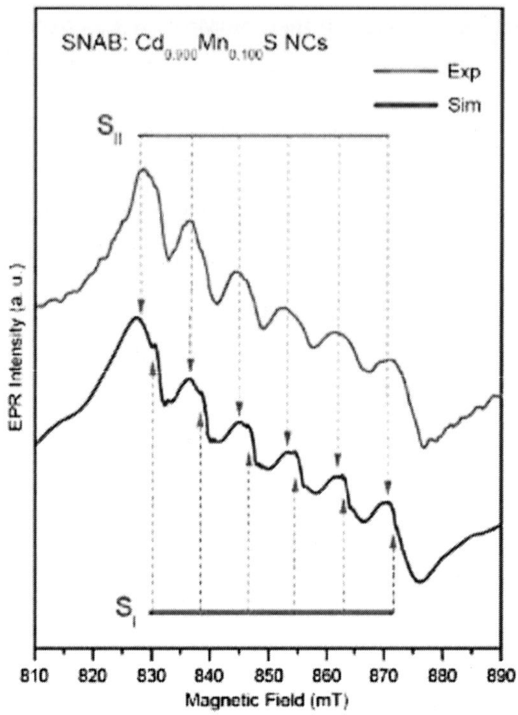

Figure 8. Room-temperature EPR spectrum in $Cd_{0.900}Mn_{0.100}S$ NCs measured in the K band. The simulated EPR transitions were obtained by the combination of a stronger spectrum associated to Mn replacing the Cd ions inside the wurtzite cell (signal S_I) and a weaker spectrum due to ions located at or near the dot surface region (signal S_{II}). The Hamiltonian model describing the fine and hyperfine magnetic interactions contributing to dipole transitions between the eigenvalues is discussed in the text.

4.3. Investigating The $Pb_{1-x}mn_xs$ Nanocrystals By Susceptibility And Magnetization Measurements

Figure 9 shows the temperature dependence of the inverse of the real part of the magnetic susceptibility (1/χ) of $Pb_{1-x}Mn_xS$ NCs growth in borosilicate glass with annealingat 500°C at 10hours, for x = 0.003 and 0.007, recorded at 100 Oe. The susceptibility data presented in Figure 9 follow the Curie-Weiss law, $\chi(T) = C/(T-\theta)$, , where θ is the Curie-Weiss temperature [25, 26]. The fitted value found for the parameter C in χ(T) allowed estimation of the isolated

Mn^{2+} ion content incorporated in the $Pb_{1-x}Mn_xS$ NCs embedded in glass matrix using:

$$x = (m_A + m_B) / [S(S+1)(g\mu_B)2N_A / (3k_B C) + m_A - m_M] \tag{2}$$

where m_A, m_B and m_M represent atomic mass for the cation (Pb), anion (S) and magnetically doped (Mn), respectively. S represents the total spin of the Mn^{2+} ion, k_B is the Boltzmann constant, and $N_A$is Avogadro's number. In this calculation g = 2 is assumed for the Mn^{2+} ion.

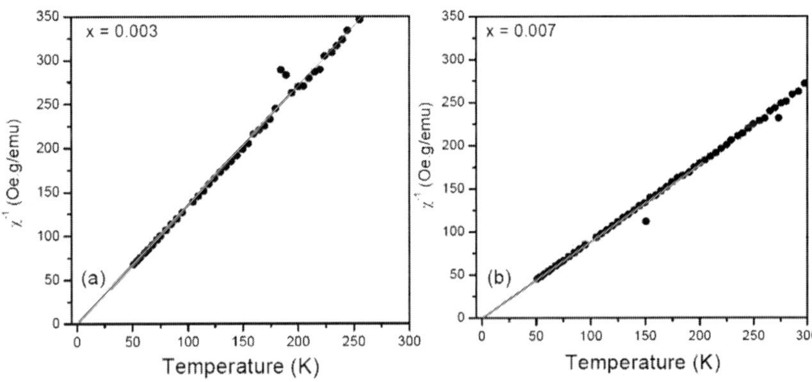

Figure 9. The temperature dependence on the inverse of susceptibility, recorded at 100 Oe, shows low temperature paramagnetic character for $Pb_{1-x}Mn_xS$ NCs embedded in borosilicate glass matrix.

Usingthe equation 02, estimates theconcentration ofMnionsincorporatedinto the crystal latticeof the nanocrystals. For thenominal concentrationsof x= 0.003and x =0.007 were estimatedreal concentrationsof x= 0.003and x =0006 respectively,resulting in theformationof$Pb_{0.997}Mn_{0.003}S$ and $Pb_{0.994}Mn_{0.006}S$NCs. Field-cooled magnetization curves of$Pb_{0.997}Mn_{0.003}S$ and $Pb_{0.994}Mn_{0.006}S$NCs growth in borosilicate glass matrix, recorded with an external applied field of 1 KOe and over a wide range of temperatures, are shown in Figures 9(a) and (b), respectively. Each inset in these figures displays the T=1.28 K field dependence of the magnetization.

The magnetic data presented in Figure 9 including the corresponding insets, reveal the dominant paramagnetic behavior of the dots.

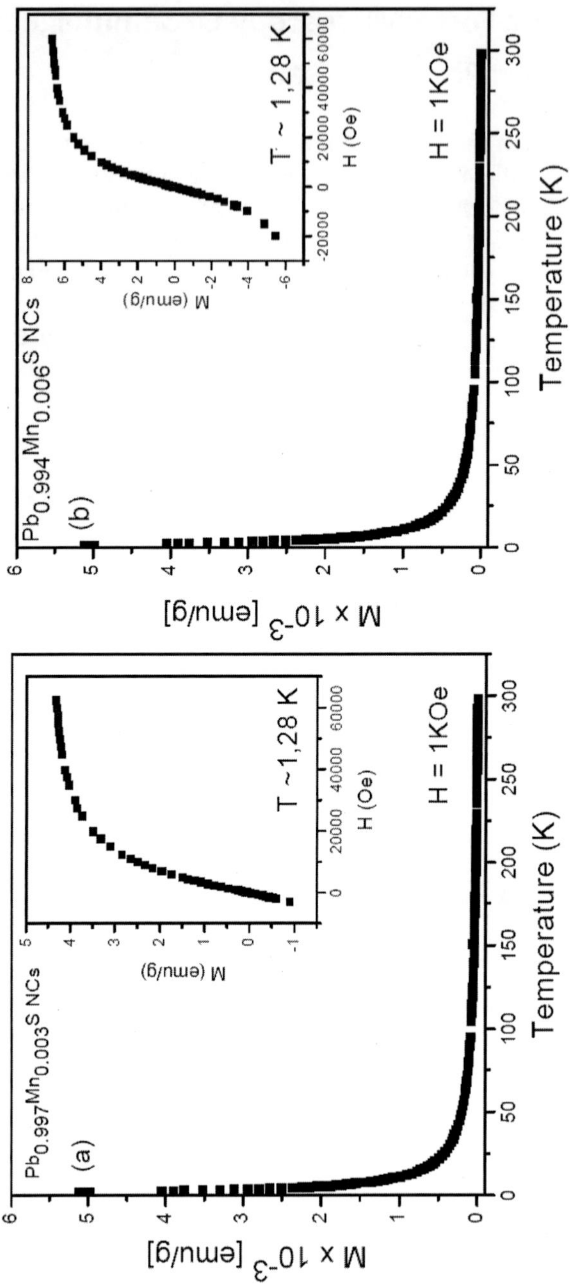

Figure 10. Magnetization as a function of temperature for the $Pb_{0.997}Mn_{0.003}S$ and $Pb_{0.994}Mn_{0.006}S$ NCs embedded in borosilicate glass matrix. Field dependence of magnetization for each sample is shown in the insets.

4.4. Magnetic Force Microscopy Of Semimagnetic Nanoparticles: $Pb_{1-x}mn_xs$ Nanocrystals

Figure 10 presents the AFM/MFM images for the $Pb_{0.993}Mn_{0.007}S$ NCs samples subjected to the thermal annealing for 10 h, where we confirmed the high density of the nanocrystals with quantum confinement properties as well as bulk-like properties. In figure 10 (a), the AFM image is also influenced by the sample topography, while in figure 10(b), in the MFM images there are only the magnetic interactions, confirming the formation of semimagnetic $Pb_{0.993}Mn_{0.007}S$ NCs, with average NCs-size of approximately 6.0 nm. In an attractive configuration, the NCs have magnetization in a parallel direction to the tip magnetization, resulting in dark areas of the MFM image. However, in a repulsive configuration, the NCs have magnetization in an antiparallel direction to the tip magnetization, resulting in bright areas of the MFM image. Therefore, the formation of the dark and bright fields related to a single spin domain in the $Pb_{0.993}Mn_{0.007}S$ NCs is shown by the clear contrast in these MFM images.

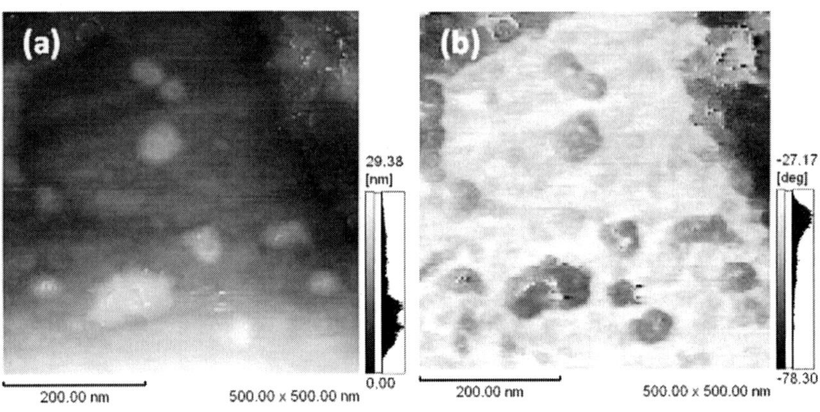

Figure 11. AFM/MFM of $Pb_{0.993}Mn_{0.007}S$ NCs growth in borosilicate glass matrix.

5. STRUCTURAL PROPERTIES OF $ZN_{1-X}MN_XO$ NANOCRYSTALS

The structural properties of $Zn_{1-x}Mn_xO$ nanocrystals were investigated by X-ray diffraction and Raman spectroscopy techniques.

5.1. Characterizing Of $Zn_{1-x}mn_xo$ Nanocrystals By X-Ray Diffraction

The XRD patterns of the $Zn_{1-x}Mn_xO$ ($x \geq 0$) NCs samples for x = 0, x = 0.0001, x = 0.0008 and x = 0.0081 are shown in Figure 11. It is noted that the typical bulk-ZnO hexagonal wurtzite crystal structure is preserved in the as-precipitated $Zn_{1-x}Mn_xO$ ($x \geq 0$) NCs samples whit treatment thermal of 500°C by 2 hours [27]. Except for the ZnO with treatment thermal the 60°C by 12 hours that is present in amorphous phase. Nevertheless, the characteristic XRD peaks shift towards lower diffraction angle values as the Mn-ion concentration in the hosting ZnO structure increases, as shown in inset for (002) peak, clearly indicating an increase of the lattice constant. Using the Cohen's method we performed the estimation of the c-axis lattice constant from the following three selected XRD peaks: (100), (002), and (101). One found that the average c-axis lattice crystal constant of the x = 0, 0.0001, 0.0008 and 0.0081 $Zn_{1-x}Mn_xO$ NC samples were 5.207, 5.208, 5.226 and 5.231 Å, respectively. The monotonic increase observed in the lattice crystal constant is attributed to the replacement of the Zn^{2+}-ion, with smaller ionic radius (0.74 Å in the hexagonal wurtzite ZnO crystal structure), by Mn^{2+}-ion with larger ionic radius (0.83 Å). These XRD findings provide evidences that the as-produced samples are high-quality and single-phased $Zn_{1-x}Mn_xO$ crystals in the nanoscale regime.

5.2.Characterizing. Of $Zn_{1-x}mn_xo$ Nanocrystals By Raman Spectroscopy

Figure 12 shows the Raman spectra of ZnO bulk and $Zn_{1-x}Mn_xO$ NCs for x = 0, 0.0008 and 0.0081. The wurtzite ZnO structure belongs to the space group C_{6v}^4 with two formula units per primitive cell. Therefore, group theory predicts that the zone-center optical phonons are described by $\Gamma_{opt} = A_1 + 2B_1 + E_1 + 2E_2$ [5]. The A_1 and E_1 modes represent Raman and infrared active polar phonons, showing frequencies for transverse-optical (TO) and longitudinal-optical (LO) modes. The E_2 mode is non-polar and is Raman active in two frequencies; the E_2 (high) associated to the oxygen anions and the E_2 (low) associated to the Zn cations in the lattice. Finally, the B_1 mode is Raman inactive. The Raman peak centered at 334 cm^{-1} is described through a multi-phonon process associated to three different modes; the dominant A_1 mode plus a weak E_2 component and an even weaker E_1component. The literature describes the frequency of this mode as the difference between the E_2 (high) and E_2 (low) modes. The A_1 (TO), E_1 (TO), A_1 (LO) and E_1 (LO) modes were observed at 382, 410, 541 and 586 cm^{-1}, respectively. The Raman peak

observed at 439 cm^{-1} represents the E_2 (high) mode associated to the oxygen anions. The Raman peaks of $Zn_{1-x}Mn_xO$ NCs show a shift to low frequencies and the peaks are asymmetry in relation to the ZnO semiconductor. With the incorporation of Mn^{2+}ions is observed a peak was observed at about 659 cm^{-1}becoming more intense with increasing of Mn-concentration in $Zn_{1-x}Mn_xO$ NCs. The 659 cm^{-1} peak is associated the two additional modes $\left[A_1(LO)+E_2(low)\right]$ originated from the precipitation phase of $ZnMn_2O_4$ as observed for x = 0.0008 and 0.0081 in the samples of $Zn_{1-x}Mn_xO$ NCs. This finding is a strong support to the picture that Zn-ions in the ZnO crystal structure are replaced by Mn-ions during the course of the chemical precipitation process.

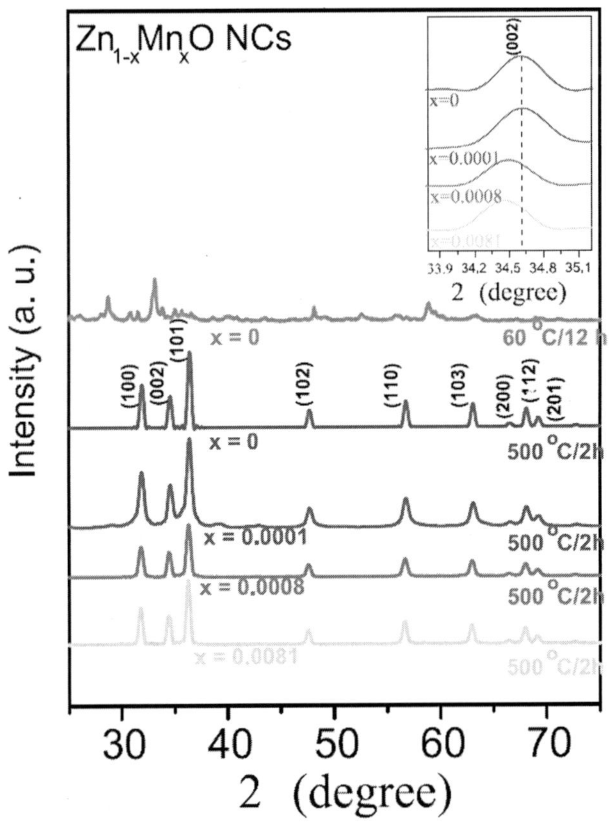

Figure 12. X-ray diffraction of ZnO bulk and $Zn_{1-x}Mn_xO$ nanocrystal samples with x = 0, 0.0008 and 0.0081. In inset shift of the (002) $Zn_{1-x}Mn_xO$ nanocrystal XRD peak as a function of Mn-ion concentration (x).

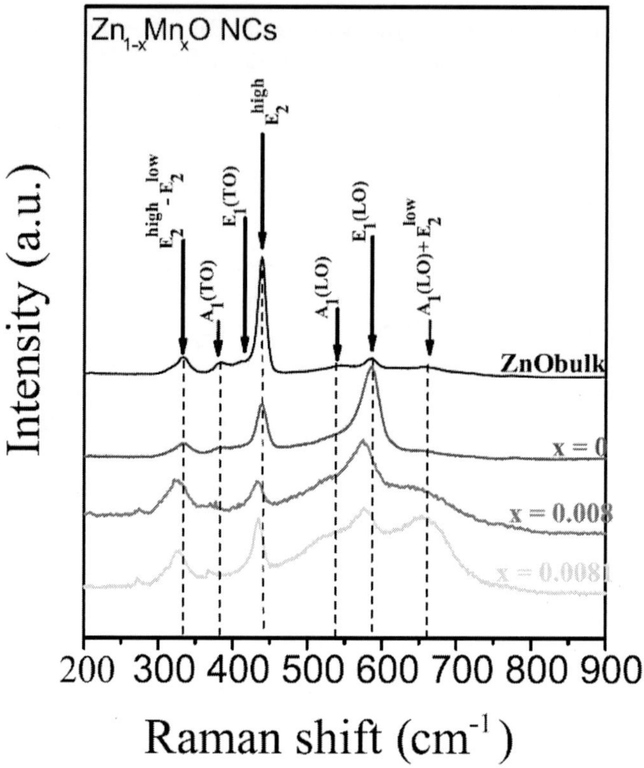

Figure 13. Room-temperature Raman spectra of ZnO bulk and $Zn_{1-x}Mn_xO$ NCs for x = 0, 0.0008 and 0.0081.

6. CONCLUSIONS

In conclusions, we report the successfully synthesis of semiconductors and semimagnetic nanocrystals by different methodologies.

PbS, $Pb_{1-x}Mn_xS$ and $Cd_{1-x}Mn_xS$ NCs were growth in borosilicate glass by fusion method and $Zn_{1-x}Mn_xO$ NCs were synthesized by co-precipitation method.

The investigation of semiconductor and semimagnetic NCs provided by experimental techniques of Optical Absorption, Photoluminescence, Electron Paramagnetic Resonance, Magnetic Force Microscopy, X-Ray Diffraction, and Raman spectroscopy, have revealed the control of optical, magnetic and structural properties and the high-quality of NCs synthesized by different methodologies.

We believe that this chapter can motivate inspire further investigation of these systems in a search for possible device applications.

ACKNOWLEDGEMENT

The authors gratefully acknowledge the financial support from the Brazilian agencies: MCT/CNPq, Capes, Fapemig and FUNEPU. We are also grateful to our collaborators: Augusto Miguel Alcalde (A. M. Alcalde), Eliane da Costa Vilela (E. C. Vilela), Felipe Chen Abrego (F. Chen), Fernando Pelegrini (F. Pelegrini), Gilmar Eugênio Marques (G. E. Marques), Henry Socrates Lavalle Sullasi (H. S. L. Sullasi), Jales Franco Ribeiro da Cunha (J. F. R. Cunha), Leonardo damigo (L. Damigo), Kely Lopes Caiado Miranda (K. L. Miranda), Marcelo de Assumpção Pereira da Silva (M. A. Pereira-da-Silva), Miguel Alexandre Novak (M. A. Novak), Patrícia Pommé Confessori Sartoratto (P. P. C. Sartoratto), Paulo César de Morais (P. C. Morais), Qu Fanyao (Qu Fanyao) and Victor Lopez Richard (V. Lopez-Richard).

REFERENCES

1. D. J. Norris, A. L. Efros, S. Erwin, 2008 (2008) Doped Nanocrystals. Science 319 17761779 .

2. N. O. Dantas, Fanyao. Qu, R. S. Silva, P. Morais, 2002 (2002) Optical, Magnetic, and Structural Properties of Semiconductor and Semimagnetic Nanocrystalshys. Chem. B. 106 74537457 .

3. Silva R. S, Morais P. C, Qu fanyai, Alcalde A. M, Dantas N. O, Sullasi H. S (2007) Synthesis process controlled magnetic properties of Pb1-xMnxS nanocrystals. App. Phy. Lett.. 90: 253114-1-253114-3.

4. E. S. F. Neto, N. O. Dantas, N. M. B. Neto, I. Guedes, Chen, 2011, Dantas N. O, Neto N. M. B, Guedes I, Chen F (2011) Optical, Magnetic, and Structural Properties of Semiconductor and Semimagnetic Nanocrystals709.

5. N. O. Dantas, L. Damigo, Fanyao. Qu, Fanyao. Qu, J. F. R. Cunha, R. S. Silva, K. L. Miranda, E. C. Vilela, P. P. C. Sartoratto, P. Morais, 2008unha J. F. R, Silva R. S, Miranda K. L, Vilela E. C, Sartoratto P. P. C, Morais P. C (2008) Raman investigation of ZnO and Zn1-xMnxO nanocrystals synthesized by precipitation method. J. Non-Cryst. Solids 354 48274829 .

6. S. V. Gaponenko, 1998 Optical, Magnetic, and Structural Properties of Semiconductor and Semimagnetic Nanocrystals, Cambridge University Press. 260 p.10.1017/CBO9780511524141

7. N. O. Dantas, E. S. F. Neto, R. S. Silva, 2010 Diluted Magnetic semiconductors in Glass matrix. In: Masuda Y. Nanocrystals. Sciyo: InTech. 143168 .

8. I. Gur, N. A. Fromer, M. L. Geier, A. Alivisatos, 2005 Air-stable all-inorganic nanocrystal solar cells processed from solution. Science 310: 462-465.

9. S. C. Erwin, L. Zu, M. I. Haftel, A. L. Efros, T.. Kennedy, D. Norris, 2005 (2005) Doping semiconductor nanocrystals. Nature: 436 9194 .

10. D. Timmerman, J. Valenta, K. Dohnalová, de , 1 W. D. A. M, Gregorkiewicz T (2011) Step-like enhancement of luminescence quantum yield of silicon nanocrystals. Nature Nanotechnology 6 710713 .Volume:

11. Vach 2011 (2011) Optical, Magnetic, and Structural Properties of Semiconductor and Semimagnetic Nanocrystals. Nano Lett. 11 54775481 .

12. J. K. Furdyna, 1988 Optical, Magnetic, and Structural Properties of Semiconductor and Semimagnetic Nanocrystals. J. Appl. Phys. 64: R29 - R64.

13. 2010, Lee D. W, Shen S, Tivakornsasithorn K, Son J. S, Park J.-H, Kim Y.-W, Hwang G. S, Dobrowolska M, Furdyna J. K, Hyeon T (2010) Giant Zeeman splitting in nucleation-controlled doped CdSe:Mn2+ quantum nanoribbons. Nature Materials 9 4753 .

14. N. O. Dantas, P. M. N. de Paula, R. S. Silva, V. López-Richard, G. Marques, 2011 (2011) Radiative versus nonradiative optical processes in PbS nanocrystals. J. Appl. Phys. 109: 024308-1- 024308-4.

15. N. O. Dantas, E. S. F. Neto, R. S. Silva, F. Chen, M. A. Pereira-da-Silva, V. López-Richard, G. Marques, 2012.S.F, Silva R. S, Chen F, Pereira-da-Silva M. A, López-Richard V, Marques G. E (2012) The migration of Mn2+ ions in Cd1-xMnxS nanocrystals: thermal annealing control. Solid State Communications 5 337340 .

16. R. S. Silva, P. C. Morais, E. Mosiniewicz-Szablewska, R. F. Cuevas, J. C. P. Campoy, F. Pelegrini, Fanyao. Qu, N. Dantas, 2008 (2008) Synthesis and Magnetic Characterization of Pb1 -xMnxS Nanocrystals in Glass Matrix. J. Phys. D: Appl. Phys. 41: 165005-1- 165005-5.

17. N. O. Dantas, F. Pelegrini, M. A. Novak, P. C. Morais, G. E. Marques, R. S. Silva, 2012 Optical, Magnetic, and Structural Properties of Semiconductor and Semimagnetic Nanocrystals: 106206-1- 106206-5.

18. N. O. Dantas, L. Damigo, Fanyao. Qu, R. S. Silva, P. P. C. Sartoratto, K. L. Miranda, E. C. Vilela, F. Pelegrini, P. C. Morais, 2008 Structural and magnetic properties of ZnO and Zn1-xMnxO nanocrystals J. Non-Cryst. Solids 354 47274729 .

19. J. Zhang, Jiang, 2008 (2008) Optical, Magnetic, and Structural Properties of Semiconductor and Semimagnetic Nanocrystals. J. Phys. Chem. B 112 95579560 .

20. A. Bagga, P. K. Chattopadhyay, Ghosh, 2006 (2006) Optical, Magnetic, and Structural Properties of Semiconductor and Semimagnetic Nanocrystals. Phys. Rev. B 74: 035341-1- 035341-7.

21. L. Brus, 1983 (1983) A Simple-Model For the Ionization-Potential, Electron-Affinity, and Aqueous Redox Potentials of Small Semiconductor Crystallites. J. of Chem. Phys. 79 55665571 .

22. N. O. Dantas, E. S. F. Neto, R. S. Silva, F. Chen, M. A. Pereira-da-Silva, V. López-Richard, G. Marques, 2012.S.F, Silva R. S, Chen F, Pereira-da-Silva M. A, López-Richard V, Marques G. E (2012) The migration of Mn2+ ions in Cd1-xMnxS nanocrystals: thermal annealing control. Solid State Communications 5 337340 .

23. D. J. Norris, A. L. Efros, S. Erwin, 2008 (2008) Doped Nanocrystals. Science 319 17761779 .

24. G. M. Dalpian, J. Chelikowsky, 2006 (2006) Optical, Magnetic, and Structural Properties of Semiconductor and Semimagnetic Nanocrystals Phys. Rev. Lett. 96, 226802 EOF -1- 226802-4.

25. N. O. Dantas, F. Pelegrini, M. A. Novak, P. C. Morais, G. E. Marques, R. S. Silva, 2012 Optical, Magnetic, and Structural Properties of Semiconductor and Semimagnetic Nanocrystals: 106206-1- 106206-5.

26. M. Górska, J. Anderson, 1988 (1988) Magnetic susceptibility and exchange in IV-VI compound diluted magnetic semiconductors. Phys. Rev. B 38 91209126 .

27. N. O. Dantas, L. Damigo, Fanyao. Qu, R. S. Silva, P. P. C. Sartoratto, K. L. Miranda, E. C. Vilela, F. Pelegrini, P. C. Morais, 2008 Structural and

magnetic properties of ZnO and Zn1-xMnxO nanocrystals. J. Non-Cryst. Solids 354 47274729 .

28. L. E. Brus, 1984 Electron-electron and electronhole interactions in small semiconductor crystallites: The size dependence of the lowest excited electronic state. J. of Chem. Phys. 80 44034409 .

Chapter 5

Surface Properties and Photocatalytic Activities of the Colloidal ZnS:Mn Nanocrystals Prepared at Various pH Conditions

Jungho Heo and Cheong-Soo Hwang *

Department of Chemistry, Institute of Nanosensor and Biotechnology, Center for Photofunctional Energy Materials (GRRC), Dankook University, 152 Yongin-si, Suji-ku, Jukjeon-ro, Gyunggi-do 448-701, South Korea

ABSTRACT

Water-dispersible ZnS:Mn nanocrystals (NC) were synthesized by capping the surface with mercaptoacetic acid (MAA) molecules at three different pH conditions. The obtained ZnS:Mn-MAA NC products were physically and optically characterized by corresponding spectroscopic methods. The UV-Visible absorption spectra and PL emission spectra showed broad peaks at 310 and 590 nm, respectively. The average particle sizes measured from the HR-TEM images were 5 nm, which were also supported by the Debye-Scherrer calculations using the X-ray diffraction (XRD) data. Moreover, the surface charges and the degrees of aggregation of the ZnS:Mn-MAA NCs were determined by electrophoretic and hydrodynamic light scattering methods, indicating formation of agglomerates in water with various sizes (50–440 nm) and different surface charge values accordingly the preparation conditions of

the NCs (−7.59 to −24.98 mV). Finally, the relative photocatalytic activities of the ZnS:Mn-MAA NCs were evaluated by measuring the degradation rate of methylene blue (MB) molecule in a pseudo first-order reaction condition under the UV-visible light irradiation. As a result, the ZnS:Mn-MAA NC prepared at the pH 7 showed the best photo-degradation efficiency of the MB molecule with the first-order rate constant (k_{obs}) of $2.0 \times 10^{-3} \cdot min^{-1}$.

KEYWORDS

ZnS:Mn; water-dispersible; MAA capping; photocatalytic activities; methylene blue degradation; aggregations

1. INTRODUCTION

Water-dispersible semiconductor nanocrystals were developed for various important applications such as luminescent labeling agents in biological systems [1,2,3], photocatalysts for the hydrogen gas production by water splitting [4,5,6], and environmental purification agents to remove pollutants from wastewater by photodegradation reactions [7,8,9]. In the syntheses of the water-dispersible nanocrystals, several organic molecules such as sulfodiisooctyl succinate [10], mercaptoethanol [11,12], and conventional aminoacids [13,14] were found as effective surfactants to solubilize the originally hydrophobic semiconductor nanocrystals in water. However, the complicated surface properties of the nanocrystals dispersed in water have not been investigated in detail. In fact, it is generally expected that nanosized semiconductor materials would exhibit much higher catalytic efficiency than their bulk counterparts due to the increased surface area-to-volume ratio [15]. However, one should also consider that there are much more complicated factors in the water-dispersed nanocrystals such as interactions between the surface capping ligands and secondary coordination by other metal ions provided during the preparation process of the nanocrystals (NCs), which majorly results in decrease of luminescence efficiency and photocatalytic activity of the NCs [16]. Therefore, it is necessary to investigate specific surface properties such as coordination modes of the capping ligands, surface charges, and degree of aggregation in water to correctly understand the observed photocatalytic activities of the NC.

In this study, water-dispersible ZnS:Mn nanocrystals were prepared by capping their surface with polar mercaptoacetic acid (MAA) molecules at three different pH conditions (pH 2, 7, and 12). The pH conditions were selected based on the

known pKa values and isoelectric point of the capping ligand ($pKa_1 = 3.64$, $pKa_2 = 10.61$, and pI = pH 7.13 for the MAA in water respectively) (CAS 68-11-1) at which different surface properties were expected to be observed for the ZnS:Mn nanocrystals. In particular, MAA was selected as the surface capping ligand, because it is a simple structured polar molecule, and has been known as an excellent stabilizer for other semiconductor nanocrystals such as CdX (X = S, Se and Te) [17] and ZnS [18] with a high solubility in water. Moreover, MAA does not have any aromatic functional group or alkyl side chain, which can further induce unexpected secondary interactions between the capping molecules on the nanocrystal surface, thus complicating the characterizations. To investigate the pH dependent nature of the ZnS:Mn-MAA nanocrystals in water, the corresponding surface charges of the nanocrystals were determined by the electrophoretic light scattering method [19]. Moreover, the degrees of aggregation of the nanocrystals in water were measured by the hydrodynamic light scattering method [20]. Finally, the relative potocatalytic activities of the ZnS:Mn-MAA nanocrystals were evaluated by measuring the degradation rate of an organic dye (methylene blue, MB) molecule under the UV light irradiation [21]. It has been known that pH is one of the most important factors affecting to the photocatalytic efficiency of traditional semiconductor nanocrystals [22]. However, especially for the ZnS:Mn-MAA NC, none of the studies have reported on the related surface properties such as surface charges and formation of aggregation together according to the different pH environments. Therefore, the main purpose of our study was to relate the pH dependent nature of the surface modified ZnS:Mn NCs and their observed relative photocatalytic activities in different environments.

2. RESULTS AND DISCUSSION

2.1. Characterizations of the ZnS:Mn-MAA NC Products Prepared at Different pH Conditions

The average particle sizes of the ZnS:Mn-MAA NCs prepared at three different pH conditions were measured from the HR-TEM images presented in Figure 1. Even though the images did not clearly show discrete individual particles, we enlarged those images as much as we could and measured about 30 identifiable particles (fringe images) to obtain the average particle sizes for the ZnS:Mn-MAA NCs. In the images, the shapes of the most particles are fairly close to a sphere, and the average of the measured particle sizes were: 3.33 nm (pH 2), 5.17 nm (pH 7), and 6.05 nm (pH 12). In the figure, little agglomerations between the particles were observed due to the evaporation of the water and alcohol mixture solvents during the sample preparation. However, the

appearance of distinct lattice planes in the fringe images with about 3.4 Å lattice spacing indicated that the obtained solids are made of single crystals rather than poly-crystalline aggregates.

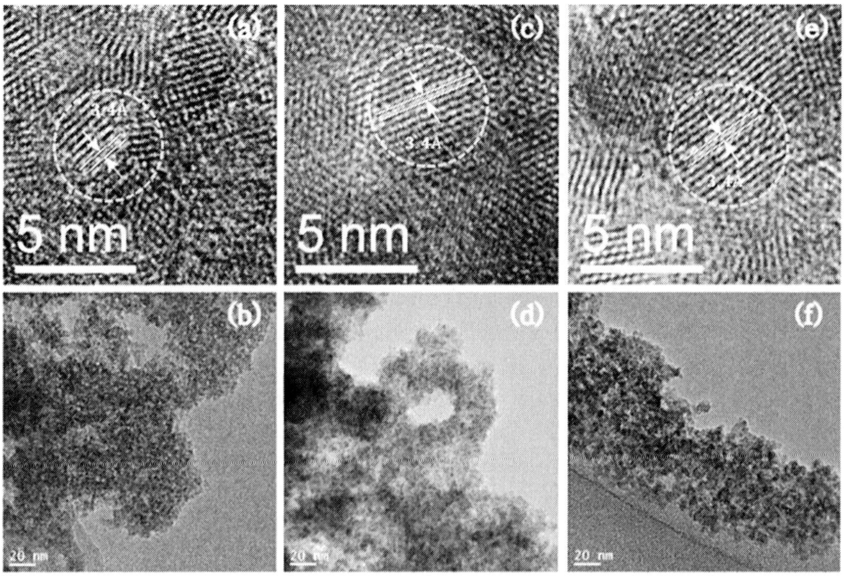

Figure 1. High-resolution transmission electron microscopy (HR-TEM) images of the ZnS:Mn-MAA nanocrystals (NCs) prepared at: (**a**,**b**) pH 2; (**c**,**d**) pH 7; and (**e**,**f**) pH 12.

To determine the elemental composition of the prepared ZnS:Mn-MAA NCs and the doping concentration of the manganese (II) ions precisely, Inductively Coupled Plasma-Atomic Emission Spectrometry (ICP-AES) analyses were performed. Three trials of the sample measurements revealed that the average elemental proportions of the Mn^{2+} ions relative to the ZnS parent crystal were 1.06% (pH 2), 1.22% (pH 7), and 1.11% (pH 12). The doping concentration of the manganese (II) ion in the ZnS parent crystal was originally aimed to be 1%–2%, which had been reported as the optimum for photoluminescence (PL) efficiency for other known ZnS:Mn nanocrystals [23].

The optical properties of the ZnS:Mn-MAA NCs were investigated by using UV-visible and photoluminescence (PL) spectroscopy, as presented in Figure 2 and Figure 3. The PL spectra showed broad emission peaks appeared at 590 nm (pH 2), 586 nm (pH 7), and 587 nm (pH 12) that were almost identical to

each other. The emission spectra were obtained by fixing the excitation wavelengths at the corresponding UV-Visible absorption peak of the NC, which were 307 nm (pH 2), 310 nm (pH 7), and 312 nm (pH 12). The dominant absorption shown in the spectrum probably resulted from the fundamental band-to-band absorption in the ZnS host [24], and the increased band gap of the ZnS:Mn nanocrystal (3.87 eV), compared to that for bulk ZnS:Mn solid (3.54 eV), is due to the well-known quantum confinement effect [25]. The yellow-orange light emissions at 590 nm were attributed to the $^4T_1$–$^6A_1$ transitions of the dopant Mn^{2+} ions [26]. In the luminescence pathway, if the surface defect states are located close to the conduction band, the direct energy transfer from the ZnS host to the Mn^{2+}activator is significantly interrupted, which can cause weakening in the emission as well as enlarging of the Stokes shift [27]. In Figure 3, the lower PL intensities for the ZnS:Mn-MAA (pH 7 and 12) NCs were probably due to the presence of sodium ions provided by the addition of NaOH to adjust the pH condition in the preparation, since a similar phenomenon has been reported that addition of sodium salts into a thiolate ligand coated CdSe/ZnS quantum dot containing colloidal solution caused a significant luminescence quenching [28]. The PL efficiencies for ZnS:Mn-MAA NCs were measured and calculated following the method originally reported by Williams *et al.* [29,30]. This method includes calculating a relative quantum yield by comparing it to that of a standard organic dye, a 0.1 M solution of quinine sulfate in H_2SO_4 (purchased from Fluka) in this case [31], of which the emission wavelength and absolute quantum efficiency are known as 550 nm and 0.546, respectively (22 °C). The excitation wavelength used for the standard was fixed at the same with the ZnS:Mn-MAA NC, which was obtained from the UV-Visible absorption spectrum of the ZnS:Mn-MAA NC. As a result, the calculated relative PL efficiencies were 4.67% (pH 2), 4.02% (pH 7), and 3.98% (pH 12).

In Figure 4, the wide-angle X-ray diffraction (XRD) pattern diagrams obtained from the ZnS:Mn-MAA NCs prepared at different pH conditions are presented. Although most of the peaks are broad, there were clearly identifiable (111), (220), and (311) peaks in the diagram, indicating that the ZnS:Mn-MAA NCs are in the cubic zinc blende phase (JCPDS 05-0566) [32]. In addition, we also calculated the particle sizes of the ZnS:Mn-MAA NCs by applying the Debye-Scherrer equation using the XRD peak data to compare with the particle sizes measured from the HR-TEM images [33,34]. From the measured full width at half maxima (FWHM) of the selected XRD peaks, we obtained the average particle sizes for ZnS:Mn-MAA NC as 3.51 nm (pH 2), 5.00 nm (pH 7), and 5.70 nm (pH 12), which showed very good agreements to that measured from the HR-TEM images.

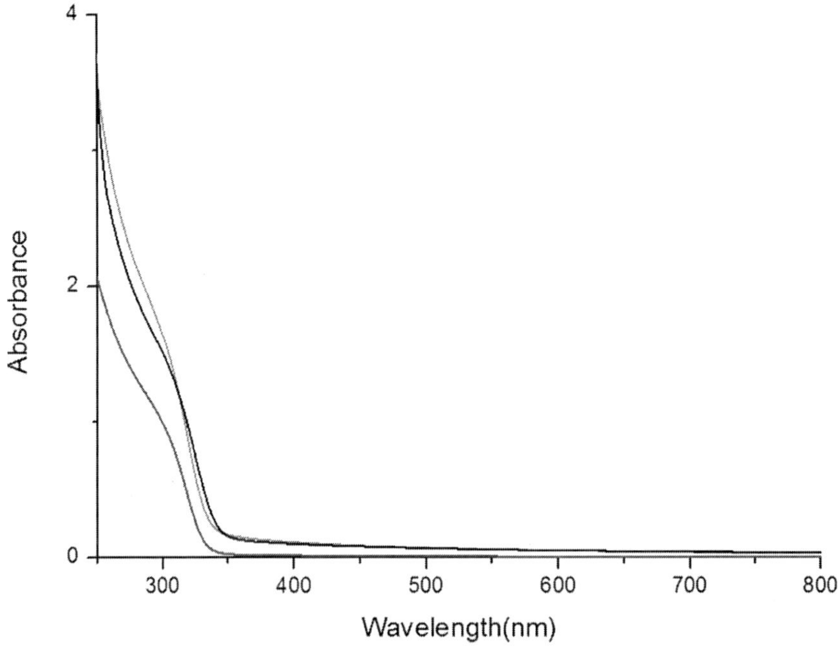

Figure 2. Ultra violet (UV)-visible absorption spectra of the ZnS:Mn-MAA NCs prepared at pH 2 (blue); pH 7 (green); and pH 12 (black).

The MAA molecules attached on the surfaces of the ZnS:Mn-MAA NC were characterized by FT-Raman spectroscopy [35].Figure 5 presents the FT-Raman spectra of ZnS:Mn-MAA NC prepared at pH 12 condition with that of free MAA molecules for comparison. The obtained peak data are listed in Table 1 and their assignments are also provided. The assignments of the specific vibrational modes were provided by comparing to previously reported paper regarding assignments of the FT-Raman spectrum of the ZnS:Mn-MAA associated with a computational calculation (DFT) method [36]. In the spectrum, the overall peaks from the capping MAA molecules were slightly shifted comparing to that of the neat MAA molecule since they were attached to much heavier transition metal ions in the nanocrystal lattices, and were immobilized to restrict some of vibrational modes [37,38]. The peaks appeared at 2575 cm^{-1} (S-H) for the free MAA molecules did not appear for the ZnS:Mn-MAA NC, since the hydrogen was removed by the addition of strong base at this pH condition. To remove any uncoordinated or unreacted MAA molecules, the centrifuged white solids were rapidly washed several times with cold alcohol/water solutions. As a result, the peaks that resulted from any free MAA molecules could be removed from the presented FT-Raman data for the ZnS:Mn-MAA NC. Finally, most

peaks appeared in the lower region, from 400 to 1000 cm^{-1} for the ZnS:Mn-MAA NC, can be assigned as mixed banding modes of the C–H, C–C, and C–O moieties of the nanocrystal coordinated MAA molecule. The peak appearing at 335 cm^{-1} was especially assigned as a longitudinal optical phonon of the ZnS lattice [39].

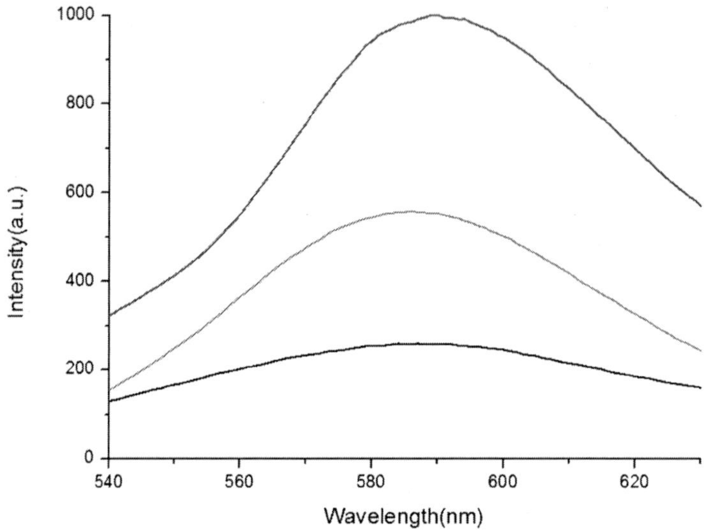

Figure 3. Room temperature solution photoluminescence (PL) emission spectra of the ZnS:Mn-MAA NCs prepared at pH 2 (blue); pH 7 (green); and pH 12 (black).

2.2. Surface Properties of the ZnS:Mn-MAA NCs

To investigate the pH dependent surface properties of the ZnS:Mn-MAA nanocrystals in water, the corresponding surface charges of the nanocrystals were determined by the electrophoretic light scattering method [40]. As a result, the zeta potentials of the nanocrystals at ambient temperature were obtained as: −24.98 mV (ZnS:Mn-MAA-pH 2), −31.67 mV (ZnS:Mn-MAA-pH 7), and −7.59 mV (ZnS:Mn-MAA-pH 12). Considering the pH condition, the negative surface charge of the ZnS:Mn-MAA-pH 2 nanocrystal can be explained that most of the protons originally belong to the MAA molecules were ionized into the water to form their conjugate base ion moieties (MAA$^-$) on the nanocrystal surface to yield the negative surface charge of the nanocrystal. For the preparation of the ZnS:Mn nanocrystal, the pH 2 condition was achieved

just by adding the MAA itself without any further modification. However, considering that the carboxyl end is not so strongly acidic like a halide acid in water, one can expect that not all of the protons in the carboxyl (–COOH) end of the MAA were ionized at this environment. Therefore, obtaining the higher negative surface charge for ZnS:Mn-MAA-pH 7 nanocrystal than that of ZnS:Mn-MAA-pH 2 nanocrystal is also relevant. The addition of sodium hydroxide (NaOH) to adjust the pH 7 condition induced additional deprotonation at the carboxyl end of the MAA to produce more negatively charged nanocrystal surface even though the pH 7 is very close to the isoelectric point of the free MAA molecule in water. In this manner, obtaining much higher negative surface charge for the ZnS:Mn-MAA-pH 12 nanocrystal was expected because the SH and COOH groups of the MAA should be completely deprotonated at this pH condition. However, much lower negative surface charge value of −7.59 mV was actually obtained. In fact, this result was explained, when an additional elemental analysis was performed for the ZnS:Mn-MAA-pH 12 nanocrystal by the ICP-AES method. It was found that 10 atom% sodium ions compared to the ZnS contents existed in the ZnS:Mn-MAA-pH 12 nanocrystal, which were probably produced by the added NaOH to adjust the pH 12 condition in the preparation. The originally produced highly negative charge of the nanocrystal surface was probably compensated by the positive charge of the surface attached Na^+ ions. Even though NaOH was also added for the ZnS:Mn-MAA-pH 7 nanocrystal, the elemental analysis revealed that less than 1 atom% of the sodium ions compared to the ZnS were found. Therefore, the amount of the attached sodium ions was not sufficient enough to cancel out the negative charge of the ZnS:Mn-MAA-pH 7 nanocrystal in this case.

Second, to investigate the pH dependent surface interactions between the water-dispersed ZnS:Mn NCs, the degree of aggregation of each nanocrystal in water was measured by the hydrodynamic light scattering method [41]. Figure 6 shows the formation of various sizes of aggregates: ZnS:Mn-MAA-pH 2 (287 nm), ZnS:Mn-MAA-pH 7 (50 nm), and ZnS:Mn-MAA-pH 12 (440 nm) nanocrystals in water. These agglomerates probably formed by the intermolecular interaction between the capping molecules on the neighboring ZnS:Mn nanocrystals, because these nanocrystals were originally found as 5 nm sized particles in the solid state according to the HR-TEM images and the Debye-Scherrer's XRD calculations. Previously, we reported very similar intermolecular attractions between the surface capping molecules on the nanocrystal surfaces for L-glycine and L-valine capped ZnS:Mn nanocrystals [42], in which the intermolecular interaction (majorly hydrogen bonding) between the aminoacid molecules caused the formation of huge aggregates of the ZnS:Mn-aminoacid nanocrystals (from 250 nm to few micrometer sizes) in water. Compared to other water-dispersed nanocrystals, the degree of aggregation for the ZnS:Mn-MAA nanocrystals was relatively low in water,

probably because of the electrostatic repulsion between the negatively charged nanocrystal surfaces [43]. The ZnS:Mn-MAA-pH 7 nanocrystal, which showed the highest negative surface charge, formed the smallest agglomerates in water, whereas the ZnS:Mn-MAA-pH 12 nanocrystal produced the largest aggregates. Moreover, the surface coordinated sodium ions probably caused the formation of sodium ion bridged complexes to form [COO–Na–OOC] moieties in water. In fact, similar phenomena have been reported previously; the addition of sodium metal ions to the 11-mercaptoundecanoic acid capped gold nanoparticles caused the formation of the metal bridged aggregates of few micrometer sizes in water [44].

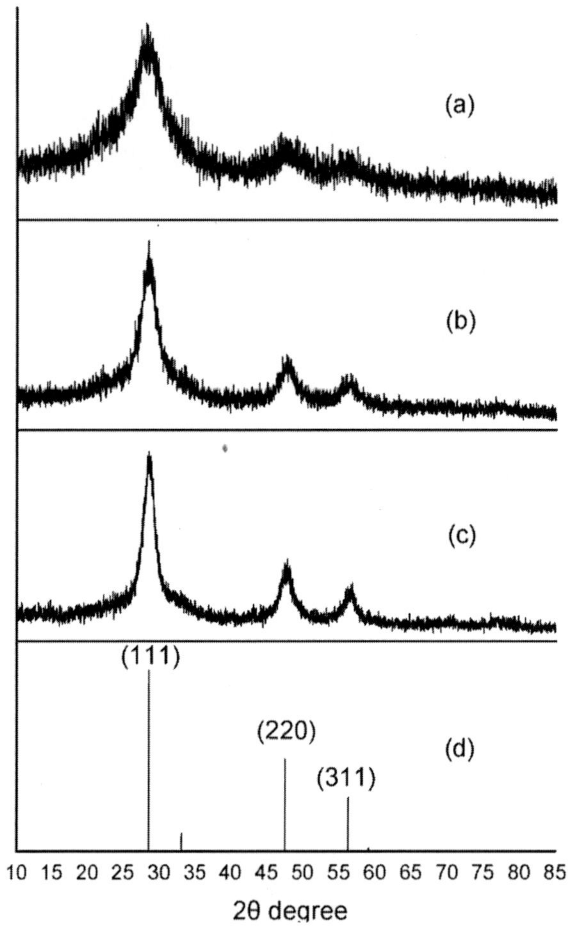

Figure 4. X-ray diffraction (XRD) pattern diagrams of the ZnS:Mn-MAA NCs prepared at: (**a**) pH 2; (**b**) pH 7; and (**c**) pH 12. The diagram (**d**) is a reference ZnS bulk solid pattern in a cubic zinc blend phase (JCPDS 05-0566).

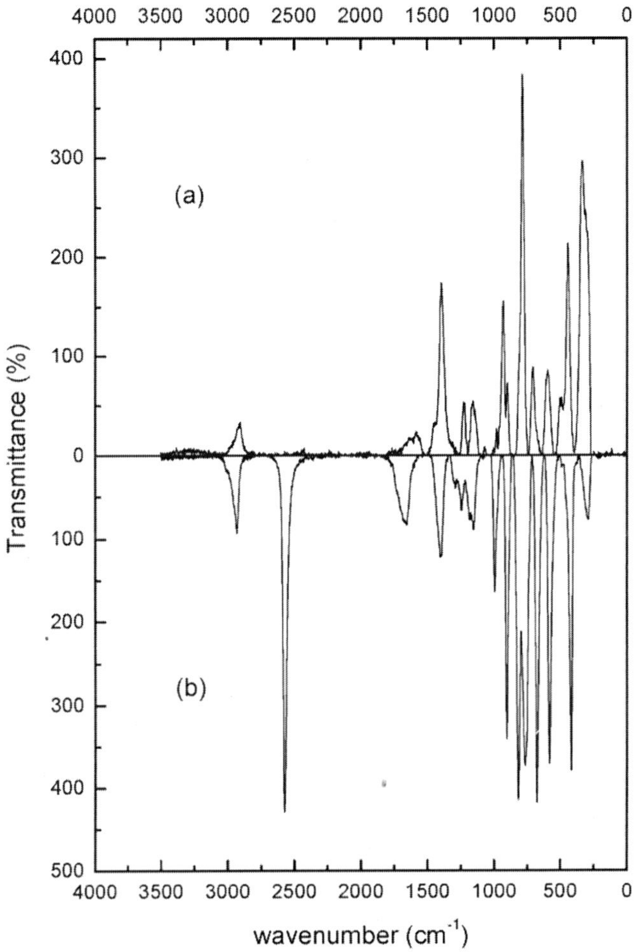

Figure 5. Fourier transform (FT) -Raman spectra of the: (**a**) ZnS:Mn-MAA NC (pH 12); and (**b**) Neat MAA molecule.

2.3. Photocatalytic Activities of the ZnS:Mn-MAA NCs

As an application, the relative photocatalytic activities of the ZnS:Mn-MAA nanocrystals were evaluated by measuring the degradation rate of an organic dye (methylene blue, MB) molecule under the UV light irradiation [45,46]. Photocatalysis is a process by which a semiconductor material absorbs energy greater than or equal to its band gap, causing excitation of the valence band (VB) electrons into the conduction band (CB). Such charge separation forms

electron hole pairs (h^+/e^-), generating free radicals in the system for redox of the organic substrate. The resulting free radicals such as hydroxyl ($HO^•$) are very efficient oxidizers of organic dye molecules such as MB [47]. The detailed photo-degradation mechanism of an MB molecule in the presence of a semiconductor nanoparticle catalyst is well known in the literature [48]. A brief description of the photo-degradation mechanism of MB molecules in the presence of the ZnS:Mn nanocrystal is as follows:

$$\text{ZnS:Mn} + h\nu \rightarrow \text{h}^+ \text{ (VB)} + \text{e}^- \text{ (CB)} \quad (1)$$

$$\text{OH}_2 + \text{h}^+ \rightarrow \text{HO}^\bullet + \text{H}^+ \text{ (VB) and } \text{O}_2 + \text{e}^- \rightarrow \text{O}_2^- \text{ (CB)} \quad (2)$$

$$\text{MB} + \text{HO}^\bullet \rightarrow \text{degradation products} \quad (3)$$

Table 1. FT-Raman data and assignments of ZnS:Mn-MAA (mercaptoacetic acid) NC (nanocrystals) prepared at pH 12 (unit in cm^{-1}).

ZnS:Mn-MAA (pH 12)	Neat MAA	Assignments
335		Zn-S phonon
444	419	δ(CCO)/ρ(OCO)
595	581	γ(CCO)/ω(OCO)
708		δ(OCO)
787	776	ν(CS)
	820	ν(CC)
934	909	ρ (CH_2)
1395	1400	ν(OCO) + δ(CH_2)
1583	1692	ν(C=O)
	2575	ν(SH)
2907	2932	ν(CH_2)

ν: stretching; ω: wagging; δ: in-plane deformation; ρ: rocking.

In this study, relative photo-degradation efficiencies of the three surface modified ZnS:Mn nanocrystals were evaluated, as shown in Figure 7 and Figure 8. These reaction rates were measured in the pseudo first-order conditions by controlling the concentration of the reactants. The pseudo first-order rate constants (k_{obs}) for the photo-degradation reaction of MB were calculated using the plots of ln(*C/Co*) *versus* irradiation time (*t*), in which *C* and *Co* are the

concentrations of MB at a certain time and the initial state respectively, and were converted from the temporal absorbance changes in the UV-visible absorption spectra. The total reaction running times were set at 300 min for all the ZnS:Mn-MAA nanocrystals, because the absorbance of the MB molecule did not show any further change after that time period. As a result, the pseudo first-order reaction constants (k_{obs}) calculated from the slopes in the fitting diagrams are as follows: 1.6 × 10^{-3} min^{-1} (ZnS:Mn-MAA-pH 2), 8.5 × 10^{-3} min^{-1} (ZnS:Mn-MAA-pH 7), and 2.0 × 10^{-3} min^{-1} (ZnS:Mn-MAA-pH 12). In the presented diagrams, all the calculated R^2 values were greater than 0.994, indicating that the data fitted well into the straight lines. Usually the pH is known as one of the most important parameters affecting the photo-oxidation process of water [49], and in general, negatively charged photo-catalysts can attract polar water molecules more efficiently to produce hydroxyl radicals than those of the neutral ones [50]. Among these nanocrystals, the ZnS:Mn-MAA-pH 7 nanocrystal, with the highest surface charge of −31.67 mV and the largest surface area (owing to the least aggregation), showed the best photo-degradation efficiency (92% after 300 min). Moreover, the photo-degradation efficiency of the ZnS:Mn-MAA-pH 7 nanocrystal was compared to a well-known commercial reference photo-catalyst, a bulk TiO_2 powder (Degussa, P-25) [51], and the calculated relative photo-degradation efficiency was 30% compared to the internal reference. Even though the ZnS:Mn-MAA NC showed lower photocatalytic efficiency than the commercial TiO_2 bulk powder, ZnS:Mn nanocrystal is still interesting as photocatalysts in a system using the sun light since the ZnS:Mn can absorb visible lights with high optical transition efficiency by tuning its band gap with a size controlling method, while TiO_2 absorbs only near UV lights which utilizes only 3%–4% of the sun light [52,53]. Moreover, ZnS:Mn nanocrystals are fairly easy to synthesize in water with a high product yield, while the preparation of TiO_2 nanocrystal in water is quite complicated and often suffers with very low product yield [54,55]. Finally, it is notable to mention that the ZnS:Mn-MAA-pH 7 NC showed much higher photocatalytic efficiency (about 10 to 20 times) than bare ZnS [56], other ligand (mercaptoetnanol) capped ZnS:Mn NC [57], and other metals doped ZnS NC such as ZnS:Fe [58] and ZnS:Cu [59] under exactly the same pH and concentration, making the ZnS:Mn-MAA NC very useful among the ZnS based photo-catalysts.

3. EXPERIMENTAL SECTION

3.1. Instrumentations

The HR-TEM images provided in this article were taken by a JEOL JEM 1210 electron microscope (JEOL, Tokyo, Japan) with a MAG mode of 1000 to 800,000

in which the accelerating voltage was 40–120 kV. For the sample preparation, dried ZnS:Mn-MAA NC powder was dispersed in methanol and placed on a carbon-coated copper grid (300 Mesh) followed by drying under vacuum for 20 h. UV-Visible absorption spectra were recorded using a Perkin Elmer Lamda 25 spectrophotometer (Waltham, MA, USA) equipped with a deuterium/tungsten lamp and the solution photoluminescence (PL) spectra were obtained by a Perkin Elmer LS-45 spectrophotometer (Waltham, MA, USA) equipped with a 500 W Xenon lamp, 0.275 m triple grating monochrometer, and PHV 400 photomultiplier tube. The powder XRD pattern diagrams were obtained using Rigaku 300 X-ray diffractometer (Rigaku, Tokyo, Japan) with Cu Kα (1.54 Å) wavelength light source. ICP-AES elemental analyses were performed by using an Optima-430 (Perkin Elmer, Waltham, MA, USA) spectrometer equipped with an Echelle optics system and segmented array charge coupled device (SCD) detector. For the surface characterization of the ZnS:Mn-MAA NC, FT-Raman spectrum was recorded by a Bruker FRA106/s spectrophotometer with a resolution of 1 cm^{-1}. Finally, the surface charges and the degrees of aggregations of the ZnS:Mn-MAA NCs in water were measured by using an ELS-8000 Zeta-PSA analyzer (Otsuka Electronics, Tokyo, Japan) equipped with a 10 mW He/Ne laser light source (KBSI).

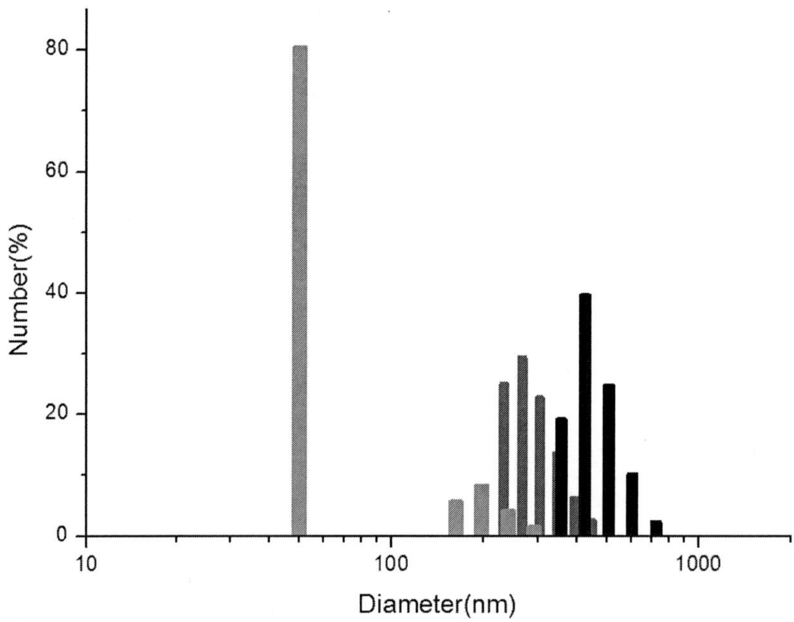

Figure 6. Particle size distribution diagrams of the ZnS:Mn-MAA NCs prepared at pH 2 (blue), pH 7 (green), and pH 12 (black).

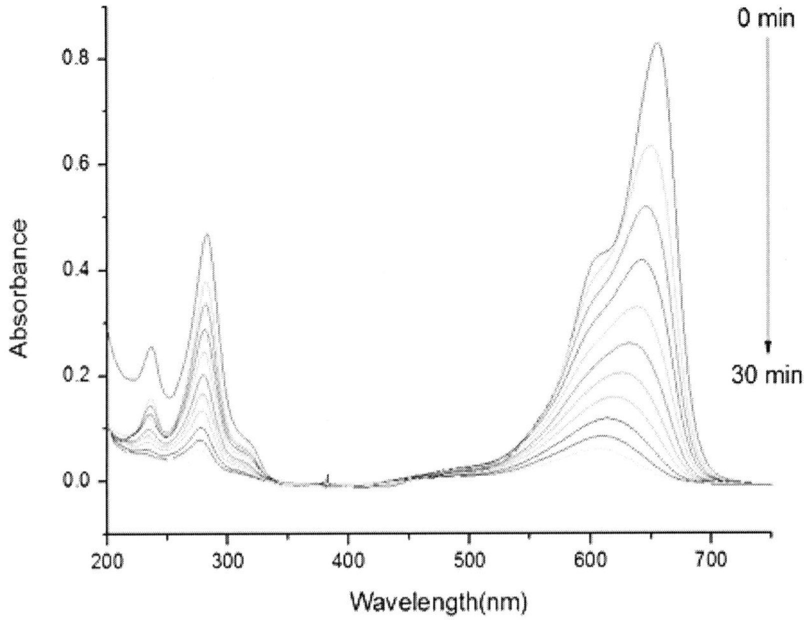

Figure 7. Temporal UV visible absorption spectral changes of methylene blue (MB) in the presence of ZnS:Mn-MAA-pH 7 NC under the UV light irradiation.

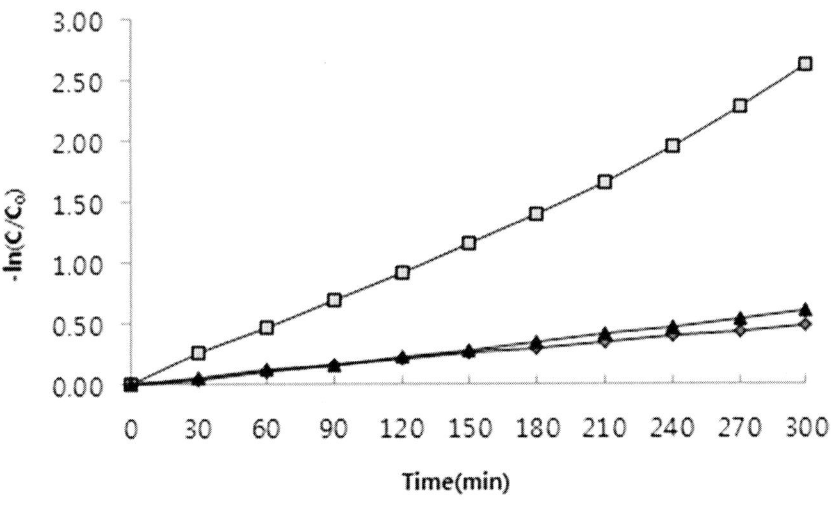

Figure 8. First-order kinetic plots of photo-degradation of MB in the presence of ZnS:Mn-MAA-pH 2 (blue), ZnS:Mn-MAA-pH 7 (green), and ZnS:Mn-MAA-pH 12 (black) NCs under the UV light irradiation.

3.2. Syntheses of the ZnS:Mn-MAA NCs

ZnS:Mn-MAA nanocrystals were synthesized following a slight modification of the previously reported aqueous synthesis of MAA capped ZnS:Mn NCs via the formation of zinc (II) ion containing reactive intermediate complexes [36]. The additional modifications were performed to adjust different pH conditions during the preparations. A 50 mL aqueous solution of $ZnSO_4{\cdot}5H_2O$ (1.44 g, 5 mmol) was slowly added to a 50 mL aqueous solution containing 10 mmol of MAA at 5 °C (ice-water bath). The solution was warmed to ambient temperature after 1 h stirring. $MnSO_4{\cdot}H_2O$ (0.02 g, 0.1 mmol) and Na_2S (0.40 g, 5 mmol) were dissolved in 20 mL DI water. The resulting mixture was subsequently transferred to the flask containing the [Zn-MAA] complexes under vigorous stirring. The subjected pH conditions were adjusted by the addition of 0.01 M NaOH aqueous solution to the mixture, measuring the pH by using a digital pH meter. The resulting solution was refluxed for 10 h. Slow cooling at ambient temperature and the addition of ethanol afforded an off-white precipitate at the bottom of the flask. Finally, the obtained solids were separated by centrifuging and decanting the supernatant. The solids were then dried for 24 h in a vacuum oven. The detailed experimental data are listed in Table 2.

Table 2. Experimental data summary of ZnS:Mn-MAA NCs prepared at different pH conditions.

Experimental Data	pH 2	pH 7	pH 12
UV/Vis absorption (λ_{max}, nm)	307	310	312
PL emission (λ_{max}, nm)	590	586	587
PL efficiencies (%)	4.67	4.02	3.98
concentration Mn dopant ICP-AES (%)	1.06	1.22	1.11
Average particle size HR-TEM (nm)	3.33	5.17	6.05
Average particle size XRD (nm)	3.51	5.00	5.70
Zeta potentials (mV)	−24.98	−31.67	−7.59
Average size of aggregates in water DLS (nm)	287	50	440

3.3. Photocatalytic Experiments

The photocatalytic reactions were performed using a 300 W xenon lamp (Newport) as the light source (emission range from 185 to 2000 nm), which was placed at the side of a 1.0 L capacity Pyrex-glass cell (Sigma-Aldrich, Seoul,

Korea) in a dark room. The glass cell was filled with 0.5 L of aqueous mixture solution containing 100 mg of MB and 5 mg of the ZnS:Mn-MAA NC as the photo-catalyst. Under this condition, the molar concentration ratio of [ZnS:Mn]:[MB] was 1:150 according to the ICP-AES elemental analyses for the ZnS:Mn NC. The resulting solution was stirred using a magnetic bar, and a magnetic stirrer was placed under the glass cell during the reaction. The degradation of MB was monitored by aliquoting 10 mL of the mixture solution at the 30 min irradiation time intervals. Subsequently, the degradation reaction rate was calculated by measuring the change in the absorbance of the MB dye solution. The absorption spectra were recorded by measuring the absorbance at 664 nm corresponding to the maximum absorption wavelength of the MB molecule with UV-Visible spectroscopy. Analogous control experiments were performed without ZnS:Mn NC (blank). Moreover, to exclude the heat factor from the calculation, generated by the UV lamp during the degradation reaction, the same reaction was performed without UV light irradiation by just heating using a hot plate (up to ~90 °C, 300 min) placed under the glass cell, confirming that no degradation reaction occurred without UV light irradiation.

4. CONCLUSIONS

Recently, low dimensional semiconductor nanocrystals have received attention as novel types of photocatalytic materials for the purification of wastewater and the hydrogen production process. Since ZnS:Mn NCs can provide a novel platform to attract polar molecules such as water, they can be used for a variety of applications especially as a photocatalyst for the degradation of pollutants in wastewater or the hydrogen production by the water splitting process. In this study, MAA capped ZnS:Mn NCs were successfully synthesized at different pH conditions, and the corresponding surface properties were investigated using spectroscopic methods. In addition, the related physical and optical properties were also investigated. The as-prepared water-dispersible MAA capped ZnS:Mn NCs indeed showed remarkable physicochemical properties for further applications in advanced photocatalysis systems. However, as demonstrated in this study, both the degree of aggregation in water and the pH dependent nature of the MAA, as the capping ligand for the ZnS:Mn NCs, are very critical factors to be considered for further applications in any commercial device.

ACKNOWLEDGMENTS

The present research was supported by the GRRC program of Dankook University (GRRC-Dankook-2014-B02).

AUTHOR CONTRIBUTIONS

Jungho Heo performed all of the experimental work and data collections in this study. Cheong-Soo Hwang supervised the experimental work, conducted data analyses, and wrote this manuscript.

CONFLICTS OF INTEREST

The authors declare no conflict of interest.

REFERENCES

1. Karakoti, A.S.; Shukla, R.; Shanker, R.; Singh, S. Surface functionalization of quantum dots for biological applications.Adv. Colloid Interface Sci. 2015, 215, 28–45.

2. Tyrakowski, C.M.; Snee, P.T. A primer on the synthesis, water-solubilization, and functionalization of quantum dots, their use as biological sensing agents, and present status. Phys. Chem. Chem. Phys. 2014, 16, 837–855.

3. Yu, W.W. Water-soluble quantum dots for biomedical applications. Biochem. Biophys. Res. Commun. 2006, 348, 781–786.

4. Ruberu, T.P.A.; Dong, Y.; Das, A.; Eisenberg, R. Photoelectrochemical Generation of Hydrogen from Water Using a CdSe Quantum Dot-Sensitized Photocathode. ACS Catal. 2015, 5, 2255–2259.

5. Chauvire, T.; Mouesca, J.-M.; Gasparutto, D.; Ravanat, J.-L.; Lebrun, C.; Gromova, M.; Jouneau, P.-H.; Chauvin, J.; Gambarelli, S.; Maurel, V. Redox Photocatalysis with Water-Soluble Core-Shell CdSe-ZnS Quantum Dots. J. Phys. Chem. C 2015, 119, 17857–17866.

6. Kudo, A.; Miseki, Y. Heterogeneous photocatalyst materials for water splitting. Chem. Soc. Rev. 2009, 38, 253–278.

7. Ray, P.Z.; Shipley, H.J. Inorganic nano-adsorbents for the removal of heavy metals and arsenic: A review. RCS Adv.2015, 5, 29885–29907.

8. Rajabi, H.R.; Farsi, M. Quantum dot based photocatalytic decolorization as an efficient and green strategy for the removal of anionic dye. Mater. Sci. Semicond. Process. 2015, 31, 478–486.

9. Ayanda, O.S.; Petrik, L.F. Nanotechnology: The Breakthrough in Water and Wastewater Treatment. Intl. J. Chem. Mater. Environ. Res. 2014, 1, 1–2.

10. Bajpai, P.K.; Yadav, S.; Tiwari, A.; Virk, H.S. Recent Advances in the Synthesis and Characterization of Chalcogenide Nanoparticles. Solid State Phenom. 2015, 222, 187–233.

11. Sharma, R.; Bisen, D.P. Thermoluminescence of mercaptoethanol capped ZnS:Mn nanoparticles. Luminescence 2015, 30, 175–181.

12. Vogel, W.; Borse, P.H.; Deshmukh, N.; Kulkarni, S.K. Structure and Stability of Monodisperse 1.4-nm ZnS Particles Stabilized by Mercaptoethanol. Langmuir 2000, 16, 2032–2037.

13. Wangoo, N.; Kaur, S.; Bajaj, M.; Jain, D.V.S.; Sharma, R.K. One pot, rapid and efficient synthesis of water dispersible gold nanoparticles using alpha-amino acids. Nanotechnology 2014, 25, 5608–5615.

14. Lee, J.H.; Kim, Y.A.; Kim, K.M.; Huh, Y.D.; Hyun, J.W.; Kim, H.S.; Noh, S.J.; Hwang, C.S. Syntheses and Optical Properties of the Water-Dispersible ZnS:Mn Nanocrystals Surface Capped by L-Aminoacid Ligands: Arginine, Cysteine, Histidine, and Methionine. Bull. Korean Chem. Soc. 2007, 28, 1091–1096.

15. Jang, H.D.; Kim, S.K.; Kim, S.J. Effect of particle size and phase composition of titanium dioxide nanoparticles on the photocatalytic properties. J. Nanopart. Res. 2001, 3, 141–147.

16. Lin, Y.T.; Weng, C.H.; Chen, F.Y. Key operating parameters affecting photocatalytic activity of visible-light-induced C-doped TiO2 catalyst for ethylene oxidation. Chem. Eng. J. 2014, 248, 175–183.

17. Farkhani, S.M.; Valizadeh, A. Review: Three synthesis methods of CdX (X = Se, S or Te) quantum dots. IET Nanobiotechnol. 2014, 8, 59–76.

18. Wageh, S.; Shu-Man, L.; You, F.T.; Xu-Rong, X. Optical properties of strongly luminescing mercaptoacetic acid capped ZnS nanoparticles. J. Lumin. 2003, 102–103, 768–773.

19. Dukhin, A.S.; Parila, S. Measuring zeta potential of protein nano-particles using electroacoustics. Col. Surf. B 2014, 121, 257–263.

20. Du, X.; Wang, X.; You, S.; Wang, Q.; Gong, X. A case study of aggregation behaviors of titanium dioxide nanoparticles in the presence of dodecylbenzene sulfonate in natural water. J. Environ. Sci. 2015, 36, 84–92.

21. Scuderi, V.; Impellizzeri, G.; Romano, L.; Scuderi, M.; Nicotra, G.; Bergum, K.; Irrera, A.; Svensson, B.G.; Privitera, V. TiO2-coated nanostructures for dye photo-degradation in water. Nanoscale Res. Lett. 2014, 9, 458–465.

22. Molea, A.; Popescu, V.; Rowson, N.A.; Dinescu, A.M. Influence of pH on the formulation of TiO2 nano-crystalline powders with high photocatalytic activity. Powder Technol. 2014, 253, 22–28.

23. Bhargava, R.N.; Gallagher, D.; Hong, X.; Nurmikko, A. Optical properties of manganese-doped nanocrystals of ZnS.Phys. Rev. Lett. 1994, 72, 416–426.

24. Kushida, T.; Tanaka, Y.; Oka, Y. Excited-state absorption spectra of ZnS:Mn Excited state absorption spectra of ZnS:Mn. Solid. State Commun. 1974, 14, 617–620.

25. Yu, I.; Isobe, T.; Senna, M. Optical properties and characteristics of ZnS nano-particles with homogeneous Mn distribution. J. Phys. Chem. Solids 1996, 57, 373–379.

26. Karar, N.; Singh, F.; Mehta, B.R. Structure and photoluminescence studies on ZnS:Mn nanoparticles. J. Appl. Phys.2004, 95, 656–660.

27. Geyer, S.M.; Scherer, J.M.; Jaworski, F.B.; Bawendi, M.G. Multispectral imaging via luminescent down-shifting with colloidal quantum dots. Opt. Mater. Exp. 2013, 3, 1167–1175.

28. Bereus, V.V.; Heyes, C.D.; Nienhaus, U. Quenching of CdSe-ZnS core-shell quantum dot luminescence by water-soluble thiolate ligands. J. Phys. Chem. C 2007, 111, 18589–18594.

29. Williams, A.T.R.; Winfield, S.A.; Miller, J.N. Relative fluorescence quantum yields using a computer-controlled luminescence spectrometer. Analyst 1983, 108, 1067–1071.

30. Chizhik, A.I.; Gregore, I.; Enderlein, J. Quantum Yield Measurement in a Multicolor Chromophore Solution Using a Nanocavity. Nano Lett. 2013, 13, 1348–1351.

31. Meech, S.R.; Philips, D. Photophysics of some common fluorescence standards. J. Photochem. 1983, 23, 193–217.

32. Hahn, T. International Tables for X-Ray Crystallography, Part III; The international union of crystallography: Chester, England, 1985; pp. 318–320.

33. Jones, F.W. The measurements of particle size by the X-ray method. Proc. R. Soc. Lond. Ser. A 1938, 166, 16–43.

34. Monshi, A.; Foroughi, M.R.; Monshi, M.R. Modified Scherrer Equation to Estimate More Accurately Nano-Crystallite Size Using XRD. World J. Nano Sci. Eng. 2012, 2, 154–160.

35. Wu, Y.; Jin, S.; Ye, J.Y.; Wang, S.; Feng, Z.; Li, C. Charge-Sensitive Surface Optical Phonon in CdS Quantum Dots Studied by Resonant Raman Spectroscopy. J. Phys. Chem. C 2014, 118, 30269–30273.

36. Kim, J.E.; Hwang, C.S.; Yoon, S. Synthesis and Surface Characterization by Raman Spectroscopy of Water-Dispersible ZnS:Mn Nanocrystals Capped with Mercaptoacetic Acid. Bull. Korean Chem. Soc. 2008, 29, 1247–1249.

37. Nakamoto, K. Infrared and Raman Spectra of Inorganic and Coordination Compounds, 5th ed.; John Wiley & Son: New York, NY, USA, 1997; pp. 350–358.

38. Ishiguro, S.; Umebayash, Y.; Kanzaki, R. Characterization of metal ions in coordinating solvent mixtures by means of Raman spectroscopy. Anal. Sci. 2004, 20, 415–421.

39. Milekhin, A.G.; Sveshnikova, L.L.; Duda, T.A.; Surovtsev, N.V.; Adichchev, S.V.; Azhniuk, Y.M.; Himcinschi, C.; Kehr, M.; Zahn, D.R.T. Resonance effects in Raman scattering of quantum dots formed by the Langmuir-Blodgett method. J. Phys. Conf. Ser. 2010, 245.

40. Xu, R. Particle Characterization: Light Scattering Methods; Kluwar Academic Publishers: Dordrecht, The Netherlands, 2001; pp. 289–340.

41. Berne, B.J.; Pecora, R. Dynamic Light Scattering; Dover Publications Inc.: New York, NY, USA, 2000; pp. 223–276.

42. Park, S.H.; Song, B.; Kong, H.Y.; Byun, J.; Hwang, C.S. Biological Toxicities and Aggregation Effects of L-Glycine and L-Alanine Capped ZnS:Mn Nanocrystals in Aqueous Solution. Bull. Korean Chem. Soc. 2014, 35, 1169–1176.

43. El Badawy, A.M.; Luxton, T.P.; Silva, R.G.; Scheckel, K.G.; Suidan, M.T.; Tolymat, T.M. Impact of Environmental Conditions (pH, Ionic Strength, and Electrolyte Type) on the Surface Charge and Aggregation of Silver Nanoparticles Suspensions. Environ. Sci. Technol. 2010, 44, 1260–1266.

44. Wang, D.; Tejerina, B.; Lagzi, I.; Kowalczyk, B.; Grzybowski, B.A. Bridging Interactions and Selective Nanoparticle Aggregation Mediated by Monovalent Cations. ACS Nano 2011, 5, 530–536.

45. Mardani, H.R.; Forouzani, M.; Ziari, M.; Biparva, P. Visible light photo-degradation of methylene blue over Fe or Cu promoted ZnO nanoparticles. Spectrochim. Acta A 2015, 141, 27–33.

46. Praus, P.; Svoboda, L.; Tokarsky, J.; Hospodkova, A.; Klemn, V. Core/shell CdS/ZnS nanoparticles: Molecular modelling and characterization by photocatalytic decomposition of Methylene Blue. Appl. Surf. Sci. 2014, 292, 813–822.

47. Soltani, N.; Saion, E.; Hussein, M.Z.; Erfani, M.; Abedini, A.; Bahmanrokh, G.; Navasery, M.; Vaziri, P. Visible Light-Induced Degradation of Methylene Blue in the Presence of Photocatalytic ZnS and CdS Nanoparticles. Int. J. Mol. Sci.2012, 13, 12242–12258.

48. Houas, A.; Lchheb, H.; Ksibi, M.; Elaloui, E.; Guillard, C.; Herrmann, J.-M. Photocatalytic degradation pathway of methylene blue in water. Appl. Catal. B Environ. 2001, 31, 145–157.

49. Guillard, C.; Lanchheb, H.; Houas, A.; Ksibi, M.; Elaloui, E.; Herrmann, J.-M. Influence of chemical structure of dyes, of pH and of inorganic salts on their photocatalytic degradation by TiO2 comparison of the efficiency of powder and supported TiO2. J. Photochem. Photobiol. A 2003, 158, 27–36.

50. Liqiang, J.; Xiaojun, S.; Jing, S.; Weimin, C.; Zili, X.; Yaoguo, D.; Honggang, F. Review of surface photovoltage spectra of nano-sized semiconductor and its applications in heterogeneous photocatalysis. Sol. Energy Mater. Sol. Cells 2003,79, 133–151.

51. Xu, N.; Shi, Z.; Fan, Y.; Dong, J.; Shi, J.; Hu, M.Z.-C. Effects of Particle Size of TiO2 on Photocatalytic Degradation of Methylene Blue in Aqueous Suspensions. Ind. Eng. Chem. Res. 1999, 38, 373–379.

52. Wold, A. Photocatalytic properties of titanium dioxide (TiO2). Chem. Mater. 1993, 5, 280–283.

53. Zhang, H.; Banfield, J. Structural Characteristics and Mechanical and Thermodynamic Properties of Nanocrystalline TiO2. Chem. Rev. 2014, 114, 9613–9644.

54. Jimmy, C.Y.; Yu, J.; Ho, W.; Zhang, L. Preparation of highly photocatalytic active nano-sized TiO2 particles viaultrasonic irradiation. Chem. Commun. 2001, 19, 1942–1943.

55. Park, H.; Park, Y.; Kim, W.; Choi, W. Surface modification of TiO2 photocatalyst for environmental applications. J. Photochem. Photobiol. C 2013, 15, 1–20.

56. Sharma, M.; Jain, T.; Sing, S.; Pandey, O.P. Photocatalytic degradation of organic dyes under UV-Visible light using capped ZnS nanoparticles. Solar Energy 2012, 86, 626–633.

57. Rajabi, H.R.; Farsi, M. Effect of transition metal ion doping on the photocatalytic activity of ZnS quantum dots: Synthesis, characterization, and application for dye decolorization. J. Mol. Catal. A 2015, 399, 53–61.

58. Chauhan, R.; Kumar, A.; Chaudhary, R.P. Photocatalytic degradation of methylene blue with Cu doped ZnS nanoparticles. J. Lumin. 2014, 145, 6–12.

59. Chauhan, R.; Kumar, A.; Chaudhary, R.P. Photocatalytic degradation of methylene blue with Fe doped ZnS nanoparticles. Spectrochim. Acta A 2013, 113, 250–256.

Chapter 6

Optical and Structural Properties of Si Nanocrystals in SiO_2 Films

Timur Nikitin and Leonid Khriachtchev *

Department of Chemistry, University of Helsinki, P.O. Box 55,
FI-00014 Helsinki, Finland

ABSTRACT

Optical and structural properties of Si nanocrystals (Si-nc) in silica films are described. For the SiO_x ($x < 2$) films annealed above 1000 °C, the Raman signal of Si-nc and the absorption coefficient are proportional to the amount of elemental Si detected by X-ray photoelectron spectroscopy. A good agreement is found between the measured refractive index and the value estimated by using the effective-medium approximation. The extinction coefficient of elemental Si is found to be between the values of crystalline and amorphous Si. Thermal annealing increases the degree of Si crystallization; however, the crystallization and the Si–SiO_2 phase separation are not complete after annealing at 1200 °C. The 1.5-eV PL quantum yield increases as the amount of elemental Si decreases; thus, this PL is probably not directly from Si-nc responsible for absorption and detected by Raman spectroscopy. Continuous-wave laser light can produce very high temperatures in the free-standing films, which changes their structural and optical properties. For relatively large laser spots, the center of the laser-annealed area is very transparent and consists of amorphous SiO_2. Large Si-nc (up to ~300 nm in diameter) are observed in the ring around the central region. These Si-nc lead to high absorption and they are typically under compressive stress, which is connected with their formation

from the liquid phase. By using strongly focused laser beams, the structural changes in the free-standing films can be made in submicron areas.

KEYWORDS

Si nanocrystal (Si-nc); SiO_2 film; Raman spectroscopy; photoluminescence; laser annealing

1. INTRODUCTION

Many limitations of modern electronic devices may be overcome by implementing photonics into electronics [1,2,3,4]. Integration of Si-based photonics with CMOS technology is a promising approach because it gives the possibility to merge electronics and photonics in the same chip [5,6,7,8,9]. Many optical functions such as, for example, light sources, amplifiers, waveguides, modulators, memory, and detectors should be achieved in order to fulfill this integration. Realization of a true monolithically fabricated injection Si laser with a small size is a particularly challenging task. The fundamental problem is the low light-emitting efficiency of bulk silicon due to its indirect band gap (~1.1 eV), which leads to long radiative times (~ms) and therefore, mostly non-radiative recombination of the excited carriers. The efficiency of the light emission can be enhanced by increasing the overlap of the wave functions of the electron and hole via, for example, spatial confinement [10,11,12,13]. The unusual properties of Si structures are observed when the size is less than the free exciton Bohr radius of 4.3 nm in bulk Si. This quantum confinement (QC) effect leads to the following main changes in the material properties: (i) an increase of the radiative probability due to delocalization of the wave functions in the momentum space, which increases the electron-hole wave function overlap and (ii) a shift in the emission wavelength to the visible region due to an increase of the band gap, which is controlled by the Si nanostructure size. Moreover, the number of sites within the volume of Si nanostructures, where non-radiative recombination can occur, decreases considerably. It should be mentioned that understanding electrical transport mechanisms in the systems considered below is very important for many applications [14]; however, this topic is outside the scope of the present review.

Porous Si is the first example of a Si-based material with enhanced luminescence efficiency. Strong room-temperature photoluminescence (PL) in visible spectral region from this material was discovered by Leigh Canham in

1990 [15]. Visible luminescence ranging from green to red in color was soon reported by Canham *et al.* [16] for other porous-Si samples and by other researchers [17,18] and ascribed to the quantum effects in Si structures with a size of ~3 nm. The blue shift of the PL and optical absorption with the increasing porosity (decreasing Si nanocluster sizes) provided the first important evidence that the QC plays an important role in light emission from porous Si. This blue shift is a result of the band gap expansion controlled by the size of Si nanostructures [13,19].

It was soon understood that oxidation of small Si crystallites has a great influence on the light emission [20,21]. The study by Wolkin *et al.* of oxidized porous Si has shown that the light-emitting centers can involve the Si=O covalent bonds located on the crystallite surface [22]. For oxygen-passivated Si crystallites, a stabilized electronic surface state is formed on a Si=O covalent bond and various recombination mechanisms can operate depending on the crystallite size. For larger sizes (>3 nm), recombination occurs via free excitons since the band gap is not wide enough to stabilize the Si=O surface state. For intermediate sizes (~2.5 nm), recombination involves a trapped electron localized on the Si atom of the Si=O bond and a free hole. For smaller sizes (<2 nm), recombination occurs via trapped excitons.

However, QC has been demonstrated for a number of Si nanomaterials. In contrast with oxygen passivation, for hydrogen-passivated porous Si, recombination occurs via free exciton states for all crystallite sizes and follows the expected QC behavior [22]. Ledoux *et al.* studied Si nanocrystals (Si-nc) prepared by pulsed CO_2 laser pyrolysis of silane in a gas flow reactor and deposited on a substrate [23]. It was observed that the PL band was blue-shifted as the size of Si-nc decreased from 8 to 2.5 nm. This dependence clearly follows the QC model. The QC mechanism operates also for alkane-terminated Si-nc. Hannah *et al.* [24] convincingly demonstrated using pressure-dependent PL studies that the PL arises from the core-states of Si-nc with indirect-gap transitions. The same PL mechanism has also been reported to operate for Si-nc in silicon-nitride films [25,26].

Another attractive Si-based material, which is the focus of the present article, is composed of Si-nc embedded in a SiO_2matrix. This material is chemically and mechanically more stable compared to porous Si and also emits light in the visible region. Si-nc in silica films can be prepared by various methods such as, for example, molecular beam deposition (MBD) [27,28], Si-ion implantation [29,30,31,32], sputtering [33,34,35,36,37,38,39], plasma enhanced chemical vapor deposition (PECVD) [33,34,40,41,42], low pressure chemical vapor deposition with subsequent thermal oxidation [43,44,45], reactive evaporation of SiO powder in oxygen atmosphere [46], and electron beam deposition [47]. Two types of architectures are most commonly prepared: (i) Si-rich silicon oxide SiO_x ($x < 2$) films [31,33,40,48,49] and (ii) SiO_x/SiO_2 [37,46,47,50,51,52,53]

and Si/SiO_2 [27,28,35,54,55,56,57] superlattices (SLs). Si-nc are formed in these materials by furnace annealing above 1000 °C. Studies of Si-nc in SiO_2 have been particularly stimulated by the observation of optical gain in this material [58,59]. In addition to PL, electroluminescence in visible region from porous Si and from Si-nc in SiO_2 has been reported [60,61].

The origin of light emission from Si-nc in a SiO_2 matrix is still controversial. It seems that the conclusions on the mechanism of the light emission from oxidized porous Si [20,21,22,62] are relevant to Si-nc in SiO_2. Similarly, the PL of Si-nc (sizes > 3 nm) in SiO_2 can be ascribed to the QC effect [63,64,65,66]. On the other hand, the importance of the Si-nc/SiO_2interface in the light-emitting properties has been stressed [31,35,67,68] and, in particular, the defect origin of the PL has been discussed [27,29,31,69,70,71]. Godefroo *et al.* have convincingly demonstrated that defects are the dominant source of the 1.5-eV PL [52]. In their experiment, an amorphous SiO/SiO_2 SL grown by reactive evaporation of SiO powders in an oxygen atmosphere was thermally annealed at 1100 °C for 1 h under N_2 atmosphere to produce Si-nc. The authors could switch the PL mechanism between the QC and defect mechanisms by passivation with hydrogen and ultraviolet illumination, respectively.

Optoelectronic applications of Si-based materials require knowledge of their optical and structural properties. For SiO_x (x< 2) films prepared by PECVD, the energy-filtered transmission electron microscopy (EFTEM) show that the Si–SiO_2 phase separation starts after annealing at 900 °C for 1 h under N_2 atmosphere when Si clusters become visible in an oxide matrix [72]. The dark-field (DF) transmission electron microscopy (TEM) technique reliably shows that the Si clusters are amorphous for annealing temperatures of 900–1000 °C, and their crystallization begins at 1100 °C. Higher annealing temperatures promote further crystallization of the Si clusters. The Si-nc size as a function of the annealing temperature and Si content has been studied for samples prepared by different deposition methods. For samples prepared by PECVD, the TEM results indicate that the Si-nc sizes increase with the increasing annealing temperature for a given Si content and with the Si content for a given annealing temperature [40]. The increase of the Si-nc sizes with the annealing temperature is also observed for SiO_x films grown by magnetron sputtering [37]. For SiO_x films prepared by ion implantation, the nucleation and pure growth stages of the nanocrystal population are almost over after 1 min of annealing at 1100 °C in N_2 [31]. For longer annealing periods, the sizes of Si-nc increase due to coalescence or Ostwald ripening process [31,73].

Several methods have been employed to study optical constants of different materials containing Si-nc. These methods include, for example, optical ellipsometry [74,75,76] and *m*-line measurements [77,78,79]. The analysis of the optical properties often uses the Bruggeman effective-medium approximation [80,81,82]. The Tauc-Lorentz model provides optical constants

of Si-nc deposited on a substrate [82]. The results show that the optical properties of Si-nc are quite different from those of amorphous and crystalline Si. It has been found that at a given wavelength, the refractive index of SiO_x films increases for larger Si-nc for Si-ion implanted samples [74] and for larger Si content for PCVD samples [75]. Chen *et al.* have developed a quantitative approach of obtaining depth profiles of the optical constants [81]. Moreno *et al.* report that the refractive index of Si-nc is lower than that of amorphous and crystalline bulk Si [80]. They also conclude that the refractive index is rather independent of the Si clusters sizes (3.6–4.6 nm) and is mostly affected by the degree of crystallinity. For SiO_x/SiO_2 SLs prepared by magnetron sputtering, negative optical birefringence (~1%) was observed, originating from the periodical set of parallel planes of two different materials constituting the SL [77,78]. At a given wavelength, the absorption coefficient of annealed SiO_x films prepared by magnetron sputtering and PECVD increases with the Si content [33,76,83]. The increase of the Si-nc mean diameter leads to an increase of absorption in porous Si [84] and in SiO_x films prepared by ion implantation [80]. For a fixed Si content, the absorption coefficient of films prepared by magnetron sputtering and PECVD decreases with the increasing annealing temperature, which is due the amorphous-to-crystalline transition of Si in the samples [33].

Another method of measuring the optical properties of SiO_x films is based on the PL-filtering effect. A silicon oxide layer containing Si-nc on a silica substrate forms a planar waveguide. Spectral filtering of the PL occurs when the PL is detected from the waveguide edge along the film surface. This effect has been found for Si/SiO_2 SLs [57] and for SiO_x films ($x < 2$) on silica substrate [79,85,86,87]. The detailed description of this effect is presented later.

The Tauc relation [88] allows one to estimate the band gap of SiO_x films containing Si-nc [31,33,36,44,74,89,90]. A study of Si-nc deposited on a silica substrate shows that the electronic band structure of Si-nc is quite different from that of bulk silicon and that the band gap increases as the Si-nc size decreases [82]. The latter observation is explained by the QC effect. The increase of the band gap for smaller Si-nc is also observed for implanted SiO_x samples [31,74]. The absorption threshold is blue-shifted for smaller Si-nc, which indicates the QC effect on the Si-nc band gap [33,36,84,90,91].

Laser annealing is an interesting approach to change the structural and optical properties of Si-based materials. The heating effect of pulsed radiation is especially strong [92]. However, it can also be substantial for continuous-wave (CW) irradiation of free-standing porous Si films [93]. A similar heating effect of CW laser radiation is observed in free-standing SiO_x films leading to a strong increase of the Raman signal of Si-nc [56,94,95,96]. Si-nc prepared by laser annealing show high compressive stress (~3 GPa), which is evidenced by an up-shift of the Raman band [56,96]. The compressive stress is formed in a solid

SiO_2 matrix when the volume of Si-nc suddenly increases after its crystallization from the liquid phase. The high stress can be relaxed by irradiating the stressed Si-nc with lower laser power [56]. Free-standing films are used to obtain the above described laser-induced effects because for films on substrates the laser-induced heat is reduced due to thermal flux to substrate [97]; however, it is still observable [98].

In agreement with the experimental observations, the theoretical simulations of isolated hydrogen-passivated Si nanostructures show that the band gap and the PL peak position change according to QC [13,19]. The behavior of these parameters is different for oxidized Si-nc [99]. Theoretical studies highlight the importance of the interface region between Si-nc and a SiO_2 matrix for the light-emitting properties [100,101]. The absorption edge tends to be red-shifted due to the formation of the Si=O or Si–O–Si bonds at the Si/SiO_2 interface [102,103,104]. The presence of oxygen atoms bonded to the surface of Si-nc as well as deformation of Si-nc are found to influence the optical band gap [105]. Luppi *et al.* have considered small Si-nc (up to 1 nm in diameter) by applying *ab initio* calculations [106]. In the case of the Si–O–Si bridge bond at the cluster surface, an emission peak at about 1.5 eV is obtained. In accord, the PL in this region is often observed experimentally from silica films containing Si-nc. The calculated emission peak is red-shifted with respect to the absorption, which is also in agreement with the experimentally observed Stokes shifts between the absorption and PL spectra.

Theoretical calculations show that the surrounding matrix can produce some strain on Si-nc [107,108,109], and this strain depends on the oxidation degree [110,111]. The magnitude of strain affects the band gap, which shows an oscillating behavior with the Si-nc size, not strictly following the QC rule [110]. Moreover, the disorder of nanoclusters has a large effect on their optoelectronic properties [112,113]. In particular, amorphization reduces the optical band gap and increases the absorption strength in the visible range.

Guerra and Ossicini have calculated the recombination rates for different classes of Si-nc in the diameter range of 0.2–1.5 nm [114]. The authors also consider different conditions of passivation, strain, and symmetry in order to find the best conditions of radiative emission. As a result, they have found that the smallest, highly oxidized, crystalline clusters are the most optically active Si/SiO_2 structures.

2. RESULTS AND DISCUSSION

2.1. Furnace-Annealed Films

In this section, we describe the properties of SiO_x ($x < 2$) films on silica substrates for different annealing temperatures (400–1200 °C) and Si contents (x from ~1.3 to ~1.98) [115,116,117,118]. The SiO_x films (thickness from ~1.5 to ~2.5 μm) are deposited on silica substrates by MBD. The MBD samples are compared with samples prepared by ion implantation [118].

2.1.1. Correlation between Optical and Structural Properties

The as-prepared MBD films (annealed at 400 °C for better mechanical stability) with substantial Si content show a broad Raman band at ~470 cm^{-1}, which is characteristic of amorphous Si (Figure 1) [27,48,55,119,120]. After annealing above 1000 °C, crystallization of the amorphous Si inclusions occurs as evidenced by the narrowing and upshift of the Raman band to ~518 cm^{-1} (Figure 1). When measured with low laser intensity, the Raman band position (518–519 cm^{-1}) is rather independent of the Si content and of the annealing temperature in the 1100–1200 °C range. According to the phonon confinement model [121,122], this result suggests that the Si-nc sizes evidenced by Raman spectroscopy do not change much under these conditions. However, growth of Si-nc with increasing annealing temperature was observed for SiO_x films prepared by other deposition methods [37,40].

For MBD SiO_x films with very small excess of Si ($x > 1.9$), the Raman bands of amorphous and crystalline Si are nearly invisible [116], in qualitative agreement with the earlier results [27]. The reason for this observation is the small amount of properly coordinated Si atoms (Si atoms bonded to four "bulk" Si atoms) for samples with low Si contents. A similar explanation was applied in a study of amorphous Si/SiO_2 SLs, where the Raman bands of amorphous Si were observed for thicker Si layers (≥2 nm) and the Raman-scattering cross section decreased for thinner Si layers [55,123].

As the annealing temperature increases, the low-frequency shoulder in the Raman spectra responsible for disordered Si and/or small Si clusters gradually decreases (Figure 1). However, even after annealing at 1200 °C, the low-frequency component of the Raman band does not disappear and its area is comparable to that of the high-frequency component. As a possibility, small Si

grains in a SiO_2 matrix may be disordered even after annealing at this temperature whereas larger Si grains are crystallized.

X-ray photoelectron spectroscopy (XPS) confirms that the structural reorganization is not complete after furnace annealing (Figure 2). The Si–SiO_2 phase separation increases with the annealing temperature, which is evidenced by the increasing amount of elemental Si and SiO_2 and the decreasing amount of SiO. However, a large proportion of suboxides (10–15 at. % for x = 1.7) is still detected after annealing at 1200 °C [116]. This result indicates incomplete Si–SiO_2 phase separation in these materials.

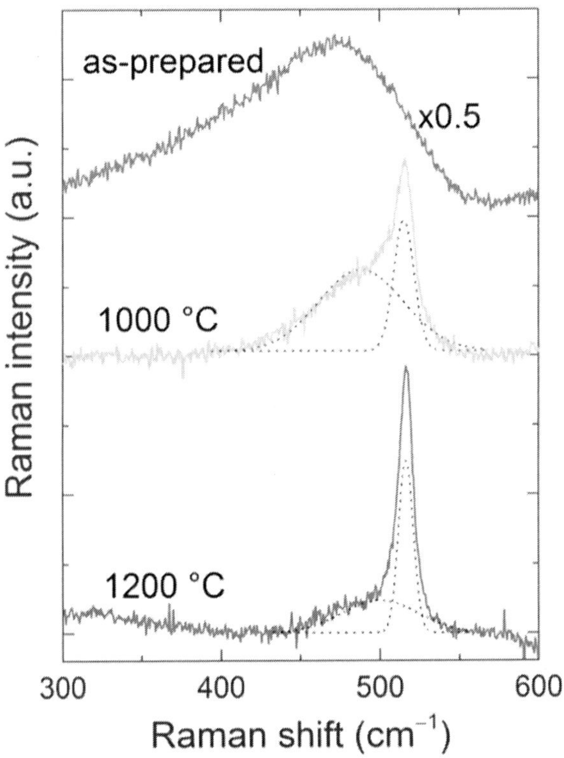

Figure 1. Raman spectra of a $SiO_{1.75}$ film as-prepared and annealed at 1000 and 1200 °C (fitted by two Gaussians). The spectra are vertically shifted for better presentation. The Raman spectra are measured with laser intensity of ~10^3 W cm^{-2} at the sample. Reproduced with permission from [116]. Copyright 2012, AIP Publishing LLC.

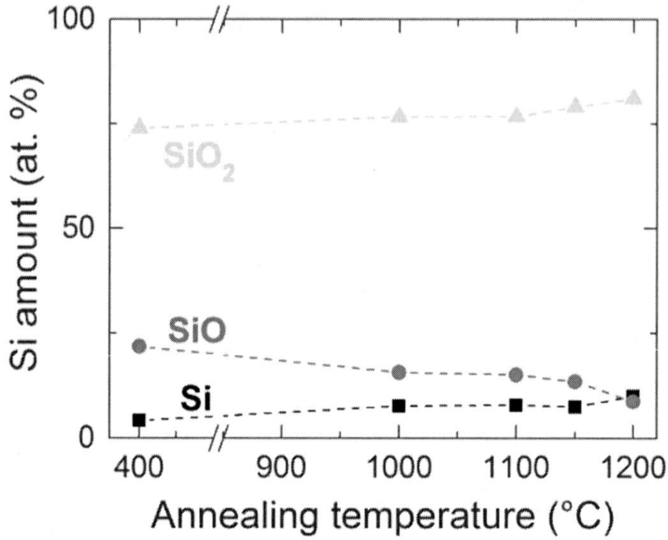

Figure 2. Composition of a $SiO_{1.7}$ film provided by X-ray photoelectron spectroscopy (XPS) as a function of the annealing temperature. Reproduced with permission from [116]. Copyright 2012, AIP Publishing LLC.

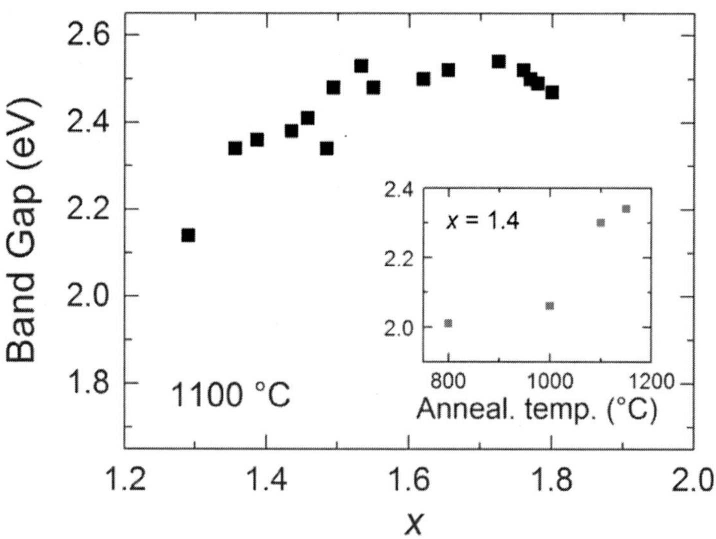

Figure 3. Band gap of SiO_x films annealed at 1100 °C as a function of *x*. The inset shows the band gap of a $SiO_{1.4}$ film as a function of the annealing temperature. Reproduced with permission from [116]. Copyright 2012, AIP Publishing LLC.

Additional information about the properties of SiO_x films can be obtained from the band gap. For a given annealing temperature (from 900 to 1200 °C), the band gap of SiO_x films increases as the Si content decreases [116]. For annealed at 1100 °C, the band gap increases from 2.2 to 2.6 eV as x increases from 1.3 to 1.8 (Figure 3), meaning that the Si-nc sizes somewhat decrease. According to the *ab initio* calculations for Si-nc with oxygen bonds at the Si/SiO_2 interface, the obtained band gap change (2.2–2.6 eV) corresponds only to a marginal change of the Si-nc sizes, the average size being ~1 nm [124]. On the other hand, the position of the maximum of the Raman bands at 518–519 cm^{-1} corresponds to larger sizes (3–4 nm) [121,122], and this value is supported by TEM measurements of similar materials [125]. These different estimates can be connected with low sensitivity of the Raman and TEM methods to very small Si grains. The low-frequency Raman scattering at 490–500 cm^{-1} can be contributed by smaller Si-nc (1–2 nm) that may be partially disordered, which would agree with the Si-nc sizes derived from the band gaps. It should also be mentioned that the band gap of Si clusters is a complex function of a number of parameters in addition to the size, such as the crystallinity, strain due to the SiO_2 matrix, and degree of oxidation [110,112]. Moreover, the Tauc law is generally applicable for amorphous semiconductors with absorption coefficients $>10^4$ cm^{-1} [88]. Thus, the band gaps obtained by applying the Tauc law to crystalline Si grains with lower absorption may be somewhat inaccurate.

The band gap of SiO_x ($x \sim 1.4$ [116] and $x \sim 1.8$ [118]) films increases with the annealing temperature (see the inset inFigure 3). This increase can be explained by amorphous-to-crystalline transitions of Si clusters rather than by a decrease of their sizes, following the discussions by Mirabella *et al.* [33]. Indeed, the Si-nc sizes should rather increase with the annealing temperature [40,42] even though this increase for the MBD samples seems to be minor as suggested by the Raman spectra.

By varying the Si content and annealing temperature, one can control the refractive index and the absorption coefficient of SiO_x films. For a given annealing temperature, the refractive index increases with the Si content (Figure 4a) [118] as also reported previously [49,79]. Extrapolation of the data to $x = 2$ yields the refractive index of silica (1.456), which is a good verification of our results. The obtained values are also in agreement with other experimental results for annealing at 1100 °C, $\lambda = 632$ nm, and $x \sim 1.38$–1.63 [80]. For a given x, the refractive index typically decreases upon annealing at temperatures from 1000 to 1100 °C but increases in the 1100–1200 °C annealing temperature range [116].

The measured refractive index is interesting to compare with the values derived from the effective medium approximation and the chemical compositions measured by XPS (Figure 4a for 1100 °C and in Table 1 for $x \sim 1.75$). The agreement between the theory and the experiment is good. The

observed small discrepancy can be due to the limitations of the used model (seeSection 3.3). It is also possible that the refractive index of nanostructures is different from that of the bulk materials. Finally, in the XPS analysis, the mixture of different suboxides constitutes the SiO component, and this can also contribute to the error.

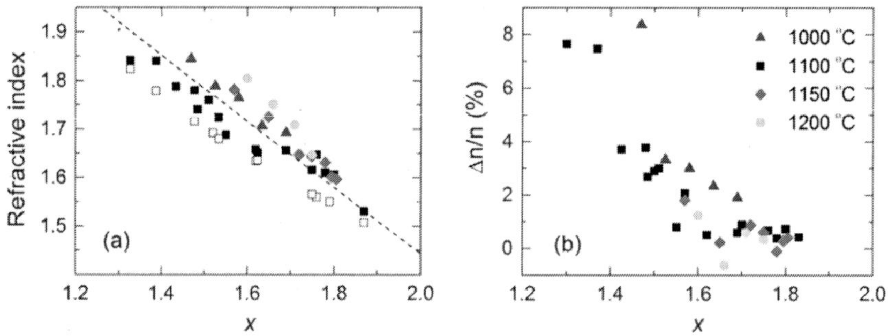

Figure 4. (**a**) TE refractive index and (**b**) birefringence of SiO_x films annealed at 1000 °C (triangles), 1100 °C (squares), 1150 °C (diamonds), and 1200 °C (circles) as a function of*x*. The open squares show the refractive index obtained using the effective medium approximation and the XPS results for annealing at 1100 °C. The dashed line is a guide for the eye. (**a**) is reproduced with permission from [116]. Copyright 2012, AIP Publishing LLC.

Table 1. Experimental refractive index n_{exp} and absorption coefficient α_{exp} of a $SiO_{1.75}$film for different annealing temperatures T_{ann} and the estimates for the refractive index n_{est} and extinction coefficient k_{est} from the effective medium approximation. Reproduced with permission from [116]. Copyright 2012, AIP Publishing LLC.

T_{ann} (°C)	n_{exp}	α_{exp} (10^4 cm^{-1})	n_{est}	k_{est}(SiO)	k_{est}(Si)
400	-	1.57	1.575	0.45	1.25
900	-	0.56	1.57	0.07	1.25
1000	1.64 *	0.134 *	1.56 *	0	0.77
1100	1.61	0.126	1.56	0	0.53
1150	1.64	0.120	1.565	0	0.535
1200	1.67	0.128	1.57	0	0.48

* Obtained by extrapolation of the data to $x = 1.75$.

The SiO_x film material prepared by MBD is found to be birefringent [79]. The classical theory of an isotropic asymmetrical planar waveguide predicts that the cut-off positions for the TE modes (with polarization parallel to the film) are at longer wavelengths than the TM cut-off positions (with polarization perpendicular to the film) [126,127]. However, the opposite order between the TE and TM peaks is found for some samples (see Figure 20a in Section 3.3). This reversed order of the TE and TM peaks is explained assuming positive optical birefringence, which is connected with the non-spherical shape of Si-nc [79]. Non spherical Si-nc have been observed in similar materials using the combination of electron tomography with plasmon-filtered microscopy [128]. It appears that that the optical birefringence is higher for SiO_x films with the higher Si content and for lower annealing temperature (Figure 4b) [79,116]. Similar values of birefringence are obtained by *m*-line measurements [115].

The absorption coefficient of SiO_x films can be analyzed using the effective medium approximation and the XPS data (Table 1). For an as-prepared $SiO_{1.75}$ film, the extracted extinction coefficient of the SiO phase (0.45) is much larger than the value known for bulk SiO (0.055) if the extinction coefficient of amorphous Si (k = 1.25) is used for elemental Si. This mismatch would be smaller if the amount of elemental Si was underestimated by XPS and/or the extinction coefficient of nanoscale amorphous Si was larger than that of bulk amorphous Si. For annealing at 900 °C, our estimates agree with the extinction coefficients of bulk SiO and amorphous Si. For films annealed above 1000 °C, the SiO component is found to be very transparent, in contrast to bulk SiO, and this difference is possibly connected with ultra-small sizes of the suboxide regions.

As the annealing temperature increases, the estimated extinction coefficient of elemental Si decreases (Table 1), which is interpreted in terms of gradual crystallization of Si [115,116]. However, this process in not complete even after annealing at 1200 °C, because the calculated extinction coefficient (0.48 for x = 1.75) is much larger than that of crystalline Si (0.079); thus, Si is partially disordered, in agreement with the earlier results [42]. This "disordered" Si should differ from "ordinary" amorphous Si because no Raman band at 470 cm^{-1} is present for the films annealed above 1000 °C. Instead, the disordered Si grains may contribute to the Raman scattering at 490–500 cm^{-1}, which is observed in Raman spectra of SiO_x films annealed up to 1200 °C (Figure 1). As a possibility, ultra-small Si grains in a SiO_2 matrix may be disordered even after the annealing whereas larger Si grains are crystallized. The presence of this disordered Si may explain the surprisingly large absorption of the annealed SiO_x films. Moreover, the optical properties of bulk and nanoscale Si may differ, and the absorption by SiO inclusions cannot be completely excluded. This explanation of the relatively large absorption coefficient of annealed SiO_xfilms

is quite speculative, and additional experimental data are required to draw more accurate conclusions.

The absorption coefficient and the integrated Raman intensity of Si-nc are nearly proportional to the amount of elemental Si obtained by XPS (see Figure 5 for absorption) [115,116], in agreement with the earlier results [48]. It follows that elemental Si determines the absorption coefficient and the Si-nc Raman signal. A similar conclusion is made for SiO_x films deposited by magnetron sputtering [129]. On the other hand, we have obtained that the Raman scattering cross-section of elemental Si in SiO_x films prepared by MBD is about three times smaller compared to that of crystalline Si [116]. This difference can be explained by several factors. Ultra-small Si-nc (diameters < 2 nm) may be invisible by Raman spectroscopy due to the low proportion of "properly" coordinated Si atoms. The interaction with a SiO_2 matrix can change the coordination of Si-nc, and this effect is stronger for smaller Si particles. In principle, the broad band gap of Si-nc may affect the Raman cross-section due to the resonance effect. Finally, the complex character of interaction of light with Si-nc embedded in SiO_2 is not well understood. The present results are not enough to give a more reliable interpretation of these observations.

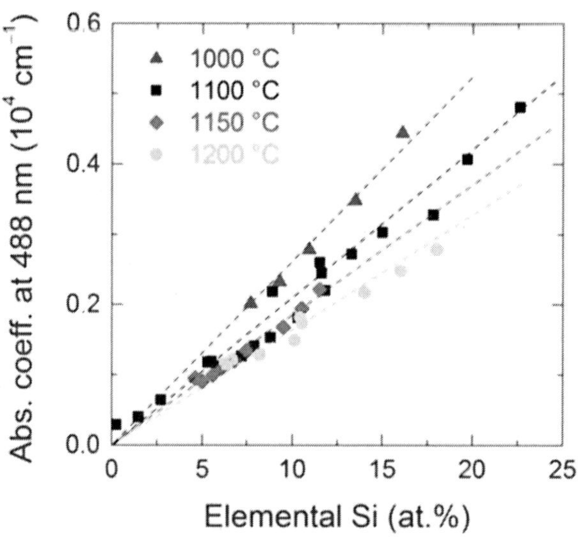

Figure 5. Absorption coefficient at 488 nm for annealing at 1000 °C (triangles), 1100 °C (squares), 1150 °C (diamonds), and 1200 °C (circles) as a function of the amount of elemental Si. The dashed line is a guide for the eye. Reproduced with permission from [116]. Copyright 2012, AIP Publishing LLC.

2.1.2. Photoluminescence of the MBD Samples

For SiO_x films annealed up to 900 °C, weak PL with the maximum at 650–750 nm is observed (Figure 6a). This light emission is usually assigned to different thermodynamically unstable radiative defects, for instance, nonbridging oxygen-hole centers or E′-type defects [31,32,130,131,132]. After annealing above 900 °C, the PL maximum shifts towards longer wavelengths (~800 nm) and the PL intensity strongly increases [116,118], which agrees with numerous literature reports [27,34,42,53,72,91]. The highest PL intensity is observed for $x \cong 1.8$–1.9 and annealing at 1100–1150 °C [116]. Annealing at 1200 °C decreases the PL intensity. A similar decrease of the PL intensity was also observed for SiO_x ($1.05 < x < 1.75$) films prepared by magnetron sputtering [34]. A slightly different behavior is observed for SiO_x ($1.05 < x < 1.75$) films prepared by PECVD. For these samples, the PL intensity increases with the annealing temperature up to 1250 °C [34,40].

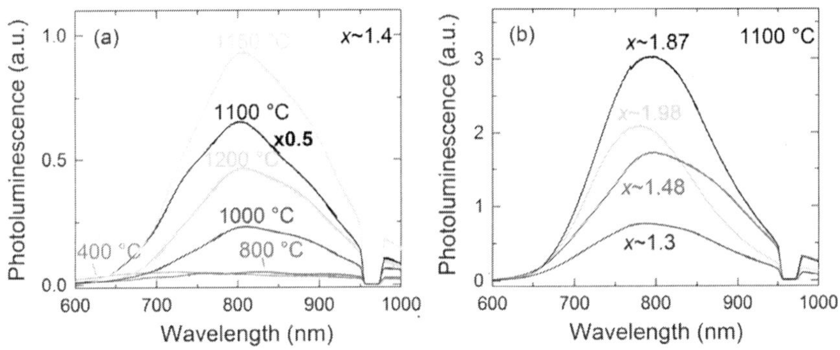

Figure 6. Photoluminescence (PL) spectra of (**a**) a $SiO_{1.4}$ film annealed at different temperatures and (**b**) SiO_x films with different x annealed at 1100 °C. The SiO_x films are prepared by molecular beam deposition (MBD). The hole in the spectra around 970 nm is due to the second order of the notch filter. Notice the multiplication factor in (**a**). (**b**) is reproduced with permission from [116]. Copyright 2012, AIP Publishing LLC.

The enhancement of the 1.5-eV PL by annealing at ≥1000 °C is accompanied by the formation of Si-nc detectable by Raman spectroscopy [40,48,49,116,118]. This correlation seemingly suggests that the PL directly originates from these Si-nc. On the other hand, the PL intensity is the highest for samples with relatively small Si contents [48,49]. Moreover, for $x > 1.9$, no Si-nc and elemental Si are evidenced by the Raman and XPS measurements, respectively,

whereas the PL is strong (Figure 6b). In fact, the PL quantum yield steadily increases as the absorption coefficient and the Si content decrease (Figure 7) [116,117]. These observations suggest that the Si-nc observed in the Raman spectra and detected by XPS as elemental Si are not direct light-emitting centers, which is in agreement with the defect origin of the 1.5-eV PL [52].

The PL position of our SiO_x films is rather stable for different annealing temperatures (above 1000 °C) and Si contents (seeFigure 6). These results seem to disagree with the studies of oxidized Si-nc prepared by magnetron sputtering [63]. Those experiments show that the PL position is size dependent for relatively large sizes (>3 nm) that are reliably measured by TEM. However, Wolkin *et al.* have shown that oxidation of Si-nc stabilizes the PL position for smaller Si-nc (<3 nm) [22].Thus, the stability of the PL position in our samples is consistent with the hypothesis that the observed PL originates from very small Si grains (<3 nm). In general, one has to distinguish different types of Si-nc samples. For example, the experiments on hydrogen-passivated Si-nc show that the PL of Si-nc with diameters smaller than 3 nm should be below 600 nm [23], which is not the case for our samples. This difference in PL positions can be explained by the decrease of the bandgap upon passivation of Si-nc with oxygen [110,133].

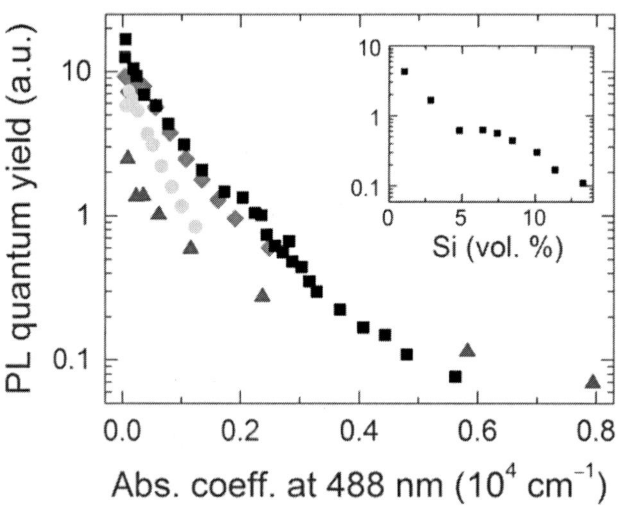

Figure 7. Relative PL quantum yield for SiO_x films prepared by MBD for annealing at 1000 °C (triangles), 1100 °C (squares), 1150 °C (diamonds), and 1200 °C (circles) as a function of the absorption coefficient at 488 nm. The inset shows the relative PL quantum yield as a function of the amount of elemental Si for annealing at 1100 °C. Reproduced with permission from [116]. Copyright 2012, AIP Publishing LLC.

Now we discuss the possibility of assigning the 1.5-eV PL to defects and/or ultra-small Si grains. By analogy with oxidized porous Si [22,62], the light-emitting centers can be related to the Si=O covalent bonds [27,48] or to the bridge Si–O–Si bonds suggested by theory [106]. Ultra-small (below ~1 nm) oxidized Si grains considered by theory [114] is another candidate to explain the red PL. Several observations are consistent with the latter interpretation. First, the stable PL position for different Si content suggests small sizes of the light-emitting clusters (<3 nm) [22]. The calculated band gaps of Si clusters with diameters ~1 nm are 2–3 eV [115], which agrees with the band gaps obtained for our samples (Figure 3) [116]. The short PL lifetimes obtained for our samples (~1 μs) [117] also suggest that the PL originates from Si-nc that are quite smaller than 3 nm, for which the PL lifetime is ~40 μs [116,134]. Moreover, the theoretical study by Luppi *et al.* suggests that Si-nc with diameters up to 1 nm and with a Si–O–Si bridge bond at the surface emit light at ~1.5 eV [106], which is observed for our samples. The decrease of the PL intensity after annealing at 1200 °C is reasonable if the amount of small optically active Si-nc decreases at this temperature.

The direct excitation of the light-emitting centers by laser light is possible. As another excitation mechanism, laser light may be absorbed by relatively large Si clusters, and then the excitation is transferred to the light-emitting centers located at the Si-nc/SiO_2 interface [68,135] or in more disordered areas [48,117]. Non-radiative defects such as dangling bonds at the Si-nc/SiO_2 interface as well as defects inside and outside the Si-nc can efficiently trap the excitation and therefore quench the PL [53]. Annealing at ~1100 °C presumably decreases the amount of these non-radiative defects and/or intensifies the energy transfer to the radiative centers, which enhances the PL. It has been shown that a considerable amount of non-paramagnetic near- and sub-gap defect states still remains after annealing, which drastically limits the PL quantum yield [136]. The amount of the light-emitting centers may increase upon annealing; however, no direct evidence of this is available.

Kusova *et al.* have suggested that compressive stress on Si-nc in a SiO_2 matrix can redshift the PL [137]. However, it is difficult to estimate stress of very small Si-nc using Raman spectroscopy, due to their low Raman scattering cross-section. Such small Si-nc can contribute to the broad low-frequency shoulder at ~500 cm^{-1} (Figure 1). On the other hand, the Raman bands at 518–519 cm^{-1} observed in our samples feature unstressed Si-nc with sizes of ~4 nm [121] that are also detected by TEM [125]. It can be speculated that no substantial stress is exerted also on smaller Si-nc.

2.1.3. Comparison of the Implanted and MBD SiO_x Samples (x ~ 1.8)

After preparation, the structural and optical properties of the MBD and implanted samples are very different [118]. Amorphous Si clusters (>2 nm) are present in the as-grown MBD film, as indicated by a broad Raman band at ~470 cm^{-1}(Figure 1) and by a large absorption coefficient (Figure 8a). For the implanted samples annealed below 900 °C, the absence of Raman bands of Si and very low absorption (Figure 8a) suggest that only very small Si clusters (presumably with sizes up to ~1 nm) may be present in the material [55,123].

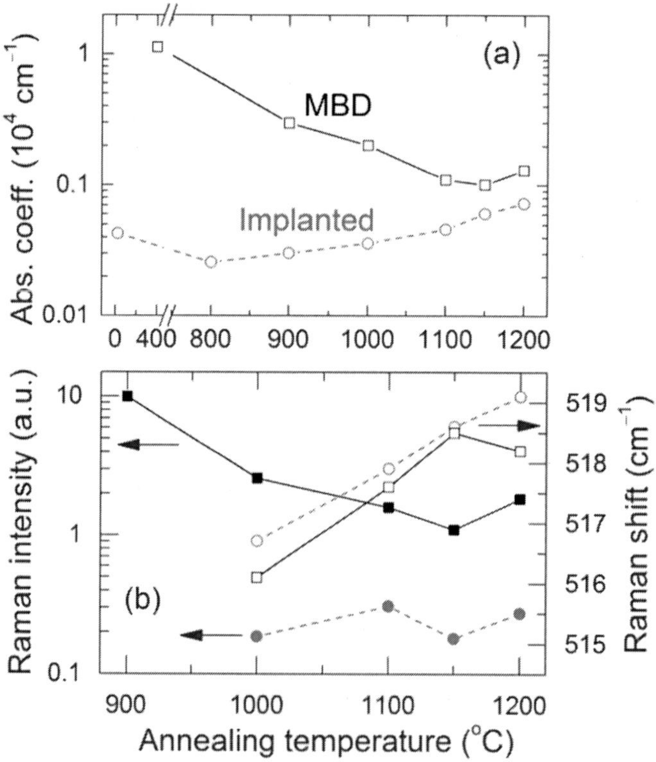

Figure 8. (**a**) Absorption coefficient at 488 nm for the MBD (squares) and implanted (circles) samples and (**b**) Normalized intensity (solid symbols) and position of the maximum (open symbols) of the Raman bands as a function of the annealing temperature. The Raman intensity is normalized by the effective film thickness (seeSection 3.2). Reproduced with permission from [118]. Copyright Wiley-VCH Verlag GmbH & Co. KGaA.

Upon annealing at 1000 °C, a narrow Raman band at ~516–517 cm^{-1} emerges for both types of samples, showing the formation of Si-nc. As compared to the as-prepared samples, the absorption coefficient of the annealed samples is increased for ion implantation but it is decreased for MBD (Figure 8a). After annealing at 1100–1200 °C, both types of samples become optically and structurally more similar. For both types of samples, the Raman band is observed at ~518 cm^{-1} and the absorption coefficients become closer (Figure 8). These trends are reasonably explained by growth and crystallization of Si clusters in the implanted sample and by crystallization of amorphous clusters in the MBD sample. Indeed, the extinction coefficient of crystalline Si is much smaller than that of amorphous Si.

After annealing (>1000 °C), the normalized Raman intensity and the absorption coefficient are larger for the MBD samples compared to the ion-implanted sample (Figure 8). This result indicates that the amount of Si-nc with diameters of ≥2 nm is greater in the MBD films. This conclusion is also supported by a smaller band gap for the MBD sample (~2.6 eV) compared to that for the implanted sample (~3.3 eV). For both types of samples, a substantial amount of suboxides (~15%) is detected by XPS even after annealing at 1200 °C, indicating that the Si–SiO_2 phase separation is not complete.

For annealing below 900 °C, the PL maximum is located at about 650–700 nm for both types of samples, and the PL intensity is much higher for the implanted sample (Figure 9). The latter fact indicates a larger amount of thermodynamically unstable radiative defects in the implanted sample [31,32,130,131,132]. Annealing at ~1100 °C enhances the 1.5-eV PL for both types of samples. After this annealing, the normalized PL is slightly stronger for the implanted sample (Figure 9), which is remarkable because the Raman band of Si-nc is weaker for this preparation method (Figure 8b). Annealing at 1200 °C weakens the PL, and this effect is more pronounced for the implanted sample. This decrease of the PL can originate from a thermal modification of the light-emitting centers, for example, an increase of the size of oxidized ultra-small Si grains. In the case of the implanted samples, this conclusion is supported by the redshift of the PL to ~900 nm and by the decrease of the band gap from 3.3 to 2.8 eV for annealing at 1150 and 1200 °C, respectively.

2.1.4. Laser Heating of Films on Substrates

Laser light is found to increase the temperature of SiO_x films on silica substrates [116]. For example, 488-nm laser power of ~130 mW focused to a ~40-μm spot (laser intensity ~10^4 W cm^{-2}) leads to a temperature of ~350 °C in a $SiO_{1.3}$ film annealed at 1100 °C. The absorption coefficient of this film increases at 350 °C

by ~40% compared to the room temperature value, which is probably due to the thermally induced shrinking of the bang gap [97]. At this laser-induced temperature, the Raman band of Si-nc is observed at ~512.5 cm^{-1}, *i.e.*, it is down-shifted by ~6 cm^{-1} from the room temperature position. The temperature dependence of the Raman shift (Figure 10a) is in excellent agreement with the calculations by Faraci *et al.* for Si-nc with sizes of 5 nm [138]. Similar temperature dependences were previously reported for crystalline Si [139] and the free-standing $SiO_{1.7}$ film [95]. The phonon confinement effect and residual stresses can explain the quantitative differences between the Raman band positions in these cases.

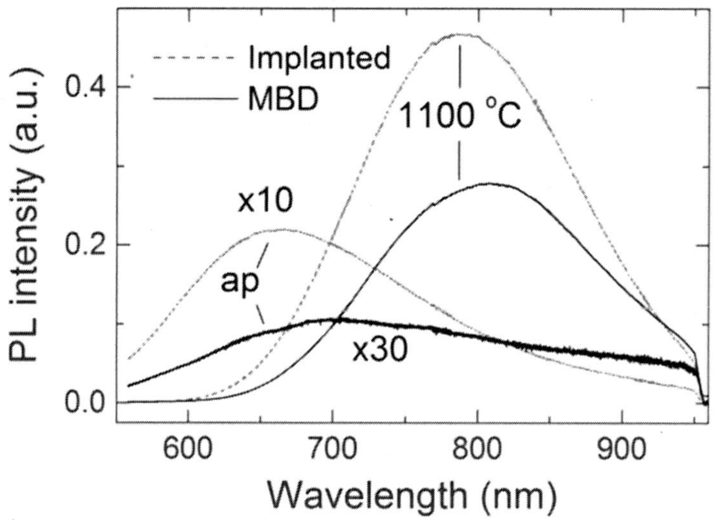

Figure 9. PL spectra of as-prepared (ap) and annealed (1100 °C) $SiO_{1.8}$ films fabricated by MBD and ion implantation. The PL spectra are normalized by the effective film thickness (see Section 3.2). Reproduced with permission from [118]. Copyright Wiley-VCH Verlag GmbH & Co. KGaA.

Figure 10b shows the laser-induced temperature for a free-standing SiO1.7 film and SiO*x* films (1.3 ≤ *x* < 1.5) on silica substrates as a function of the laser power absorbed by a unit volume, which is proportional to (1−eαd)Pd, where *P* is the laser power, α is the absorption coefficient, and *d* is the film thickness [95]. The linear fit of the temperature *versus* the absorbed power yields the room temperature at zero power, which verifies the results. For the same absorbed power, the laser-induced temperature is much higher in the

free-standing films compared to the films on substrates, which is due to the thermal flow to the substrate in the latter case.

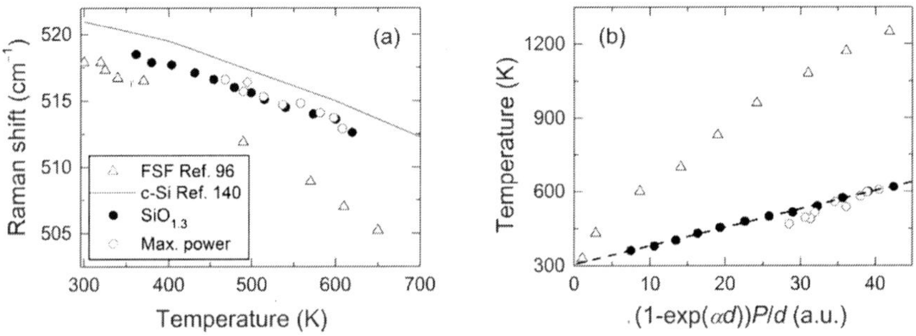

Figure 10. (**a**) Raman shift as a function of the laser-induced temperature and (**b**) Laser-induced temperature as a function of the absorbed laser power. The measurements are made for a $SiO_{1.3}$ film on a silica substrate annealed at 1100 °C using various 488-nm laser intensities (solid circles) and for SiO_x films ($1.3 < x < 1.5$) at the maximum laser intensity of ~10^4 W cm^{-2} (open circles) [116]. The data for a free-standing $SiO_{1.7}$ film [95] (FSF, triangles) and for crystalline Si [139] (solid line) are shown for comparison. The dashed line in panel b is a linear fit. Reproduced with permission from [116]. Copyright 2012, AIP Publishing LLC.

In Raman measurements of Si-nc, relatively high laser powers are usually employed to obtain a good signal-to-noise ratio. It follows that the laser-induced heat should always be carefully considered even for films on substrates (especially when using Raman microscopes) to obtain reliable information of the material structure. In general, laser heat, local stress, and phonon confinement are overlapping effects with respect to the Raman spectra.

2.2. Laser Annealing of Free-Standing Films

2.2.1. Laser-Annealed Areas

We have studied CW laser annealing (488 nm) of free-standing SiO_x films (0.5 and 2 µm-thick) and Si/SiO_2 SL (0.5-µm-thick). After preparation (Section 3.1), the free-standing films presumably contain small Si-nc (up to 4 nm in diameter)

[95,125]. The laser annealing is usually performed for ~1 s in air atmosphere (if different is not stated). An optical photograph of a typical laser-annealed area on a free-standing film and a Raman map from the same area are presented inFigure 11. Three regions can be distinguished: (i) central region, (ii) ring around the central region, and (iii) pristine film outside the irradiated area [140,141,142].

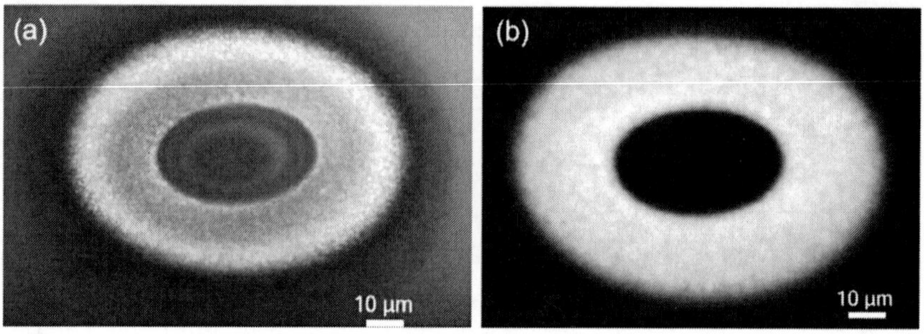

Figure 11. (**a**) Optical microscope photograph and (**b**) Raman map of the Si-nc band. A 2-μm-thick free-standing SiO_x film (x ~ 1.7–1.8) is exposed to 488-nm light (~110 mW) through a lens (focal length of 150 mm). In the optical photograph, the sample is illuminated through the objective. The oval shape is due to a non-perpendicular direction of the annealing laser beam. Brighter regions in the Raman map correspond to stronger Raman signals. The Raman signal is integrated in the 510–540 cm^{-1} spectral range. Reproduced with permission from [140]. Copyright 2010, AIP Publishing LLC.

In the central region, the material is essentially amorphous SiO_2. For 0.5-μm-thick SiO_x films (x ~ 1.7–1.8), this composition is directly shown by XPS and electron energy loss spectroscopy (EELS) [140]. In accord, the TEM images and selected area diffraction (SAD) patterns indicate a very homogeneous and amorphous structure of this region, and the Fourier-transform infrared spectra (FTIR) measured with high spatial resolution are similar to those of thermal silica. In this region, no Si-related peak is detected by Raman spectroscopy, the absorption is very low, and no PL is observed (Figure 12). In agreement, for 0.5-μm-thick Si/SiO_2 SL (2-nm-thick Si and SiO_2 layers), no Si clusters are detected in bright-field scanning TEM (BF-STEM) images, and energy-dispersive X-ray spectroscopy (EDS) indicates an increase of x in the central region compared to the pristine film [141,142]. Quantitative analysis based on the EDS measurements is not possible because quantification using low-energy X-ray lines such as oxygen K alpha is not feasible. Small Si clusters (<1 nm) invisible by these microscopic methods are in principle possible in the central region.

However, very small Si clusters are thermodynamically unfavorable [143], and their formation seems to be improbable at high temperatures. Furthermore, these very small Si grains are presumably luminescent; however, practically no PL is detected from the central region.

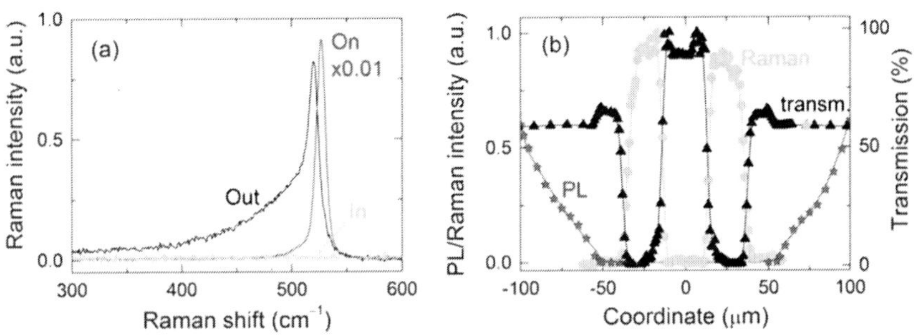

Figure 12. (**a**) Raman spectra of a laser-annealed 2-μm-thick SiO_x (x ~ 1.7–1.8) film from regions in the center (in), on the ring (on), and outside the laser-annealed area (out). (**b**) Raman (circles), PL (stars), and transmission (triangles) cross-sections of the laser-annealed area. The cross-sections are measured in the vertical direction through the center of the laser-annealed area shown in Figure 11. Reproduced with permission from [140]. Copyright 2010, AIP Publishing LLC.

In the ring around the central region, the EFTEM, BF-STEM, and TEM images evidence large Si-nc embedded in amorphous SiO_2 (Figure 13) [140,141,142]. The maximum Si-nc diameters (~200 nm) for the 0.5-μm-thick Si/SiO_2 SLs [141] are on average two times larger than those detected for the 0.5-μm-thick free-standing SiO_x films (x ~ 1.7–1.8) [140]. This difference is reasonably explained by a larger Si content in Si/SiO_2 SLs (x ~ 0.7 from the deposition conditions) compared to the SiO_x films (x ~ 1.7–1.8). The crystallinity of the large Si clusters in the ring region is confirmed by selected-area diffraction measurements (for 0.5-μm-thick SiO_x film) and Raman spectroscopy. The Raman band is quite narrow and does not have a shoulder at 450–500 cm^{-1}, indicating the absence of amorphous phase and therefore suggesting efficient Si crystallization (see Figure 12a). The EELS spectra show the presence of the Si and SiO_2 phases. In accord, the FTIR spectra correspond to stoichiometric SiO_2 perturbed by the Si–SiO_2 interface. The Raman bands from the ring region is stronger by about 2–3 orders of magnitude compared to the pristine film, which is explained by the resonance size effect for the large Si-nc [144]. The Raman bands from Si-nc in the ring region are typically shifted up to ~530 cm^{-1} (from the value of 521 cm^{-1} for crystalline Si) indicating a compressive stress on Si-nc. According to the

mechanism proposed earlier [56], these large Si-nc are formed from the liquid phase, and the stress appears when their volume suddenly increased in the solid SiO_2 matrix. Thus, the temperature in the ring region exceeds the melting point of Si (1685 K). Indeed, a temperature of ~2000 K was estimated by fitting the strong light-emission background upon irradiation with high laser intensities (~16 kW cm^{-2} at the sample) [94]. The bright thermal radiation is also seen by the naked eye. High processing temperatures in the central region are also confirmed by the decrease of the film roughness detected by AFM. For 0.5-μm-thick Si/SiO_2 SLs, scanning electron microscopy (SEM), TEM, and atomic force microscopy (AFM) show several large holes (up to 200 nm) in the inner part of the ring region (Figure 13a,b).

These results demonstrate a Si–SiO_2 phase separation on a macroscopic scale. The central region (up to 30 μm) is cleaned of the Si excess. On the other hand, in the ring region, large Si-nc are formed and the Si excess is enhanced as found for 0.5-μm-thick SiO_x film by XPS and EELS [140]. The macroscopic phase separation can be explained as follows [140]. In the beginning of laser irradiation, the film structure is homogeneous so that the temperature gradient approximately corresponds to the intensity gradient in the Gaussian beam. The temperature gradient gives rise to a concentration gradient due to different diffusion properties of the mixture constituents. For example, Si, O, and SiO can diffuse in silica [145], which changes the relative concentrations. Diffusion in temperature gradient is known as thermodiffusion or Ludwig-Soret effect [146,147]. As detected by EELS, XPS, and AFM, the central region is typically thinner than the pristine film by about 30%–40% and ~50% for 0.5-μm-thick Si/SiO_2 SLs and SiO_x films, respectively. Moreover, AFM shows an increase of the film thickness in the ring region compared to the pristine film [142]. These results are also indicative of macroscopic movement of Si excess from the central region.

2.2.2. Surface Structure

For laser-annealed 0.5-μm-thick Si/SiO_2 SLs, surface analysis (SEM and AFM) has been used to discriminate Si-nc located near the film surface [141]. In the tilt TEM images (Figure 13c), these "surface Si-nc" have an unusual pear-like shape with the thinner part presumably sticking out of the laser-illuminated surface. The AFM and SEM methods estimate that the thinner part above the surface (surface feature) has a typical size of 60–70 nm, in agreement with the TEM measurements. The subsurface part has a diameter of ~200 nm, as shown by TEM and BF-STEM images.

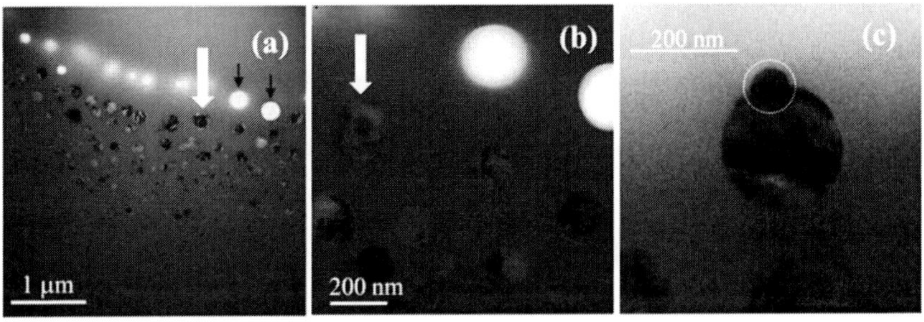

Figure 13. TEM images of a 0.5-μm-thick Si/SiO_2 SL showing a Si particle located near the film surface (marked with a white arrow). The images in (**a**), (**b**), and (**c**) are taken with different magnification. (**c**) shows this particle in the tilt geometry, the surface subparticle being indicated by a dotted circle. The bright spots are holes in the film; the holes observed in (**b**) are marked by black arrows in (**a**). Figure courtesy of Simona Boninelli and Fabio Iacona. Reproduced with permission from [141]. Copyright 2012, AIP Publishing LLC.

The formation of the surface features is explained by the following mechanism. Since most of large Si-nc formed by laser annealing are under high compressive stress, it is possible that the laser-pressurized silicon can erupt from the film if it is separated from the ambient atmosphere by a very thin silica layer. The erupted Si is seen as the subparticle in Figure 13c. The volume of the subparticle (~70 nm in diameter) is ~5% of the volume of the larger subsurface part of the nanocrystal (~200 nm in diameter). Eruption of this volume would lead to a substantial decrease of the compressive stress of the parent particle [56]. The fact that no eruptions are detected on the backside of the film seems to be consistent with this mechanism. Indeed, the temperature should be somewhat higher on the front (laser-illuminated) side of the film; hence, the covering silica film is softer here, which favours Si eruption.

Raman measurements confirm the proposed eruption model. The surface Si nanocrystal shown in the TEM images inFigure 13 is located in position 1 in Figure 14a. The Raman shift obtained in this position is ~520 cm^{-1} (spectrum 1, Figure 14b), which indicates a low stress. It is reasonable that Si eruption can relax stress in the vicinity of surface Si-nc. The compressive stress is partially relaxed at a distance of about 1 μm from the surface feature (spectrum 2). The Raman spectrum measured at position 3, which is far away from the surface feature, shows a higher compressive stress (spectrum 3 in Figure 14b).

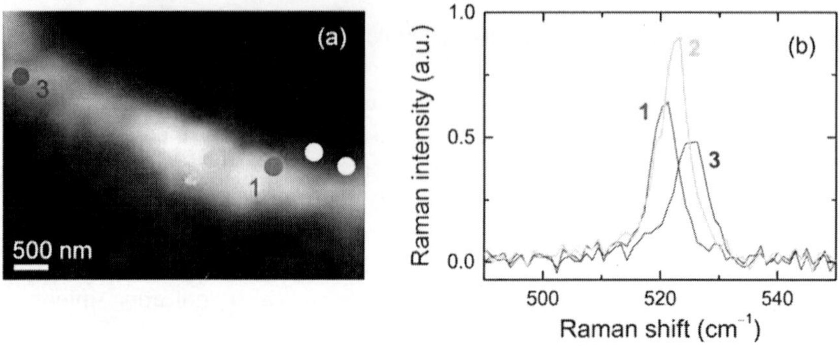

Figure 14. Raman characterization of the area shown in Figure 13a. (**a**) Raman map of Si-nc (integration from 515 to 530 cm^{-1}) obtained with a 100× objective and steps of 0.2 μm. Brighter areas correspond to stronger Raman signals. The white circles indicate the positions of two holes used as the reference points. (**b**) Raman spectra in positions 1, 2, and 3 indicated in (**a**). Position 1 corresponds to the surface Si-nc shown in Figure 13. Reproduced with permission from [141]. Copyright 2012, AIP Publishing LLC.

2.2.3. Effect of the Annealing Atmosphere

The studies described above have been performed in the atmosphere of air (containing ~20% of oxygen), which can cause some oxidation of Si during laser annealing. Indeed, oxygen can diffuse through the film oxide to Si-nc [148,149,150]. This process may decrease the excess of Si, especially in the central region, where the temperature in the beginning of laser annealing is the highest. To evaluate this effect, we compare laser annealing in atmospheres containing different amounts of oxygen (Ar, N_2, O_2, and air). Detailed measurements are performed for 0.5-μm-thick Si/SiO_2 SL and 1-s exposure [142].

For all annealing atmospheres, the central region of laser-annealed areas most probably consists of amorphous silica without Si excess. This composition is indicated by the absence of the Raman signal of Si in this region, an increase of *x* compared to the pristine film detected by EDS, and the absence of Si clusters in the BF-STEM images. In the ring region, large Si-nc (up to 150–300 nm) are formed, and the film thickness tends to increase compared to that of the pristine film. The change of the film thickness was estimated using the AFM images taken from both sides of the film. These results support the concept of macroscopic Si–SiO_2 phase separation as a result of laser annealing.

The effect of oxygen can be analyzed by comparing the film thickness in the central region obtained for different atmospheres. Indeed, two competitive processes, thermodiffusion and Si oxidation, affect the film thickness in the opposite directions, tending, respectively, to decrease and to increase it. For Ar atmospheres, the central region is thinner by ~150 nm than the pristine film, *i.e.*, the thickness is about 350 nm. This observation supports the concept of the macroscopic phase separation because Si oxidation does not operate in the inert atmospheres. A similar change of the thickness occurs in air (the thickness of the central region is 300–350 nm) also indicating minor Si oxidation.

In contrast, laser annealing in pure O_2 increases the film thickness in the central region by ~20 nm compared to the pristine film. Thus, both thermodiffusion and Si oxidation seem to operate in this case, practically compensating each other. It seems that a significant part of Si is oxidized upon laser annealing in pure oxygen. The difference between the film thickness of the central region for inert and O_2 atmospheres is ~200 nm (larger in oxygen). 200 nm of SiO_2 can be formed by oxidation of about 100-nm Si layer, which corresponds to ~40% of all amount of elemental Si in the pristine SL (125 2-nm-thick Si layer). According to the model suggested by Deal and Grove for crystalline Si, 1 bar of dry O_2 at 1200 °C produces ~20-nm oxide film during 1 s, whereas a ~33-nm oxide is formed for 100 s [149], which is in approximate agreement with our estimates. It should be reminded that the temperature of laser annealing is sufficiently higher [94]. In addition, the extent of oxidation can differ from these values because our pristine material is very different from crystalline Si.

The structure of the ring region is also affected by the amount of O_2 in the ambient atmosphere. For inert atmospheres, the AFM and BF-STEM images show that the ring region is very inhomogeneous. It has areas with large Si-nc (up to 150 nm in diameter), with large non-spherical holes up to 300 nm, and with smaller Si-nc (up to 50 nm in diameter). No surface Si-nc are observed in this case, as shown by AFM and SEM. Raman spectroscopy indicates Si-nc with different stresses (Raman shifts from 518 to 530 cm^{-1}) and regions with very different intensities of Raman bands of Si-nc. The BF-STEM images show a "belt" of nano-channels (or pores) around the ring region. The size of these nano-channels can be as low as several nanometers.

For laser annealing in air, the ring region contains Si-nc with diameters of up to 200 μm. A chain of holes with sizes up to 200 μm is seen at the inner part of the ring of large Si-nc (Figure 15a). A number of surface features are observed as discussed above.

For pure O_2 atmosphere, in addition to Si-nc inside the film, SEM and AFM show a regular chain of surface features that are located in the inner part of the ring at the front side of the film (Figure 15b). The undersurface part of the

surface Si-nc can be up to 300 nm in diameter. In agreement with the suggested eruption model, Raman spectroscopy shows a relaxed stress for these surface Si-nc (Raman shift of ~520 cm^{-1}) whereas the Si-nc at the outer part are still under compressive stress (Raman shift up to ~530 cm^{-1}). No holes in the film are observed after laser annealing in O_2 atmosphere.

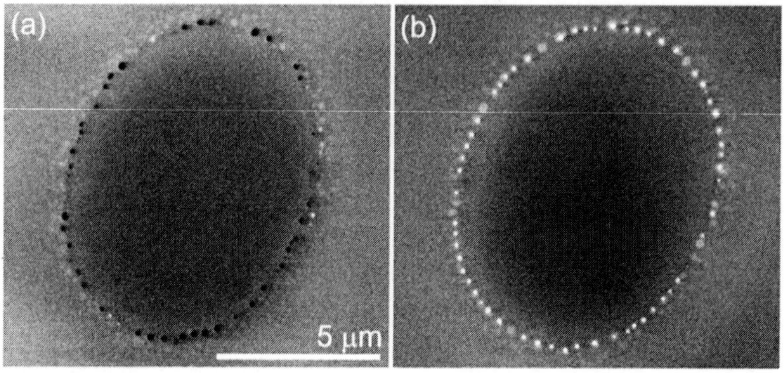

Figure 15. Secondary electron (SE) images of the front side of laser-annealed areas produced in (**a**) air and (**b**) O_2 atmospheres during 1-s exposure. Surface features are white, holes are black. The scale is the same in both panels. Figure courtesy of Marianna Kemell. Reprinted with permission from [142]. Copyright 2014, American Scientific Publishers.

Based on these results, it can be suggested that the holes produced in inert and air atmospheres are formed via removal of surface Si-nc from the film. For laser annealing in air, the holes appear in the positions where a chain of large surface Si-nc is observed in the case of O_2 atmosphere (see Figure 15). The laser-annealed areas with larger number of holes contain less surface Si-nc, and the presence of oxygen increases the amount of surface Si-nc. This observation suggests that oxygen decreases the removal of large Si-nc by the formation of a thin oxide layer covering surface Si-nc, which "glues" them to the film. The large irregular holes obtained in the inert atmospheres also presumably originate from large Si-nc removed from the film. The suggested mechanism of hole formation is consistent with the dependence of the maximum Si-nc sizes on the annealing atmosphere. For inert atmospheres, the largest Si-nc visible in the BF-STEM images are "only" ~150 nm in diameter because bigger Si-nc are removed from the films forming holes with sizes of up to 300 nm. For O_2 atmosphere, the largest Si-nc (up to 300 nm) are detected because they are protected by thin oxide layers. Annealing in air represents an intermediate case where the largest Si-nc are ~200 nm in diameter, and some Si-nc are removed

forming holes. In accord, with the proposed mechanism, holes are not observed in a 0.5-µm-thick SiO_x film (x ~ 1.7–1.8) annealed in air [140]. For this film, Si-nc have diameters "only" up to 100 nm, which are substantially smaller than the film thickness. These Si-nc are mostly located below the film surface and, hence cannot be removed to form holes.

Figure 16 shows the cross-sections of the Raman signal measured through characteristic ring regions of the laser-annealed areas prepared in inert (Ar) and O_2 atmospheres. For inhomogeneous ring regions formed in Ar atmosphere, two different characteristic cross-sections are observed (Figure 16a,b) whereas for O_2 and air atmospheres, the laser-annealed area is more uniform having similar Raman cross-sections in different ring regions (see Figure 16c for O_2). For 488-nm laser light, the Raman enhancement is known to be the strongest for spherical Si particles with diameters of ~112.5 nm [144]. In agreement, Si-nc with sizes ~100 nm are observed in the areas with the strongest Raman signal (at coordinate = 0). At coordinates >5 µm, the Raman spectra are similar to that of the pristine film. In Figure 16b, the Raman intensity is relatively weak in the ring region because the Si-nc diameters in this case do not exceed ~50 nm, which is smaller than the resonance value of ~112.5 nm.

For inert atmospheres (Figure 16a,b), the Raman signal is very weak at coordinates 2–3 µm. For air and O_2, this decrease of Raman intensity is less pronounced but still observable (Figure 16c). We call this effect the secondary Si depletion, and it supports the model of macroscopic phase separation. We tentatively explain the formation of this interval with low Raman intensity in the following way. When the large Si-nc are formed in the ring region during laser annealing, the absorption coefficient in this area strongly increases resulting in an increase of the temperature. Due to the thermal conductivity, the temperature of the vicinity also increases, which leads to a temperature gradient and additional macroscopic phase separation outside the ring region. This process seems to be more efficient for the inert compared to oxygen-containing atmospheres. The position of this spatial interval roughly coincides with the position of the belt of nano-pores and nano-channels observed by SEM in the inert atmospheres. The exact mechanism of the formation of these pores is unclear.

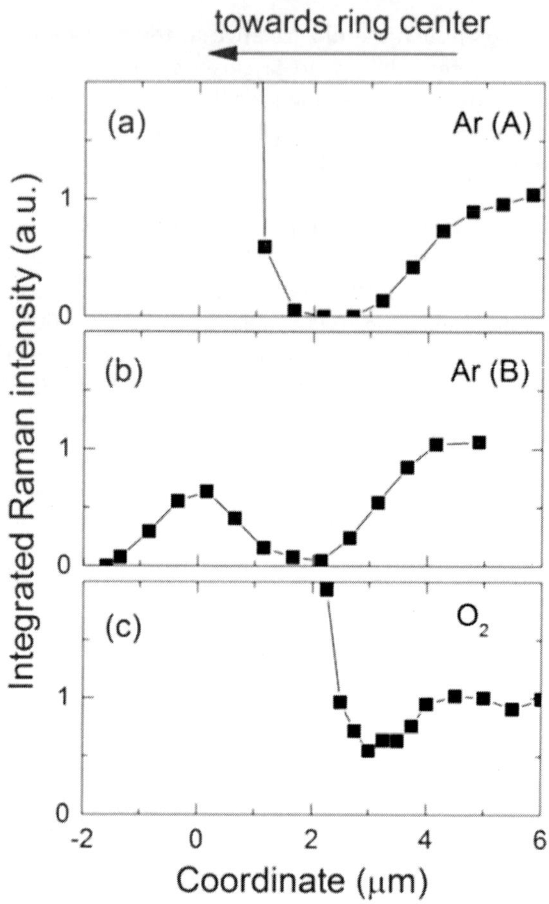

Figure 16. Characteristic Raman cross-sections of the ring region. Two characteristic cross-sections (**a**) and (**b**) are shown for Ar atmosphere, and the cross-sections (**c**) are similar for O_2. For air, the cross-sections are similar to the case of O_2. The Raman signal is integrated in the 230–585 cm^{-1} range. The data are normalized by the Raman intensity measured from the pristine film. Reprinted with permission from [142]. Copyright 2014, American Scientific Publishers.

2.2.4. Effect of the Exposure Period

Laser annealing with different exposures (10 ms and 100 s, in addition to 1 s) has been studied for 0.5-μm-thick Si/SiO_2 SL mostly in air atmosphere [142]. For the central area, the results are similar for all exposures: the EDS

measurements show an increase of *x* (relative to the pristine film), no Raman signal from Si-nc is detected, and no Si clusters are visible in the BF-STEM images. Thus, 10-ms exposure is enough for the removal of excess Si from the central region. In other words, the formation of the central region occurs in a millisecond time scale.

On the other hand, the ring region differs for different exposures. For 10-ms exposures, no holes and surface features are found and Si-nc have diameters of up to 20–30 nm. For 1-s and 100-s laser exposures, some holes and surface features are formed and Si-nc with diameters up to 200 and 250 nm, respectively, are observed. Thus, longer laser exposure increases the Si-nc size. Coalescence of Si-nc can be suggested for this process [73], because the number of Si-nc decreases for longer exposure periods. The increase of the Si-nc sizes with the exposure period can probably explain the greater amount of surface Si-nc observed for 100-s compared to 1-s exposure. For other atmospheres, the prolonged laser annealing (100 s) also decreases the amount of small Si-nc and increases the average sizes of Si-nc compared to 1-s exposure.

The Si-nc diameters for 10-ms exposure (20–30 nm) are smaller than the resonance size of 112.5 nm. This explains why no Raman enhancement is observed from these Si-nc (Figure 17). In contrast, for 1-s and 100-s exposures, the Raman intensity is strongly enhanced compared to that from the pristine film, which is explained by the presence of Si-nc with diameters of ~100 nm. The Raman band observed after 10-ms exposure has a smaller low-frequency shoulder compared to the pristine film, meaning that the amount of disordered Si and/or small Si clusters is reduced. As the exposure increases, the Raman band narrows indicating that the structure of Si-nc becomes more regular. For 100-s exposure, the width of the Si-nc Raman band is as narrow as that of a crystalline Si wafer, which indicates an almost perfect crystalline structure.

For 10-ms and 1-s exposures, the Si-nc in the ring region are under compressive stress because the Raman shifts are larger than that of crystalline Si (Figure 17); thus, these Si-nc are formed from the liquid phase. On the contrary, no stress on Si-nc is detected for 100-s exposure for all studied annealing atmospheres. This fact may show that for the long exposure, the temperature of large Si-nc becomes at some moment below the melting point. The decrease of the temperature for long exposure may be explained by the decrease of the absorption when the Si-nc become much larger than the resonance size of 112.5 nm [144]. Annealing at temperatures below the melting point relaxes the stress [56,96].

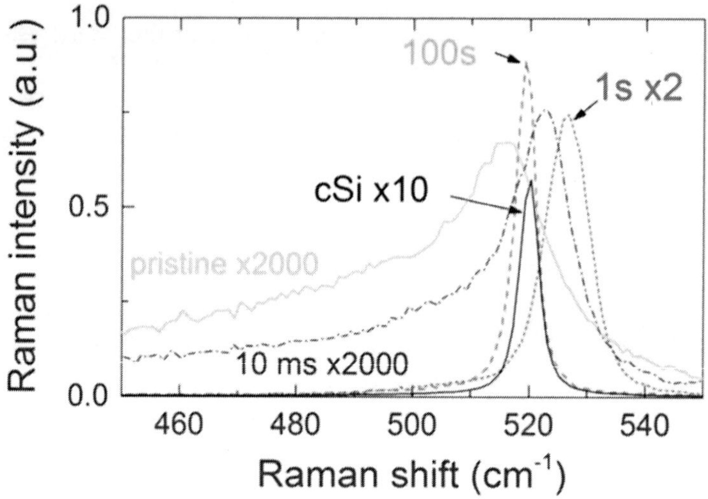

Figure 17. Typical Raman spectra of the ring region for 10-ms, 1-s, and 100-s exposures in air. Raman spectra of the pristine film and crystalline Si wafer (cSi) are also shown. Reprinted with permission from [142]. Copyright 2014, American Scientific Publishers.

2.2.5. Optical Memory

We describe two concepts of optical memory based on laser annealing of Si nanostructures in free-standing Si/SiO_2 SLs [151]. The first concept is based on the enhancement of Raman signal. The sample is irradiated by a 488 nm laser beam focused by a 100× objective of a Raman microscope, which produces laser-annealed areas with sizes below 1 µm. These areas are clearly visible in camera photographs (Figure 18a), and this indicates a change in the reflection and/or absorption coefficient of the film. The Raman, PL, and transmission profiles through a typical laser-annealed spot are presented inFigure 18b.The Raman signal is strongly enhanced whereas the transmission is decreased due to temperature-induced Si–SiO_2 phase separation and ordering and growth of Si-nc [140,141,142].

The observed laser effect is a case of non-volatile memory. Binary digits can take values of either 0 or 1 as demonstrated inFigure 18c. The smaller Raman intensity corresponds to 0, whereas the higher Raman intensity represents 1. Similar data reading can be obtained by transmission or reflection measurements. The estimated data density for this method is ~1 bit/µm^2, which can be improved by using stronger objectives and shorter laser wavelength. The nature of the used material provides high thermal stability of

the described optical memory. The thermal stability of the laser-crystallized spots was tested by furnace annealing for 1 h at 1200 °C in a nitrogen atmosphere. As expected, Raman and transmission signals remained unchanged. This approach does not allow erasing and rewriting information.

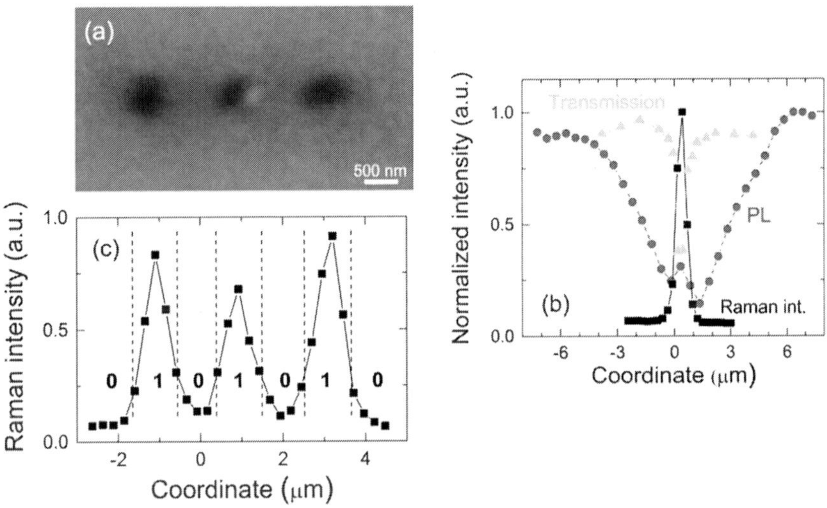

Figure 18. Writing digital data on a free-standing 2-μm-thick Si/SiO_2 SL. (**a**) Camera photograph of three spots produced by 488-nm laser annealing for ~10 ms. (**b**) Cross-section of the Raman, transmission, and 1.5-eV PL signals through a typical laser spot. (**c**) Cross-section of the Raman intensity through the spots presented in panel a, which illustrates digital data points. The Raman intensity is integrated in the 500–540 cm^{-1}range. Reproduced with permission from [151]. Copyright 2009, AIP Publishing LLC.

The second concept is based on Si-nc stress. First, a free-standing Si/SiO_2 SL is shortly irradiated with a relatively large laser spot (10–40 μm in diameter), which produces an extended area with a compressive stress of ~3 GPa (see Figure 11). This stress is evidenced by Raman bands at ~527 cm^{-1} (Figure 19a). Next, the stressed area is irradiated by a highly focused laser beam using a 100× objective and reduced laser power. As a result of laser annealing below the Si melting temperature, the stress is released in the irradiated spot and the Raman band shifts down to ~518 cm^{-1}. The typical size of the spot with relaxed stress is ~1 μm (Figure 19b). The optical constants are practically unchanged upon the stress relaxation, and this effect is invisible in optical photographs. The laser-induced decrease of the Raman shift writes data pixels on the film

similarly to the concept based on the Raman enhancement. The data can be erased by inducing stress using a broad laser beam and then re-written again. The retention time of the stress-based memory is estimated to be about 1 year at ~300 °C [56].

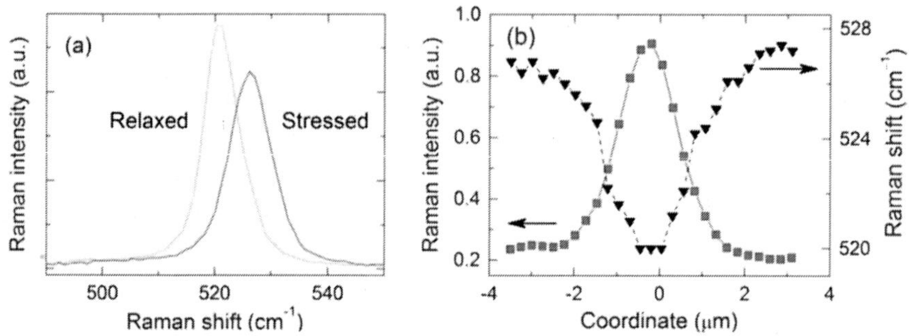

Figure 19. (**a**) Raman spectra of Si-nc with stress of ~3 GPa and after laser-induced relaxation (for 2-μm-thick Si/SiO_2 SL). (**b**) Cross-sections of the Raman band position and Raman intensity integrated in the 510–520 cm^{-1} region. Reproduced with permission from [151]. Copyright 2009, AIP Publishing LLC.

3. EXPERIMENTAL SECTION

3.1. Samples

Most of the samples used in this study have been prepared by MBD in the Aalto University (Finland) by S. Novikov. The SiO_x films are deposited on silica wafers. Some of the samples have a gradient of the Si content in the lateral direction, and they are called optical wedges. Due to the deposition conditions, these films also have a varying thickness (1.5–2.5 μm). The advantage of the optical wedges is that the areas with different Si contents are annealed under the same conditions and the O/Si concentration ratio *x* can be continuously changed by selecting the film area. The wedged samples are cut in slices (5–7 cm long and ~1 cm wide) along the Si content gradient and annealed in a furnace at temperatures up to 1200 °C for 1 h in nitrogen atmosphere. In [116], three samples covering *x* from ~1.3 to ~1.98 are studied.

The free-standing films are prepared from the SiO_x films and Si/SiO_2 SL sdeposited on Si wafers and annealed in a furnace at 1100 °C in nitrogen atmosphere. The free-standing films are made by chemical etching of the Si substrate [125,140,141,142]. The Si/SiO_2 SLs are constituted of 125 or 500 pairs

of 2-nm-thick Si and SiO_2 layers, yielding the total film thicknesses 0.5 or 2 μm, respectively. The thicknesses of the free-standing SiO_x films (x ~ 1.7–1.8) are 0.5 and 2 μm.

One sample is made by Si-ion implantation into a silica substrate in the Accelerator Laboratory (University of Helsinki) [118]. The implantation of Si ions into a silica plate was done with different energies (30, 56, 98, 160, 250, and 400 keV) and doses to achieve a nearly constant Si excess in the SiO_x layer with a thickness of ~750 nm. The O/Si concentration ratio $x$estimated by XPS is ~1.8. The implanted sample is annealed at various temperatures up to 1200 °C for 1 h in a nitrogen atmosphere.

3.2. Equipment

The main analytical techniques used in our laboratory are Raman, PL, and UV-visible absorption methods. Two setups have been used to record the Raman and PL spectra. The first setup consists of an argon-ion laser (488 nm, Melles Griot 543-AP-A01), a spectrometer (Acton SpectraPro 500I), and a charge-coupled device camera (Andor InstaSpec IV). The laser beam is focused with a lens to a spot with a diameter of ~40 μm. The second setup allows us to perform micro-Raman and micro-PL measurements. It consists of a confocal Raman microscope (LabRAM HR800, Horiba Jobin Yvon) equipped with an argon-ion laser (488 nm). The size of the laser spot in the microscopic measurements is down to ~1 μm. For both setups, the spectral resolution is ~2 cm^{-1} and 10 cm^{-1} for Raman and PL measurements, respectively. The Raman bands of the films annealed at ≥1000 °C are fitted with two or three Gaussians. The position of the Raman band of Si-nc is obtained from the position of the higher-frequency Gaussian (~518–530 cm^{-1}). The PL spectra are corrected for the spectral sensitivity of the apparatus. The Raman and PL intensities are normalized by the effective thickness $\int_0^d e^{(-2\alpha x)}dx$ and $\int_0^d e^{(-2\alpha x)}dx$ respectively, where the same absorption coefficients α is assumed for the laser and Raman light, the absorption for the PL is neglected, and d is the film thickness [55].

Laser annealing of free-standing films is performed using a TEM_{00} beam of an argon-ion laser (488 nm, power up to 110 mW at the sample) focused to spots with diameters ranging from ~1 to ~40 μm. The exposure time is mainly ~1 s, but 10-ms and 100-s exposures is also used in some experiments. Most of laser-annealing studies are performed in air; however, a number of measurements are done in O_2, N_2, and Ar atmospheres.

The absorption coefficient is obtained from the transmission and reflection spectra [152], recorded with a fiber-optics spectrometer (SD2000, Ocean Optics) and a broadband light source (DH-2000, Top Sensor Systems) or by measuring transmitted and reflected laser light using a power meter (NOVA II, OPHIR). The micro-transmission measurements of the 488-nm light are carried out in the LabRAM microscope with a large-area photodiode located behind the sample.

Additional measurements by other methods have been available from cooperation: XPS [115,116,117,118] including measurements with high spatial resolution [140], FTIR spectroscopy with high spatial resolution [140], TEM including bright-field scanning (BF), STEM and EFTEM [140,141], EELS [140], SEM and EDS [141,142], AFM [141,142], and *m*-line measurements [115]. The equipment used in these methods is described in the given references.

3.3. PL-Filtering Effect

The method to measure the optical properties of the films utilizes the PL-filtering effect. The PL filtering is observed when the PL spectra are measured in a direction along the film surface. In this case, narrow and polarized spectral peaks are detected (Figure 20a). The PL filtering has been observed in various silica films containing Si-nc [49,57,59,79,85,86,87,94].

The PL-filtering effect can be interpreted in terms of delocalization of guided modes near the cut-off frequency [49,57,79,85]. The minimal losses occur near the cut-off wavelengths, for which the delocalized light travels mostly in the transparent substrate rather than in the absorbing film. Thus, the narrow peaks seen in the PL spectra obtained in the waveguiding direction correspond to the cut-off wavelengths. The mode localization is a function of the generalized frequency parameter written for an asymmetrical planar waveguide in the form:

$$V = 2\pi \sqrt{(n_1^2 - n_2^2)}\, d/\lambda \qquad (1)$$

where n1 and n2 are the refractive indexes of the film and substrate, respectively, *d* is the film thickness, and λ is the wavelength. This equation is obtained assuming weak-guiding approximation $(n1-n2) \ll 1$ [127]. For the cut-off condition$V=(2m+1)\pi/2$, Equation (1) yields the cut-off wavelengths.

Figure 20b shows a typical transmission spectrum of an annealed $SiO_{1.5}$ film on a silica substrate. The spacing of the interference patterns (in cm^{-1}) can be written as:

$$\delta\nu = 1/2\, n_1 d \tag{2}$$

which provides the optical thickness (*nd*). The numerical information can be extracted from the transmission spectrum by fitting it, for example, with the following empirical function:

$$T(\lambda) = P_7 + P_1\left(1 - \frac{P_3}{\lambda - P_2}\right)\left[1 - P_4 \sin\left(\frac{P_5}{\lambda} + P_6\right)\right] \tag{3}$$

where *T* is the transmittance, and P_i are the fitting parameters [49]. In particular, the P_5 parameter is connected with the period of the oscillations, *i.e.*, with the optical thickness. The TE refractive index is relevant to transmission spectra. By using Equation (1) for the spacing between the TE peaks and Equation (2), the TE refractive index and the film thickness can be extracted [85]. Next, the obtained layer thickness together with the spacing between the TM peak positions is used to calculate the TM refractive index. The obtained thickness is also used to calculate the absorption (extinction) coefficient from the transmission and reflection measurements.

The model described above is called a *simple model* because it uses a few simplifications: (i) Equation (1) corresponds to the weak-guiding approximation $(n1-n2) \ll 1$; (ii) the procedure uses only the interval between the cut-off peaks disregarding their actual positions; (iii) the material dispersion is neglected; and (iv) a step-like profile is assumed for the refractive index.

To take the dispersion of the material into account, the transmission spectra can be fitted with a function similar to Equation (3) but containing an additional term describing the wavelength dependence of the spacing $\delta\nu$ [115]. Equations (1) and (2) are then solved taking the dispersion correction into account. The results show that the dispersion corrections for the refractive index and the film thickness are within experimental errors (Figure 21).

The refractive index and the film thickness can be also calculated using the model of reflection from a multilayer system [153], which does not limit the refractive index (exact equations). The analysis includes both spacing and positions of the cut-off peaks. Equation (3) with an additional term describing the material dispersion is also used in this case [115]. Again, the differences from the simple model are within the experimental errors ("exact" in Figure 21). Using the XPS data for a $SiO_{1.55}$ film annealed at 1100 °C, it has been concluded that the use of smooth index profile of the refractive index is not

crucial to obtain good description of the delocalized cut-off modes. These results suggest that the simple model is sufficient to achieve adequate estimates of the refractive index and film thickness. The *m*-line measurements lead to the results very similar to those obtained by the PL-filtering effect.

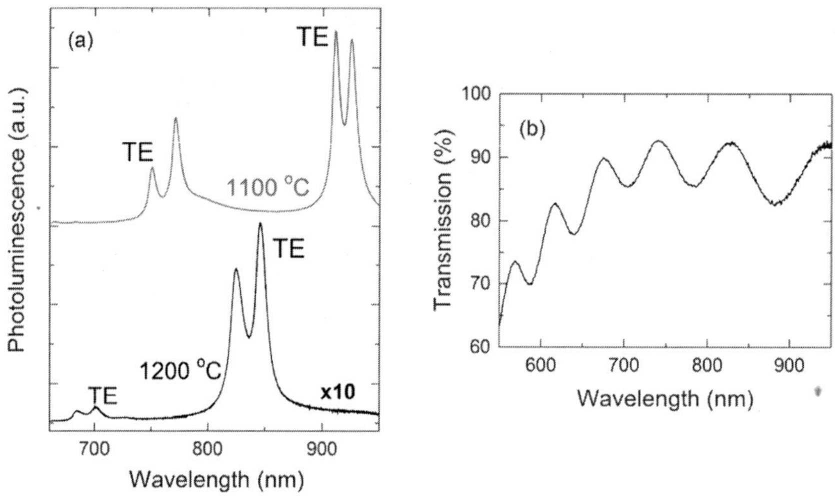

Figure 20. (**a**) Cut-off spectra of a $SiO_{1.5}$ film annealed at 1100 and 1200 °C, measured from the sample edge along the film. The excitation spot was at a distance of several millimeters from the sample edge. The positions of the TE peaks are marked and the other bands have the TM polarization. The TE and TM peaks have polarization parallel and perpendicular to the film surface, respectively. (**b**) Transmission spectrum of the $SiO_{1.5}$ film annealed at 1100 °C. Reproduced with permission from [115]. Copyright 2008, AIP Publishing LLC.

PL-filtering effect has also been interpreted in terms of substrate leaky (or radiation) modes of a planar waveguide [86,87]. In this model, the regular guided modes experience greater loss propagating over a macroscopic distance from the place of creation towards the sample edge. The most likely source of the loss is absorption and scattering by nanocrystals, as well as scattering at the "interface" between the core and cladding layers. In contrast, the beams responsible for generating the substrate modes travel shorter distances through the core region, and the substrate mode undergoes virtually no loss in traveling to the substrate facet. There seems to be no contradiction between the descriptions based on mode delocalization and substrate leaky modes. In particular, both of these models explain the PL-filtering effect based on a smaller absorption in the substrate compared to the film containing Si-nc.

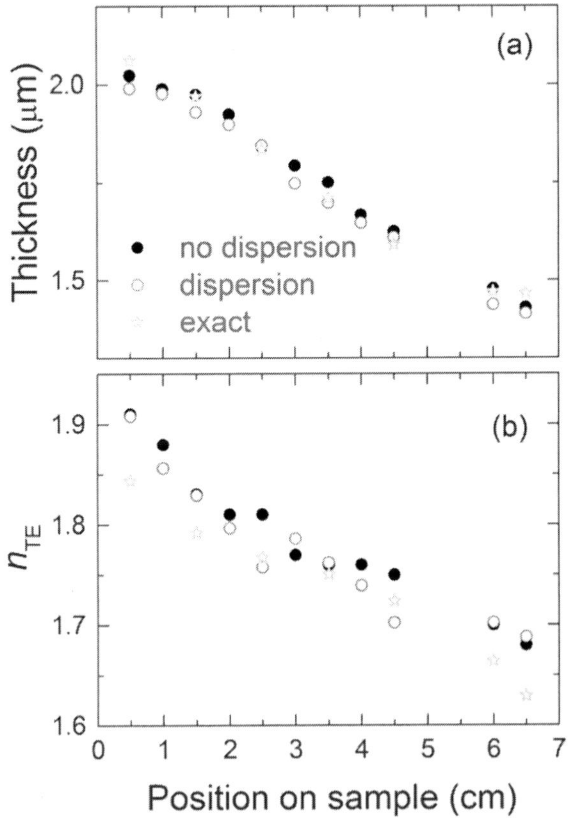

Figure 21. Optical parameters of a SiO_x optical wedge ($1.5 < x < 1.8$) obtained with different models: simple model (solid circles), taking dispersion into account (open circles), and using the exact equations [153] (stars). Reproduced with permission from [115]. Copyright 2008, AIP Publishing LLC.

3.4. X-Ray Photoelectron Spectroscopy

XPS measurements provide detailed information about the elemental and chemical composition of the material. Since only a thin layer of the sample (≤10 nm) is investigated with this method, the sample is sputtered with Ar^+ ions to obtain information about the bulk material.

In the case of SiO_x films, the Si 2*p* spectrum is fitted by several peaks to obtain the amounts of Si in different oxidation states. Three peaks are often used to describe Si atoms bonded as in crystalline Si (elemental Si), "SiO" (mixture of various suboxides), and SiO_2 materials (Figure 22). More complicated fitting

with five peaks can be used to obtain the five oxidation states of Si (elemental Si, Si_2O, SiO, Si_2O_3, and SiO_2) [154,155,156]. For our samples, the fittings with three and five peaks give similar amounts of elemental Si and SiO_2 [116]. Moreover, the use of five fitting curves is problematic for low Si content.

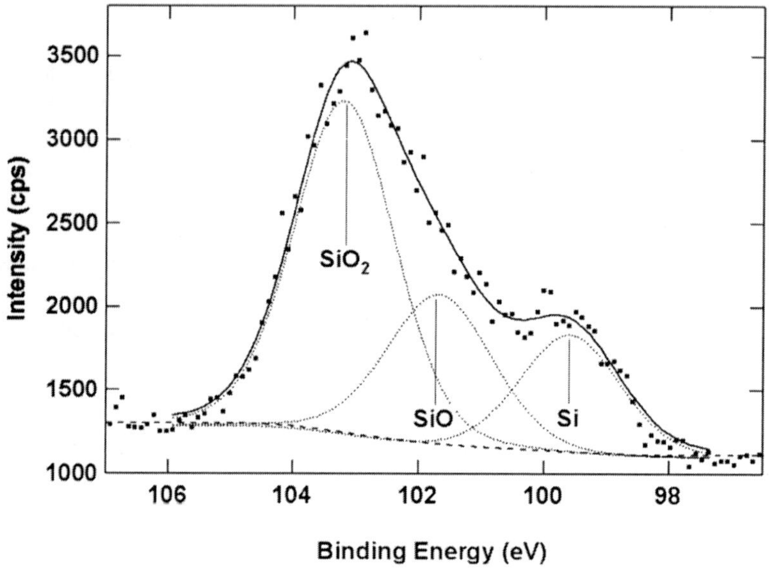

Figure 22. Typical XPS spectrum of a SiO_x film fitted with three Gaussians corresponding to Si bonded as in elemental Si, SiO (mixture of suboxides), and SiO_2. Figure courtesy of Jouko Lahtinen. Reproduced with permission from [115]. Copyright 2008, AIP Publishing LLC.

3.5. Effective Medium Approximation

The effective medium approximation (Bruggeman theory) connects the effective dielectric constant with the dielectric constants of the three structural components and their volumetric fractions [157]:

$$a\left(\frac{\varepsilon_a - \varepsilon_m}{\varepsilon_a + 2\varepsilon_m}\right) + b\left(\frac{\varepsilon_b - \varepsilon_m}{\varepsilon_b + 2\varepsilon_m}\right) + c\left(\frac{\varepsilon_c - \varepsilon_m}{\varepsilon_c + 2\varepsilon_m}\right) = 0 \tag{4}$$

where *a*, *b*, and *c* are the volumetric fractions of each component, $\varepsilon_{a,b,c}$ are the corresponding dielectric constants, and ε_m is the effective dielectric constant. The effective medium approximation is employed to calculate the complex

refractive index$n^* = n + ik$ (where k is the extinction coefficient) using the chemical compositions (Si, SiO, and SiO_2) obtained from the XPS measurements.

3.6. Temperature Measurements

Laser-induced temperature (T) can be measured *in-situ* using the intensities of the Stokes (I_S) and anti-Stokes (I_{AS}) Raman bands (Figure 23) and the relation:

$$\frac{I_{AS}}{I_S} = Ae^{\left(-E_R/kT\right)} \tag{5}$$

where E_R is the phonon energy and A is a correction coefficient [95,139]. The coefficient $A \sim 0.95$ is obtained by measuring the laser-induced temperature of a Si wafer as a function of the laser power and assuming room temperature in the irradiated volume at zero laser power.

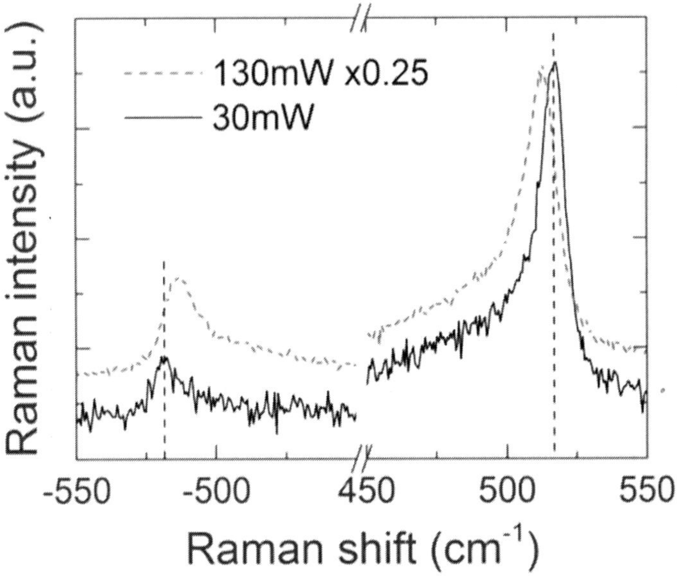

Figure 23. Raman spectra of a $SiO_{1.3}$ film on a silica substrate measured with two laser powers (~30 and ~130 mW focused to a ~40 µm spot). The spectra are vertically shifted and scaled for better presentation.

3.7. Optical Absorption

The film absorption is obtained by measuring transmission and reflection in ultraviolet and visible regions using a broadband light source. The band gap of semiconductors can be obtained from the absorption spectra using the Tauc relation, according to which, the absorption spectrum obeys the following equation [88,158]:

$$\sqrt{\alpha \cdot h\nu} = A(h\nu - E_{Tauc}) \qquad (6)$$

where A is a correction coefficient. Function $\sqrt{\alpha \cdot h\nu}$ is plotted *versus* hν and fitted with a straight line. The value of the optical gap (E_{Tauc}) is obtained from the intercept of the linear fit with the abscissa.

4. CONCLUSIONS

Optical and structural properties of Si-nc in silica films have been described. In order to prepare Si-nc, SiO_x ($x < 2$) films are annealed in a furnace at temperatures up to 1200 °C. For different Si contents, the Si-nc Raman signal and the absorption coefficient are proportional to the amount of elemental Si detected by XPS. On the other hand, the Raman scattering cross-section of elemental Si is about three times smaller than that of crystalline Si. This difference can be explained by the presence of small Si-nc (<2 nm) and/or by the different properties of bulk and nanoscale Si.

The measured optical properties of SiO_x films are compared with the values estimated by the effective medium approximation using the XPS results. A good agreement is found between the measured and calculated refractive index. The results for absorption suggest high transparency of nanoscale suboxide in the annealed samples. Thermal annealing increases the degree of Si crystallization; however, the crystallization is not complete after annealing at 1200 °C. The extinction coefficient of elemental Si is found to be between the values of crystalline and amorphous Si.

The PL quantum yield increases as the amount of elemental Si decreases. It follows that the 1.5-eV PL from the SiO_x films annealed above 1000 °C is probably not directly from Si-nc responsible for absorption and detected by Raman spectroscopy. According to these results, the light-emitting centers may be small (~1 nm) oxidized Si grains or oxygen-related defects, which are not detectable by Raman spectroscopy. This conclusion is in agreement with the theoretical simulations [114].

The as-prepared SiO_x films deposited by MBD and ion implantation are structurally and optically very different. Relatively large amorphous clusters (>2 nm) are present in the MBD films whereas they are absent in the implanted samples as suggested by absorption and Raman spectroscopy. After annealing at ≥1100 °C, these two kinds of samples become similar and possess comparable amounts of Si-nc.

Laser-induced thermal effects are found for SiO_x films on silica substrates when illuminated by focused laser light. In our experiments, a temperature of ~350 °C was estimated for SiO_x films (x ~ 1.3) on a silica substrate for 488-nm laser intensity of ~10^4 W cm^{-2}, which decreases the Raman shift by ~6 cm^{-1}. This effect should be taken into account in optical measurements using focused laser beams, especially in Raman microscopy.

CW laser light can produce very high temperatures in free-standing SiO_x films and Si/SiO_2 SLs, which strongly changes their structure and optical properties. The center of a laser-annealed area is very transparent and presumably consists of amorphous SiO_2. Large Si-nc (up to 300 nm) are observed in the ring around the central region. These Si-nc possess high absorption of visible light and they are typically under compressive stress, which is connected with the crystallization from the liquid phase. Some of these large Si-nc exhibit surface features that are formed by eruption of pressurized Si from the film. Some of these "surface" Si-nc are removed from the film leading to holes of similar sizes. The presence of oxygen in the laser-annealing atmosphere decreases the amount of removed Si particles most probably due to oxidation of their surface.

The structure of laser-annealed areas is explained by thermodiffusion in temperature gradient, which means the macroscopic Si–SiO_2 phase separation. Comparison of the structure of central regions for laser annealing in oxygen, air, and inert atmospheres excludes the dominating effect of Si oxidation, thus, supporting the mechanism of the macroscopic Si–SiO_2 phase separation.

By using a strongly focused laser beam, the structural changes in the free-standing films can be made in areas of submicron sizes. A concept of high-density non-volatile optical memory with superior thermal stability is demonstrated where the information can be read, rewritten and erased by optical means.

ACKNOWLEDGMENTS

The authors thank all colleagues who contributed to the results described in this article and whose names appear in the joint publications.

AUTHOR CONTRIBUTIONS

TN and LK have written this review.

CONFLICTS OF INTEREST

The authors declare no conflict of interest.

REFERENCES

1. Silicon Nanophotonics: Basic Principles, Present Status and Perspectives; Khriachtchev, L., Ed.; Pan Stanford: Singapore, Singapore, 2008.

2. Silicon Photonics II: Components and Integration; Lockwood, D.J., Pavesi, L., Eds.; Springer-Verlag: Berlin, Germany, 2011.

3. Nanostructures in Electronics and Photonics; Rahman, F., Ed.; Pan Stanford: Singapore, Singapore, 2008.

4. Yariv, A.; Yeh, P. Photonics: Optical Electronics in Modern Communications; Oxford University Press: Oxford, UK, 2007.

5. Jalali, B.; Fathpour, S. Silicon photonics. J. Lightwave Technol. 2006, 24, 4600–4615.

6. Soref, R. The past, present, and future of silicon photonics. IEEE J. Sel. Top. Quant. Electron. 2006, 12, 1678–1687.

7. Daldosso, N.; Pavesi, L. Nanosilicon photonics. Laser Photonics Rev. 2009, 3, 508–534.

8. Silicon Photonics: The State of the Art; Reed, G.T., Ed.; John Wiley & Sons, Ltd: Chichester, UK, 2008.

9. Pavesi, L. Will silicon be the photonic material of the third millennium? J. Phys. Cond. Matter 2003, 15, R1169–R1196.

10. Lockwood, D.J. Light emission in silicon nanostructures. NATO Asi Ser. Ser. E 1998, 348, 185–209.

11. Lockwood, D.J.; Pavesi, L. Silicon fundamentals for photonics applications. Top. Appl. Phys. 2004, 94, 1–50.

12. Delley, B.; Steigmeier, E.F. Quantum confinement in Si nanocrystals. Phys. Rev. B 1993, 47, 1397–1400.

13. Ogut, S.; Chelikowsky, J.R.; Louie, S.G. Quantum confinement and optical gaps in Si nanocrystals. Phys. Rev. Lett. 1997,79, 1770–1773.

14. Balberg, I. Electrical transport mechanisms in three dimensional ensembles of silicon quantum dots. J. App. Phys. 2011,110, 061301.

15. Canham, L.T. Silicon quantum wire array fabrication by electrochemical and chemical dissolution of wafers. Appl. Phys. Lett. 1990, 57, 1046–1048.

16. Cullis, A.G.; Canham, L.T. Visible-light emission due to quantum size effects in highly porous crystalline silicon.Nature 1991, 353, 335–338.

17. Gardelis, S.; Rimmer, J.S.; Dawson, P.; Hamilton, B.; Kubiak, R.A.; Whall, T.E.; Parker, E.H.C. Evidence for quantum confinement in the photoluminescence of porous Si and SiGe. Appl. Phys. Lett. 1991, 59, 2118–2120.

18. Bsiesy, A.; Vial, J.C.; Gaspard, F.; Herino, R.; Ligeon, M.; Muller, F.; Romestain, R.; Wasiela, A.; Halimaoui, A.; Bomchil, G. Photoluminescence of high porosity and of electrochemically oxidized porous silicon layers. Surf. Sci.1991, 254, 195–200.

19. Delerue, C.; Allan, G.; Lannoo, M. Theoretical aspects of the luminescence of porous silicon. Phys. Rev. B 1993, 48, 11024–11036.

20. Prokes, S.M. Light emission in thermally oxidized porous silicon: Evidence for oxide-related luminescence. Appl. Phys. Lett. 1993, 62, 3244–3246.

21. Gole, J.L.; Dudel, F.P.; Grantier, D.; Dixon, D.A. Origin of porous silicon photoluminescence: Evidence for a surface bound oxyhydride-like emitter. Phys. Rev. B 1997, 56, 2137–2153.

22. Wolkin, M.V.; Jorne, J.; Fauchet, P.M.; Allan, G.; Delerue, C. Electronic states and luminescence in porous silicon quantum dots: The role of oxygen. Phys. Rev. Lett. 1999, 82, 197–200.

23. Ledoux, G.; Gong, J.; Huisken, F.; Guillois, O.; Reynaud, C. Photoluminescence of size-separated silicon nanocrystals: Confirmation of quantum confinement. Appl. Phys. Lett. 2002, 80, 4834–4836.

24. Hannah, D.C.; Yang, J.H.; Podsiadlo, P.; Chan, M.K.Y.; Demortiere, A.; Gosztola, D.J.; Prakapenka, V.B.; Schatz, G.C.; Kortshagen, U.; Schaller, R.D. On the origin of photoluminescence in silicon nanocrystals: Pressure-dependent structural and optical studies. Nano Lett. 2012, 12, 4200–4205.

25. Kim, T.Y.; Park, N.M.; Kim, K.H.; Sung, G.Y.; Ok, Y.W.; Seong, T.Y.; Choi, C.J. Quantum confinement effect of silicon nanocrystals in situ grown in silicon nitride films. Appl. Phys. Lett. 2004, 85, 5355–5357.

26. Rezgui, B.; Sibai, A.; Nychyporuk, T.; Lemiti, M.; Bremond, G. Photoluminescence and optical absorption properties of silicon quantum dots embedded in Si-rich silicon nitride matrices. J. Lumin. 2009, 129, 1744–1746.

27. Khriachtchev, L.; Novikov, S.; Lahtinen, J. Thermal annealing of Si/SiO2 materials: Modification of structural and photoluminescence emission properties. J. Appl. Phys. 2002, 92, 5856–5862.

28. Lu, Z.H.; Lockwood, D.J.; Baribeau, J.M. Quantum confinement and light emission in SiO2/Si superlattices. Nature1995, 378, 258–260.

29. Zhuravlev, K.S.; Gilinsky, A.M.; Kobitsky, A.Y. Mechanism of photoluminescence of Si nanocrystals fabricated in a SiO2 matrix. Appl. Phys. Lett. 1998, 73, 2962–2964.

30. Dhara, S.; Lu, C.Y.; Nair, K.G.M.; Chen, K.H.; Chen, C.P.; Huang, Y.F.; David, C.; Chen, L.C.; Raj, B. Mechanism of bright red emission in Si nanoclusters. Nanotechnology 2008, 19, 395401.

31. Garrido Fernandez, B.; Lopez, M.; Garcia, C.; Perez-Rodriguez, A.; Morante, J.R.; Bonafos, C.; Carrada, M.; Claverie, A. Influence of average size and interface passivation on the spectral emission of Si nanocrystals embedded in SiO2. J. Appl. Phys. 2002, 91, 798–807.

32. Guha, S. Characterization of Si+ ion-implanted SiO2 films and silica glasses. J. Appl. Phys. 1998, 84, 5210–5217.

33. Mirabella, S.; Agosta, R.; Franzo, G.; Crupi, I.; Miritello, M.; Lo Savio, R.; Di Stefano, M.A.; Di Marco, S.; Simone, F.; Terrasi, A. Light absorption in silicon quantum dots embedded in silica. J. Appl. Phys. 2009, 106, 103505.

34. Franzo, G.; Miritello, M.; Boninelli, S.; Lo Savio, R.; Grimaldi, M.G.; Priolo, F.; Iacona, F.; Nicotra, G.; Spinella, C.; Coffa, S. Microstructural evolution of SiOx films and its effect on the luminescence of Si nanoclusters. J. Appl. Phys.2008, 104, 094306.

35. Ternon, C.; Dufour, C.; Gourbilleau, F.; Rizk, R. Roles of interfaces in nanostructured silicon luminescence. Eur. Phys. J. B 2004, 41, 325–332.

36. Alonso, M.I.; Marcus, I.C.; Garriga, M.; Goñi, A.R.; Jedrzejewski, J.; Balberg, I. Evidence of quantum confinement effects on interband optical transitions in Si nanocrystals. Phys. Rev. B 2010, 82, 045302.

37. Roussel, M.; Talbot, E.; Pareige, P.; Gourbilleau, F. Influence of the supersaturation on Si diffusion and growth of Si nanoparticles in silicon-rich silica. J. Appl. Phys. 2013, 113, 063519.

38. Antonova, I.V.; Gulyaev, M.; Savir, E.; Jedrzejewski, J.; Balberg, I. Charge storage, photoluminescence, and cluster statistics in ensembles of Si quantum dots. Phys. Rev. B 2008, 77, 125318.

39. Balberg, I.; Savir, E.; Jedrzejewski, J.; Nassiopoulou, A.G.; Gardelis, S. Fundamental transport processes in ensembles of silicon quantum dots. Phys. Rev. B 2007, 75, 235329.

40. Iacona, F.; Franzo, G.; Spinella, C. Correlation between luminescence and structural properties of Si nanocrystals. J. Appl. Phys. 2000, 87, 1295–1303.

41. Trwoga, P.F.; Kenyon, A.J.; Pitt, C.W. Modeling the contribution of quantum confinement to luminescence from silicon nanoclusters. J. Appl. Phys. 1998, 83, 3789–3794.

42. Boninelli, S.; Iacona, F.; Franzo, G.; Bongiorno, C.; Spinella, C.; Priolo, F. Formation, evolution and photoluminescence properties of Si nanoclusters. J. Phys.: Condens. Matter 2007, 19, 225003.

43. Nassiopoulou, A.G. Encyclopedia of Nanoscience and Nanotechnology; Nalwa, H.S., Ed.; American Scientific: Valencia, Spain, 2004; Volume 9, pp. 793–813.

44. Gardelis, S.; Nassiopoulou, A.G.; Manousiadis, P.; Milita, S.; Gkanatsiou, A.; Frangis, N.; Lioutas, C.B. Structural and optical characterization of two-

dimensional arrays of Si nanocrystals embedded in SiO2 for photovoltaic applications.J. Appl. Phys. 2012, 111, 083536.

45. Gardelis, S.; Nassiopoulou, A.G.; Vouroutzis, N.; Frangis, N. Effect of exciton migration on the light emission properties in silicon nanocrystal ensembles. J. Appl. Phys. 2009, 105, 113509.

46. Hiller, D.; Goetze, S.; Munnik, F.; Jivanescu, M.; Gerlach, J.W.; Vogt, J.; Pippel, E.; Zakharov, N.; Stesmans, A.; Zacharias, M. Nitrogen at the Si-nanocrystal/SiO2 interface and its influence on luminescence and interface defects.Phys. Rev. B 2010, 82, 195401.

47. Seguini, G.; Castro, C.; Schamm-Chardon, S.; BenAssayag, G.; Pellegrino, P.; Perego, M. Scaling size of the interplay between quantum confinement and surface related effects in nanostructured silicon. Appl. Phys. Lett. 2013, 103, 023103.

48. Khriachtchev, L.; Räsänen, M.; Novikov, S.; Pavesi, L. Systematic correlation between Raman spectra, photoluminescence intensity, and absorption coefficient of silica layers containing Si nanocrystals. Appl. Phys. Lett.2004, 85, 1511–1513.

49. Khriachtchev, L.; Räsänen, M.; Novikov, S.; Lahtinen, J. Tunable wavelength-selective waveguiding of photoluminescence in Si-rich silica optical wedges. J. Appl. Phys. 2004, 95, 7592–7601.

50. Zacharias, M.; Blasing, J.; Veit, P.; Tsybeskov, L.; Hirschman, K.; Fauchet, P.M. Thermal crystallization of amorphous Si/SiO2 superlattices. Appl. Phys. Lett. 1999, 74, 2614–2616.

51. Zacharias, M.; Heitmann, J.; Scholz, R.; Kahler, U.; Schmidt, M.; Blasing, J. Size-controlled highly luminescent silicon nanocrystals: A SiO/SiO2 superlattice approach. Appl. Phys. Lett. 2002, 80, 661–663.

52. Godefroo, S.; Hayne, M.; Jivanescu, M.; Stesmans, A.; Zacharias, M.; Lebedev, O.I.; van Tendeloo, G.; Moshchalkov, V.V. Classification and control of the origin of photoluminescence from Si nanocrystals. Nat. Nanotechnol. 2008, 3, 174–178.

53. Hiller, D.; Jivanescu, M.; Stesmans, A.; Zacharias, M. P-b(0) centers at the Si-nanocrystal/SiO2 interface as the dominant photoluminescence quenching defect. J. Appl. Phys. 2010, 107, 084309.

54. Novikov, S.V.; Sinkkonen, J.; Kilpela, O.; Gastev, S.V. Visible light emission from MBD-grown Si/SiO2 superlattices. J. Cryst. Growth 1997, 175, 514–518.

55. Khriachtchev, L.; Räsänen, M.; Novikov, S.; Kilpela, O.; Sinkkonen, J. Raman scattering from very thin Si layers of Si/SiO2 superlattices: Experimental evidence of structural modification in the 0.8-3.5 nm thickness region. J. Appl. Phys. 1999, 86, 5601–5608.

56. Khriachtchev, L.; Räsänen, M.; Novikov, S. Laser-controlled stress of Si nanocrystals in a free-standing Si/SiO2superlattice. Appl. Phys. Lett. 2006, 88, 013102.

57. Khriachtchev, L.; Novikov, S.; Lahtinen, J.; Räsänen, M. Wavelength-selective optical waveguiding of photoluminescence in a thermally annealed Si/SiO2 superlattice. J. Phys.: Condens. Matter 2004, 16, 3219–3228.

58. Pavesi, L.; Dal Negro, L.; Mazzoleni, C.; Franzo, G.; Priolo, F. Optical gain in silicon nanocrystals. Nature 2000, 408, 440–444.

59. Khriachtchev, L.; Räsänen, M.; Novikov, S.; Sinkkonen, J. Optical gain in Si/SiO2 lattice: Experimental evidence with nanosecond pulses. Appl. Phys. Lett. 2001, 79, 1249–1251.

60. Koshida, N.; Koyama, H. Visible electroluminescence from porous silicon. Appl. Phys. Lett. 1992, 60, 347–349.

61. Franzo, G.; Irrera, A.; Moreira, E.C.; Miritello, M.; Iacona, F.; Sanfilippo, D.; Di Stefano, G.; Fallica, P.G.; Priolo, F. Electroluminescence of silicon nanocrystals in MOS structures. Appl. Phys. A 2002, 74, 1–5.

62. Zhou, F.L.; Head, J.D. Role of Si=O in the photoluminescence of porous silicon. J. Phys. Chem. B 2000, 104, 9981–9986.

63. Takeoka, S.; Fujii, M.; Hayashi, S. Size-dependent photoluminescence from surface-oxidized Si nanocrystals in a weak confinement regime. Phys. Rev. B 2000, 62, 16820–16825.

64. Nesheva, D.; Raptis, C.; Perakis, A.; Bineva, I.; Aneva, Z.; Levi, Z.; Alexandrova, S.; Hofmeister, H. Raman scattering and photoluminescence from Si nanoparticles in annealed SiOx thin films. J. Appl. Phys. 2002, 92, 4678–4683.

65. Lioudakis, E.; Othonos, A.; Hadjisavvas, G.C.; Kelires, P.C.; Nassiopoulou, A.G. Quantum confinement and interface structure of Si nanocrystals of sizes 3–5 nm embedded in a-SiO2. Phys. E 2007, 38, 128–134.

66. Garcia, C.; Garrido, B.; Pellegrino, P.; Ferre, R.; Moreno, J.A.; Morante, J.R.; Pavesi, L.; Cazzanelli, M. Size dependence of lifetime and absorption cross section of Si nanocrystals embedded in SiO2. Appl. Phys. Lett. 2003, 82, 1595–1597.

67. Kanemitsu, Y.; Ogawa, T.; Shiraishi, K.; Takeda, K. Visible photoluminescence from oxidized Si nanometer-sized spheres: Exciton confinement on a spherical shell. Phys. Rev. B 1993, 48, 4883–4886.

68. Kanemitsu, Y.; Uto, H.; Masumoto, Y.; Matsumoto, T.; Futagi, T.; Mimura, H. Microstructure and optical properties of free-standing porous silicon films: Size dependence of absorption spectra in Si nanometer-sized srystallites. Phys. Rev. B 1993, 48, 2827–2830.

69. Garoufalis, C.S.; Zdetsis, A.D. High accuracy calculations of the optical gap and absorption spectrum of oxygen contaminated Si nanocrystals. Phys. Chem. Chem. Phys. 2006, 8, 808–813.

70. Dohnalova, K.; Kusova, K.; Pelant, I. Time-resolved photoluminescence spectroscopy of the initial oxidation stage of small silicon nanocrystals. Appl. Phys. Lett. 2009, 94, 211903.

71. Averboukh, B.; Huber, R.; Cheah, K.W.; Shen, Y.R.; Qin, G.G.; Ma, Z.C.; Zong, W.H. Luminescence studies of a Si/SiO2superlattice. J. Appl. Phys. 2002, 92, 3564–3568.

72. Iacona, F.; Bongiorno, C.; Spinella, C.; Boninelli, S.; Priolo, F. Formation and evolution of luminescent Si nanoclusters produced by thermal annealing of SiOx films. J. Appl. Phys. 2004, 95, 3723–3732.

73. Wang, Y.Q.; Smirani, R.; Ross, G.G. The formation mechanism of Si nanocrystals in SiO2. J. Cryst. Growth 2006, 294, 486–489.

74. Naciri, A.E.; Mansour, M.; Johann, L.; Grob, J.J.; Eckert, C. Optical study of Si nanocrystals in Si/SiO2 layers by spectroscopic ellipsometry. Nucl. Instrum. Meth. B 2004, 216, 167–172.

75. Falcony, C.; Calleja, W.; Aceves, M.; Siqueiros, J.M.; Machorro, R.; CotaAraiza, L.; Soto, G.; Farias, M.H. Characterization of excess Si in nonstoichiometric SiO2 films by optical and surface analysis techniques. J. Electrochem. Soc. 1997, 144, 379–383.

76. Charvet, S.; Madelon, R.; Gourbilleau, F.; Rizk, R. Ellipsometric spectroscopy study of photoluminescent Si/SiO2systems obtained by magnetron co-sputtering. J. Lumin. 1998, 80, 257–261.

77. Daldosso, N.; Melchiorri, M.; Pavesi, L.; Pucker, G.; Gourbilleau, F.; Chausserie, S.; Belarouci, A.; Portier, X.; Dufour, C. Optical losses and absorption cross-section of silicon nanocrystals. J. Lumin. 2006, 121, 344–348.

78. Navarro-Urrios, D.; Riboli, F.; Cazzanelli, M.; Chiasera, A.; Daldosso, N.; Pavesi, L.; Oton, C.J.; Heitmann, J.; Yi, L.X.; Scholz, R.; et al. Birefringence characterization of mono-dispersed silicon nanocrystals planar waveguides. Opt. Mater.2005, 27, 763–768.

79. Khriachtchev, L.; Navarro-Urrios, D.; Pavesi, L.; Oton, C.J.; Capuj, N.E.; Novikov, S. Spectroscopy of silica layers containing Si nanocrystals: Experimental evidence of optical birefringence. J. Appl. Phys. 2007, 101, 044310.

80. Moreno, J.A.; Garrido, B.; Pellegrino, P.; Garcia, C.; Arbiol, J.; Morante, J.R.; Marie, P.; Gourbilleau, F.; Rizk, R. Size dependence of refractive index of Si nanoclusters embedded in SiO2. J. Appl. Phys. 2005, 98, 013523.

81. Chen, T.P.; Liu, Y.; Tse, M.S.; Fung, S.; Dong, G. Profile of optical constants of SiO2 thin films containing Si nanocrystals. J. Appl. Phys. 2004, 95, 8481–8483.

82. Amans, D.; Callard, S.; Gagnaire, A.; Joseph, J.; Ledoux, G.; Huisken, F. Ellipsometric study of silicon nanocrystal optical constants. J. Appl. Phys. 2003, 93, 4173–4179.

83. Mishra, P.; Jain, K.P. Raman, photoluminescence and optical absorption studies on nanocrystalline silicon. Mat. Sci. Eng. B 2002, 95, 202–213.

84. von Behren, J.; van Buuren, T.; Zacharias, M.; Chimowitz, E.H.; Fauchet, P.M. Quantum confinement in nanoscale silicon: The correlation of size with bandgap and luminescence. Solid State Commun. 1998, 105, 317–322.

85. Khriachtchev, L.; Räsänen, M.; Novikov, S. Efficient wavelength-selective optical waveguiding in a silica layer containing Si nanocrystals. Appl. Phys. Lett. 2003, 83, 3018–3020.

86. Valenta, J.; Ostatnicky, T.; Pelant, I.; Elliman, R.G.; Linnros, J.; Honerlage, B. Microcavity-like leaky mode emission from a planar optical waveguide

made of luminescent silicon nanocrystals. J. Appl. Phys. 2004, 96, 5222–5225.

87. Pelant, I.; Ostatnicky, T.; Valenta, J.; Luterova, K.; Skopalova, E.; Mates, T.; Elliman, R.G. Waveguide cores containing silicon nanocrystals as active spectral filters for silicon-based photonics. Appl. Phys. B 2006, 83, 87–91.

88. Tauc, J.; Grigorov, R.; Vancu, A. Optical properties and electronic structure of amorphous germanium. Phys. Status Solidi B 1966, 15, 627–637.

89. BenChorin, M.; Averboukh, B.; Kovalev, D.; Polisski, G.; Koch, F. Influence of quantum confinement on the critical points of the band structure of Si. Phys. Rev. Lett. 1996, 77, 763–766.

90. Podhorodecki, A.; Zatryb, G.; Misiewicz, J.; Wojcik, J.; Mascher, P. Influence of the annealing temperature and silicon concentration on the absorption and emission properties of Si nanocrystals. J. Appl. Phys. 2007, 102, 043104.

91. Inokuma, T.; Wakayama, Y.; Muramoto, T.; Aoki, R.; Kurata, Y.; Hasegawa, S. Optical properties of Si clusters and Si nanocrystallites in high-temperature annealed SiOx films. J. Appl. Phys. 1998, 83, 2228–2234.

92. Thompson, M.O.; Galvin, G.J.; Mayer, J.W.; Peercy, P.S.; Poate, J.M.; Jacobson, D.C.; Cullis, A.G.; Chew, N.G. Melting temperature and explosive crystallization of amorphous silicon during pulsed laser irradiation. Phys. Rev. Lett. 1984,52, 2360–2363.

93. Koyama, H.; Fauchet, P.M. Laser-induced thermal effects on the optical properties of free-standing porous silicon films. J. Appl. Phys. 2000, 87, 1788–1794.

94. Khriachtchev, L.; Räsänen, M.; Novikov, S. Free-standing silica film containing Si nanocrystals: Photoluminescence, Raman scattering, optical waveguiding, and laser-induced thermal effects. Appl. Phys. Lett. 2005, 86, 141911.

95. Khriachtchev, L.; Räsänen, M.; Novikov, S. Continuous-wave laser annealing of free-standing Si/SiO2 superlattice: Modification of optical, structural, and light-emitting properties. J. Appl. Phys. 2006, 100, 053502.

96. Khriachtchev, L.; Novikov, S. Laser-induced thermal effects on Si/SiO2 free-standing superlattices. Appl. Phys. A 2007,87, 761–766.

97. Koyama, H.; Fauchet, P.M. Very large continuous-wave-laser-induced optical absorption in porous silicon films: Evidence for thermal effects. Appl. Phys. Lett. 1998, 73, 3259–3261.

98. Mchedlidze, T.; Arguirov, T.; Kouteva-Arguirova, S.; Kittler, M. Light induced solid-phase crystallization of Si nanolayers in Si/SiO2 multiple quantum wells. J. Appl. Phys. 2010, 107, 124302.

99. Khriachtchev, L.; Ossicini, S.; Iacona, F.; Gourbilleau, F. Silicon nanoscale materials: From theoretical simulations to photonic applications. Int. J. Photoenergy 2012, 2012, 872576.

100. Daldosso, N.; Luppi, M.; Ossicini, S.; Degoli, E.; Magri, R.; Dalba, G.; Fornasini, P.; Grisenti, R.; Rocca, F.; Pavesi, L.; et al. Role of the interface region on the optoelectronic properties of silicon nanocrystals embedded in SiO2. Phys. Rev. B2003, 68, 085327.

101. Djurabekova, F.; Nordlund, K. Atomistic simulation of the interface structure of Si nanocrystals embedded in amorphous silica. Phys. Rev. B 2008, 77, 115325.

102. Luppi, M.; Ossicini, S. Multiple SiO bonds at the silicon cluster surface. J. Appl. Phys. 2003, 94, 2130–2132.

103. Vasiliev, I.; Chelikowsky, J.R.; Martin, R.M. Surface oxidation effects on the optical properties of silicon nanocrystals.Phys. Rev. B 2002, 65, 121302.

104. Puzder, A.; Williamson, A.J.; Grossman, J.C.; Galli, G. Computational studies of the optical emission of silicon nanocrystals. J. Am. Chem. Soc 2003, 125, 2786–2791.

105. Pennycook, T.J.; Hadjisavvas, G.; Idrobo, J.C.; Kelires, P.C.; Pantelides, S.T. Optical gaps of free and embedded Si nanoclusters: Density functional theory calculations. Phys. Rev. B 2010, 82, 125310.

106. Luppi, E.; Iori, F.; Magri, R.; Pulci, O.; Ossicini, S.; Degoli, E.; Olevano, V. Excitons in silicon nanocrystallites: The nature of luminescence. Phys. Rev. B 2007, 75, 033303.

107. Kleovoulou, K.; Kelires, P.C. Stress state of embedded Si nanocrystals. Phys. Rev. B 2013, 88, 085424.

108. Hadjisavvas, G.; Kelires, P.C. Structure and energetics of Si nanocrystals embedded in a-SiO2. Phys. Rev. Lett. 2004, 93, 226104.

109. Kleovoulou, K.; Kelires, P.C. Local rigidity and physical trends in embedded Si nanocrystals. Phys. Rev. B 2013, 88, 245202.

110. Guerra, R.; Degoli, E.; Ossicini, S. Size, oxidation, and strain in small Si/SiO2 nanocrystals. Phys. Rev. B 2009, 80, 155332.

111. Watanabe, T.; Tatsumura, K.; Ohdomari, I. SiO2/Si interface structure and its formation studied by large-scale molecular dynamics simulation. Appl. Surf. Sci. 2004, 237, 125–133.

112. Guerra, R.; Marri, I.; Magri, R.; Martin-Samos, L.; Pulci, O.; Degoli, E.; Ossicini, S. Silicon nanocrystallites in a SiO2matrix: Role of disorder and size. Phys. Rev. B 2009, 79, 155320.

113. Guerra, R.; Marri, I.; Magri, R.; Martin-Samos, L.; Pulci, O.; Degoli, E.; Ossicini, S. Optical properties of silicon nanocrystallites in SiO2 matrix: Crystalline vs. amorphous case. Superlattices Microstruct. 2009, 46, 246–253.

114. Guerra, R.; Ossicini, S. High luminescence in small SiO/SiO2 nanocrystals: A theoretical study. Phys. Rev. B 2010, 81, 245307.

115. Khriachtchev, L.; Nikitin, T.; Oton, C.J.; Velagapudi, R.; Sainio, J.; Lahtinen, J.; Novikov, S. Optical properties of silicon nanocrystals in silica: Results from spectral filtering effect, m-line technique, and X-ray photoelectron spectroscopy. J. Appl. Phys. 2008, 104, 104316.

116. Nikitin, T.; Velagapudi, R.; Sainio, J.; Lahtinen, J.; Räsänen, M.; Novikov, S.; Khriachtchev, L. Optical and structural properties of SiOx films grown by molecular beam deposition: Effect of the Si concentration and annealing temperature. J. Appl. Phys. 2012, 112, 094316.

117. Khriachtchev, L.; Nikitin, T.; Velagapudi, R.; Lahtinen, J.; Novikov, S. Light-emission mechanism of thermally annealed silicon-rich silicon oxide revisited: What is the role of silicon nanocrystals? Appl. Phys. Lett. 2009, 94, 043115.

118. Nikitin, T.; Aitola, K.; Novikov, S.; Räsänen, M.; Velagapudi, R.; Sainio, J.; Lahtinen, J.; Mizohata, K.; Ahlgren, T.; Khriachtchev, L. Optical and structural properties of silicon-rich silicon oxide films: Comparison of ion implantation and molecular beam deposition methods. Phys. Status Solidi A 2011, 208, 2176–2181.

119. Hartstein, A.; Tsang, J.C.; Dimaria, D.J.; Dong, D.W. Observation of amorphous silicon regions in silicon-rich silicon dioxide films. Appl. Phys. Lett. 1980, 36, 836–837.

120. Khriachtchev, L.; Kilpela, O.; Karirinne, S.; Keranen, J.; Lepisto, T. Substrate-dependent crystallization and enhancement of visible photoluminescence in thermal annealing of Si/SiO2 superlattices. Appl. Phys. Lett. 2001, 78, 323–325.

121. Faraci, G.; Gibilisco, S.; Russo, P.; Pennisi, A.R.; La Rosa, S. Modified Raman confinement model for Si nanocrystals.Phys. Rev. B 2006, 73, 033307.

122. Campbell, I.H.; Fauchet, P.M. The effects of microcrystal size and shape on the one phonon Raman spectra of crystalline semiconductors. Solid State Commun. 1986, 58, 739–741.

123. Khriachtchev, L.; Novikov, S.; Kilpela, O. Optics of Si/SiO2 superlattices: Application to Raman scattering and photoluminescence measurements. J. Appl. Phys. 2000, 87, 7805–7813.

124. Seino, K.; Bechstedt, F.; Kroll, P. Influence of SiO2 matrix on electronic and optical properties of Si nanocrystals.Nanotechnology 2009, 20, 135702.

125. Novikov, S.; Sinkkonen, J.; Nikitin, T.; Khriachtchev, L.; Räsänen, M.; Haimi, E. Free-standing SiO2 films containing Si nanocrystals directly suitable for transmission electron microscopy. Microelectron. J. 2008, 39, 518–522.

126. Kogelnik, H.; Ramaswamy, V. Scaling rules for thin-film optical waveguides. Appl. Opt. 1974, 13, 1857–1862.

127. Unger, H.G. Planar Optical Waveguides and Fibres; Oxford University Press: Oxford, UK, 1977.

128. Yurtsever, A.; Weyland, M.; Muller, D.A. Three-dimensional imaging of nonspherical silicon nanoparticles embedded in silicon oxide by plasmon tomography. Appl. Phys. Lett. 2006, 89, 151920.

129. Zatryb, G.; Podhorodecki, A.; Misiewicz, J.; Cardin, J.; Gourbilleau, F. Correlation between matrix structural order and compressive stress exerted on silicon nanocrystals embedded in silicon-rich silicon oxide. Nanoscale Res. Lett. 2013, 8, 40.

130. Munekuni, S.; Yamanaka, T.; Shimogaichi, Y.; Tohmon, R.; Ohki, Y.; Nagasawa, K.; Hama, Y. Various types of nonbridging oxygen hole center in highpurity silica glass. J. Appl. Phys. 1990, 68, 1212–1217.

131. Glinka, Y.D.; Lin, S.H.; Hwang, L.P.; Chen, Y.T. Photoluminescence from mesoporous silica: Similarity of properties to porous silicon. Appl. Phys. Lett. 2000, 77, 3968–3970.

132. Borsella, E.; D'Amato, R.; Fabbri, F.; Falconieri, M.; Trave, E.; Bello, V.; Mattei, G.; Nie, Y.R.; Wang, D.Y. On the role of non-bridging oxygen centers in the red luminescence emission from silicon nanocrystals. Phys. Status Solidi C 2011, 8, 974–978.

133. Koponen, L.; Tunturivuori, L.O.; Puska, M.J.; Nieminen, R.M. Effect of the surrounding oxide on the photoabsorption spectra of Si nanocrystals. Phys. Rev. B 2009, 79, 235332.

134. Garcia, C.; Garrido, B.; Pellegrino, P.; Ferre, R.; Moreno, J.A.; Pavesi, L.; Cazzanelli, M.; Morante, J.R. Absorption cross-sections and lifetimes as a function of size in Si nanocrystals embedded in SiO2. Phys. E 2003, 16, 429–433.

135. Shimizu-Iwayama, T.; Kurumado, N.; Hole, D.E.; Townsend, P.D. Optical properties of silicon nanoclusters fabricated by ion implantation. J. Appl. Phys. 1998, 83, 6018–6022.

136. Lee, B.G.; Hiller, D.; Luo, J.W.; Semonin, O.E.; Beard, M.C.; Zacharias, M.; Stradins, P. Strained interface defects in silicon nanocrystals. Adv. Funct. Mater. 2012, 22, 3223–3232.

137. Kusova, K.; Ondic, L.; Klimesova, E.; Herynkova, K.; Pelant, I.; Danis, S.; Valenta, J.; Gallart, M.; Ziegler, M.; Honerlage, B.; et al. Luminescence of free-standing versus matrix-embedded oxide-passivated silicon nanocrystals: The role of matrix-induced strain. Appl. Phys. Lett. 2012, 101, 143101.

138. Faraci, G.; Gibilisco, S.; Pennisi, A.R. Quantum confinement and thermal effects on the Raman spectra of Si nanocrystals. Phys. Rev. B 2009, 80, 193410.

139. Balkanski, M.; Wallis, R.F.; Haro, E. Anharmonic effects in light scattering due to optical phonons in silicon. Phys. Rev. B 1983, 28, 1928–1934.

140. Khriachtchev, L.; Nikitin, T.; Räsänen, M.; Domanskaya, A.; Boninelli, S.; Iacona, F.; Engdahl, A.; Juhanoja, J.; Novikov, S. Continuous-wave laser annealing of Si-rich oxide: A microscopic picture of macroscopic Si-SiO2 phase separation. J. Appl. Phys. 2010, 108, 124301.

141. Nikitin, T.; Kemell, M.; Puukilainen, E.; Boninelli, S.; Iacona, F.; Räsänen, M.; Ritala, M.; Novikov, S.; Khriachtchev, L. Surface fingerprints of individual silicon nanocrystals in laser-annealed Si/SiO2 superlattice: Evidence of nanoeruptions of laser-pressurized silicon. J. Appl. Phys. 2012, 111, 124302.

142. Nikitin, T.; Kemell, M.; Puukilainen, E.; Räsänen, M.; Ritala, M.; Novikov, S.; Khriachtchev, L. Continuous-wave laser annealing of a Si/SiO2 superlattice: Effect of the ambient atmosphere and exposure period. Sci. Adv. Mater. 2014, 6, 1000–1010.

143. Bachels, T.; Schafer, R. Binding energies of neutral silicon clusters. Chem. Phys. Lett. 2000, 324, 365–372.

144. Murphy, D.V.; Brueck, S.R.J. Enhanced raman-scattering from silicon microstructures. Opt. Lett. 1983, 8, 494–496.

145. Doremus, R.H. Viscosity of silica. J. Appl. Phys. 2002, 92, 7619–7629.

146. Wada, K.; Suzuki, A.; Sato, H.; Kikuchi, R. Soret effect in solids. J. Phys. Chem. Solids 1985, 46, 1195–1205.

147. Zhang, K.J.; Briggs, M.E.; Gammon, R.W.; Sengers, J.V. Optical measurement of the Soret coefficient and the diffusion coefficient of liquid mixtures. J. Chem. Phys. 1996, 104, 6881–6892.

148. Susa, M.; Nagata, K. Thermal oxidation of silicon substrates through oxygen diffusion. Mat. Sci. Eng. A 1991, 146, 51–62.

149. Deal, B.E.; Grove, A.S. General relationship for thermal oxidation of silicon. J. App. Phys. 1965, 36, 3770–3778.

150. Lukes, F.; Schmidt, E. Oxidation of silicon in dry oxygen. J. Phys. Chem. Solids 1965, 26, 1353–1357.

151. Nikitin, T.; Khriachtchev, L.; Räsänen, M.; Novikov, S. Optical memory of silicon nanocrystals with submicron spatial resolution and very high thermal stability. Appl. Phys. Lett. 2009, 94, 173116.

152. Khriachtchev, L. Comment on Optical absorption measurements of silica containing Si nanocrystals produced by ion implantation and thermal annealing. Appl. Phys. Lett. 2002, 81, 1357–1358.

153. Oton, C.J.; Ghulinyan, M.; Gaburro, Z.; Bettotti, P.; Pavesi, L.; Pancheri, L.; Gialanella, S.; Capuj, N.E. Scattering rings as a tool for birefringence measurements in porous silicon. J. Appl. Phys. 2003, 94, 6334–6340.

154. Renault, O.; Marlier, R.; Gely, M.; De Salvo, B.; Baron, T.; Hansson, M.; Barrett, N.T. Synchrotron radiation X-ray photoelectron spectroscopy of Si nanocrystals grown onto Al2O3/Si surfaces. Appl. Phys. Lett. 2005, 87, 163119.

155. Kim, S.; Kim, M.C.; Choi, S.H.; Kim, K.J.; Hwang, H.N.; Hwang, C.C. Size dependence of Si 2p core-level shift at Si nanocrystal/SiO2 interfaces. Appl. Phys. Lett. 2007, 91, 103113.

156. Barbagiovanni, E.G.; Goncharova, L.V.; Simpson, P.J. Electronic structure study of ion-implanted Si quantum dots in a SiO2 matrix: Analysis of quantum confinement theories. Phys. Rev. B 2011, 83, 035112.

157. Bruggeman, D.A.G. Berechnung verschiedener physikalischer Konstanten von heterogenen Substanzen. I. Dielektrizitätskonstanten und Leitfähigkeiten der Mischkörper aus isotropen Substanzen. Ann. Phys. 1935, 416, 636–664.

158. Knief, S.; von Niessen, W. Disorder, defects, and optical absorption in a-Si and a-Si:H. Phys. Rev. B 1999, 59, 12940–12946.

Chapter 7

Nanomaterials for biosensing applications: A review

Michael Holzinger[*], **Alan Le Goff** and **Serge Cosnier**

Département de Chimîe Moléculaire UMR 5250, Biosystèmes Electrochimique and Analytiques, CNRS, University of Grenoble Alpes, Grenoble, France

A biosensor device is defined by its biological, or bioinspired receptor unit with unique specificities toward corresponding analytes. These analytes are often of biological origin like DNAs of bacteria or viruses, or proteins which are generated from the immune system (antibodies, antigens) of infected or contaminated living organisms. Such analytes can also be simple molecules like glucose or pollutants when a biological receptor unit with particular specificity is available. One of many other challenges in biosensor development is the efficient signal capture of the biological recognition event (transduction). Such transducers translate the interaction of the analyte with the biological element into electrochemical, electrochemiluminescent, magnetic, gravimetric, or optical signals. In order to increase sensitivities and to lower detection limits down to even individual molecules, nanomaterials are promising candidates due to the possibility to immobilize an enhanced quantity of bioreceptor units at reduced volumes and even to act itself as transduction element. Among such nanomaterials, gold nanoparticles, semi-conductor quantum dots, polymer nanoparticles, carbon nanotubes, nanodiamonds, and graphene are intensively studied. Due to the vast evolution of this research field, this review summarizes in a non-exhaustive way the advantages of nanomaterials by focusing on nano-objects which provide further beneficial properties than "just" an enhanced surface area.

1. INTRODUCTION

As in many different technological sections, nanomaterials have demonstrated their appropriateness for biosensing applications. The intelligent use of such nano-objects led to clearly enhanced performances with increased sensitivities and lowered detection limits of several orders of magnitudes. One general advantage of all nanomaterials is the high specific surface thus already enabling the immobilization of an enhanced amount of bioreceptor units. However, one of the constant challenges is the immobilization strategy used to conjugate intimately the bio-specific entity onto such nanomaterials. Therefore, the technique used to immobilize the enzyme is one of the key factors in developing a reliable biosensor.

Efficient methods for the biofunctionalization of nanomaterials are summarized in reference (Putzbach and Ronkainen, 2013). Briefly, non-covalent approaches representing electrostatic interaction, π–π stacking, entrapment in polymers, or van der Waals forces between the nanomaterial and the biological entity. These principles preserve all specific properties of both, nanomaterial and biomolecule.

Covalent binding: the strategy to attached covalently biomolecules to nanomaterials has an advantage in terms of stability and reproducibility of the surface functionalization and lowers unspecific physisorption. Covalent links can be formed, e.g., by classic amide coupling reactions, cross-linking, or click chemistry. One drawback is the uncontrolled anchoring of the biomolecule which can affect the domain which is responsible for the recognition event.

The immobilization of biomolecules via supramolecular or coordinative interactions: this technique has achieved wide acceptance in recent years in binding biological species to surfaces. The most famous example used in the field of biosensor engineering is the biotin/avidin (or streptavidin) system (Wilchek and Bayer, 1988). Biotinylated biomolecules can be attached to biotinylated substrates via avidin (or streptavidin) bridges. Other affinity systems have been reported like the nitrilotriacetic acid (NTA)/Cu^{2+}/histidine complex (Haddour et al., 2005) or the host-guest system adamantane/β-cyclodextrin (Holzinger et al., 2009). The advantage of such systems, compared to the other immobilization methods, is the reversibility, enabling the possibility to regenerate the transducer element. Furthermore, all components like the functionalized transducer surface and the modified bio-receptor can be characterized individually assuring the reproducibility of the constructed biosensor.

According to the chemical composition, almost all nanomaterials can be equipped with appropriate functions via direct functionalization (in some cases

already during synthesis), or via coating with functional polymers without affecting their specific properties (Biju, 2014). Such functionalization not only allows the reproducible immobilization of bioreceptor units but can also increase the biocompatibility of these materials.

One particular issue in biosensing devices is that the recognition event is not directly detectable by the used transduction technique. This is the case for affinity biosensors like the immunoreaction between an antigen and its antibody or the hybridization of corresponding DNA strands. Here, further biospecific components (secondary antibodies or DNA strands) modified with labels for optical or electrochemical transduction have to be used. The specific properties of some nanomaterials clearly contributed to the development of "label free" transduction techniques or contribute to clear signal amplifications when used as labels.

2. GOLD NANOPARTICLES

Within the group of noble metal nanoparticles, gold nanoparticles are mostly used for biosensor application (Li et al., 2010a) due to their biocompatibility, their optical and electronic properties, and their relatively simple production and modification (Biju, 2014).

Particular interesting is the optical behavior of gold surfaces where irradiation with light of one specific wavelength causes an oscillation of the electrons in the conduction band, called resonant surface plasmons. When the particle size is much smaller than the incident wavelength, the oscillating electrons cannot propagate along the surface as it is for classic surface plasmon resonance (SPR) setups. The electron density is then polarized on one side of the particle where the plasmons oscillate in resonance with the light frequency (Figure 1). This phenomenon was described applying the Mie theory (Mulvaney, 1996; Hao et al., 2004) and is strongly dependent on the size, shape of the nanoparticle and the dielectric constant of its environment (Kelly et al., 2002). This environmental dependency represents a great advantage for (bio)-analytics since the recognition event can result in a change of the oscillation frequency and therefore to a color change of the gold nanoparticles observable with bare eye. In this context, a wide series of efficient colorimetric biosensors were developed for DNA or oligonucleotide detection, or immunosensors (Reynolds et al., 2000; Oldenburg et al., 2002; Liu and Lu, 2004; Xu et al., 2009).

Gold nanoparticles have also shown their ability to form a powerful transduction platform for single molecule detection. By refractive index sensing of localized surface plasmon resonance (LSPR) coupled with enzyme linked

immunosorbent assay (ELISA) using isolated gold nanoparticles of 60 nm sizes. The model enzyme horseradish peroxidase (HRP) was immobilized on these gold nanoparticles via biotin streptavidin linkage. HRP is widely used as label in biosensing applications since it can form colored, fluorescent, or redox active molecules while reducing hydrogen peroxide to water (Veitch, 2004). In the work of Chen et al., HRP oxidized the soluble monomer 3, 3'-Diaminobenzidine (DAB) to insoluble colored polybenzimidazole which aggregates around the enzyme. This aggregation led to an additional shift of the LSPR scattering wavelength enabling the detection of few or even single HRP proteins attached to the gold particles (Chen et al., 2011). The beneficial effects of gold nanoparticles could be validated for high sensitive biosensing using surface enhanced Raman spectroscopy (SERS). Based on surface plasmon assisted signal amplification (Moskovits, 1978) of the vibrational spectrum of adsorbed or immobilized compounds, detection limits also down to the single molecule level could be reached (Nie and Emory, 1997; Hossain et al., 2009; Lim et al., 2011; Saha et al., 2012).

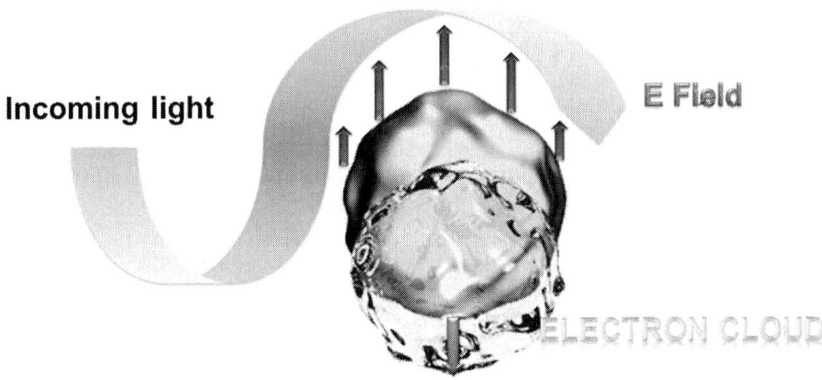

Figure 1. Schematic presentation of the polarization of the electron density at resonant excitation wavelength. At particle sizes smaller than the excitation wavelength, the oscillating electrons (surface plasmons) cannot propagate along the gold surface leading to a polarization of the electron cloud at one side of the particle.

Gold nanoparticles have also demonstrated their advantages in bioanalysis using SPR transduction. This method is usually based on the change of the dielectric constant of propagating surface plasmons' environment of gold films where the detection of the analyte can be recorded in different ways like the changes of the angle, intensity, or phase of the reflected light (Wijaya et al.,

2011; Guo, 2012). Beside the use of a pure gold nanoparticle based SPR transduction replacing the gold film (Pedersen and Duncan, 2005), a clear SPR signal enhancement can be obtained when gold films and gold nanoparticles are used in a sandwich configuration. In fact, the surface plasmons on gold nanoparticles provoke a perturbation of the evanescent field of the gold film in addition to the immobilized bioreceptor unit and the recognized analyte. The optimal configuration of this approach was determined for gold nanoparticles smaller than 40 nm at a distance to the gold film surface of 5 nm (Zeng et al., 2013) as illustrated in Figure 2. In this case, gold nanoparticles serve as labels when attached to secondary antibodies or DNA strands. Even when such labeling needs further preparative steps than label less detection, the signal enhancement is of several orders of magnitudes which is a convincing argument for such an approach.

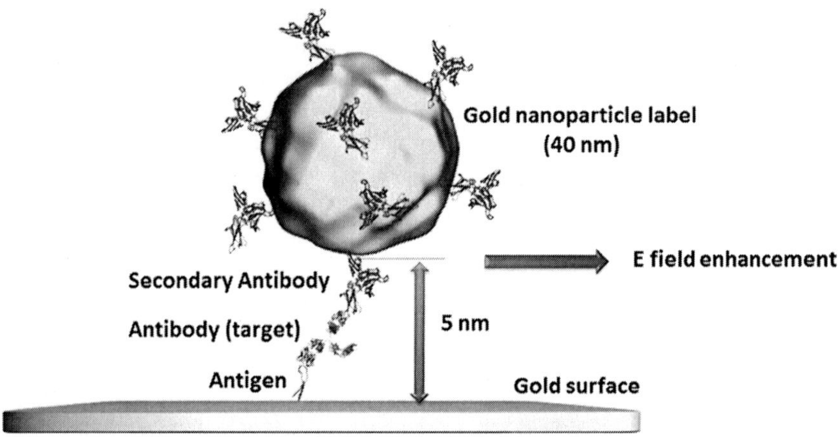

Figure 2. Illustration of the perturbation of propagating surface plasmons on gold surfaces provoked by gold nanoparticles of defined size and distance leading to an additional change of the evanescent field and therefore to an enhanced signal.

Besides the outstanding optical properties, gold nanoparticles also have the ability to transfer electrons between a wide range of electroactive biological species and the electrode. This principle is principally used for redox enzyme biosensing where the bioreceptor unit catalyzes the oxidation or reduction of the analyte. In classical electrochemical enzyme biosensors setups, the formed species are oxidized or reduced by the electrode, giving the electrochemical signal. The disadvantage of this approach is that the detectable molecules have to diffuse to the electrode where a non-negligible amount is lost in solution.

Gold nanoparticles can act as electron shuttles, i.e., the gold nanoparticles can approach to the redox center of the enzyme regenerating this biocatalyst by transferring the electrons involved in the redox reaction to the electrode. Since such wiring of enzymes prevents the need of enzymatically formed redox species which have to reach the electrode surface, a clear increase of the electrochemical signal is to expect. Particular efficient wiring could be obtained using enzymes where metal ions are involved in the catalytic redox process as it is for HRP (Xu et al., 2006). Nonetheless, the holy grail of enzymes to wire is glucose oxidase (GOx) since its active center is deeply embedded inside the protein structure and contains no metal ion which would facilitate electron transfer (Mena et al., 2005; Willner et al., 2006; Pingarrón et al, 2008). Wiring of GOx is not only a great challenge for glucose sensing applications but also for the design of bioanodes for glucose fuel cells. This is a steady expanding research field since such glucose biofuel cells could generate power to supply implantable medical devices out of body liquids (Cosnier et al., 2014). GOx catalyzes the oxidation of glucose to gluconolactone by releasing 2 electrons. These electrons are generally used to reduce oxygen to hydrogen peroxide. The involved natural electron shuttle is the co-factor flavin adenine dinucleotide (FAD) which is after glucose oxidation in reduced form ($FADH_2$). The goal is therefore to provide an electron carrier which regenerates FAD with favored kinetics to avoid the oxygen reduction process. Xiao et al. proposed an original approach by the reconstitution of apo-GOx (GOx without the cofactor FAD) with a FAD modified gold nanoparticle (Figure 3). This work even demonstrated that the as reactivated enzyme showed a 7-fold improved electron transfer turnover rate compared to the natural electron transfer rate to oxygen (Xiao et al., 2003).

These outstanding properties of gold nanoparticles made them promising candidates not only for bioanalytics but also for many other research fields. The particular properties of such gold nanoparticles can be tuned and adjusted. Whatever the desired application, almost any desired shape or size can be obtained using the appropriate synthesis technique. These different morphologies result in different optical, catalytic, and electronic behavior of these gold nanoparticles (Eustis and El-Sayed, 2006).

3. QUANTUM DOTS

Another prominent example of nanomaterials used for bioanalytics are luminescent semiconducting nanocrystals called quantum dots (QDs). The most studied colloidal QDs are based on cadmium chalcogenides (S, Se, Te) (Murray et al., 1993; Park et al., 2007; Reiss et al., 2009) which provide a very large absorption spectrum with a size-dependent narrow emission spectrum. This

phenomenon is due to the varying band gaps of the semiconductor material for different nanocrystal sizes (the bigger the particle the lower the band gap) which leads to distinct emission wavelengths from the recombination of the electron-hole exiton (Weller, 1993). The availability of a large range of emission wavelengths of QDs with different sizes enables efficient multiplexed analysis using classic optical transduction (Geißler et al., 2010;Petryayeva and Algar, 2014). However, structural defects in the crystal lattice can trap the exited electrons or holes leading to non-radiative relaxation (Murphy, 2002). To overcome this issue, core/shell composited with another semiconductor material with a wider band gap range (generally ZnS) were realized to passivate these surface defects and to enhance quantum yields and photo-stability (Dabbousi et al., 1997; Jaiswal et al., 2003). Due to this high photochemical stability of core/shell QDs, this material became a promising alternative to organic fluorophores (Resch-Genger et al., 2008).

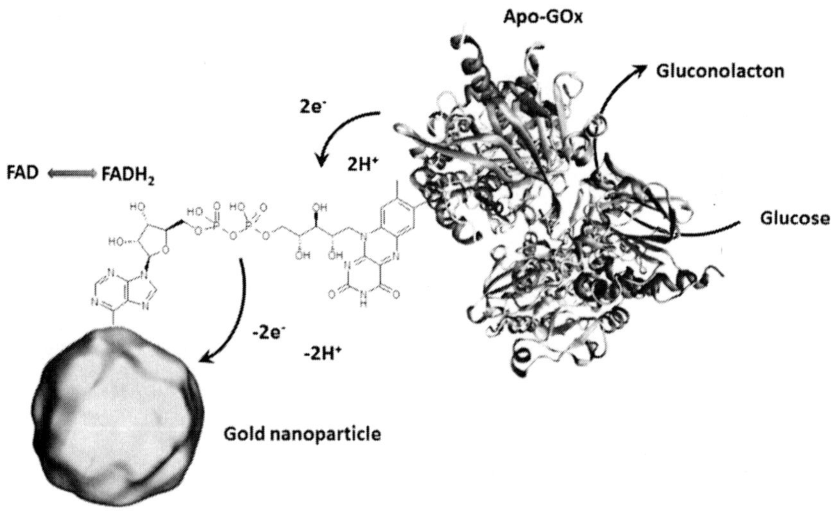

Figure 3. Schematic presentation of the reconstitution of apo-GOx with FAD modified gold nanoparticles enabling the transfer of electron released by the electrocatalytic oxidation of glucose. The collected electrons are then directly transferred to the external circuit giving an electrochemical signal.

In order to provide functional groups for bioreceptor immobilization and also to anticipate possible toxicity issues, QDs are nowadays available with inert or biocompatible coatings (Biju et al., 2010). Thus, almost any kind of biomolecule

can be attached to these nanocrystals as long as the photophysical recombination event is not affected.

Such so-called fluorescence quenching of QDs, described by the Förster (Fluorescence) Resonance Energy Transfer (FRET) (Clapp et al., 2006), is based on a non-radiative energy transfer between the exited QD (donor) and a quencher (acceptor) such as organic fluorophores. Such FRET quenching revealed original approaches used for optical transduction since QD fluorescence reappears when the quencher is removed. This strategy is particularly used for optical DNA and oligonucleotide sensors (Zhang et al., 2005;Freeman et al., 2013). The challenge is to bring the quenching dye in close contact to the QD and to release it after the recognition event. One example deals with the hybridization of a QD tagged receptor DNA and a short corresponding DNA sequence equipped with a gold nanoparticle. As already described, gold nanoparticles are excellent acceptors and are highly efficient QD quenchers. Due to the more favorable hybridization kinetics of the analyte DNA, the short sequence with the gold nanoparticle is released and the QDs' fluorescence reappears where its intensity is correlated to the analyte concentration (Dyadyusha et al., 2005; Dai et al., 2007) as shown in Figure 4.

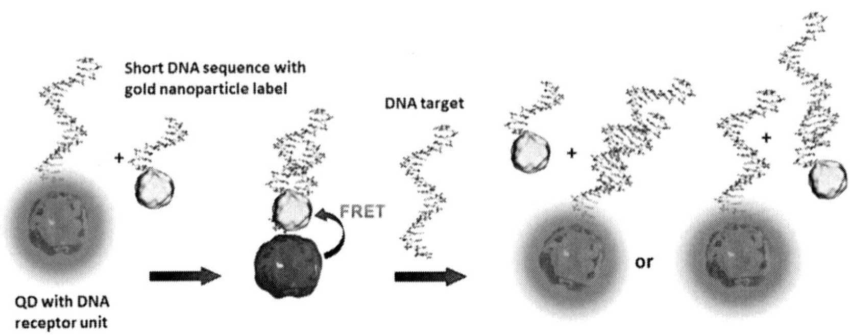

Figure 4. Illustration of the FRET principle between QDs and gold nanoparticles and the reestablishing of fluorescence after the recognition event. The target DNA can hybridize either with the receptor DNA, the short sequence of the gold nanoparticle label, or with both. These statistical hybridization events have to be considered for the quantification of the analyte.

Another use of non-radiative energy transfer provoking QD fluorescence is called Bioluminescence Resonance Energy Transfer (BRET) (So et al., 2006). Here, a light-emitting protein label transfers the energy to QDs and eliminates

the necessity of an external excitation light source. FRET and BRET, together with charge transfer quenching and chemiluminescence resonance energy transfer (CRET) (Huang et al., 2006) are the most common strategies in biosensing applications using QDs as optical transducers (Frasco and Chaniotakis, 2009; Algar et al., 2010; Petryayeva et al., 2013).

For all cases, the distance between the quencher and QD is one of the key factors in the sensitivity of such detection principles. For instance, an enhancement of QD fluorescence was observed when gold nanoparticles are localized at around 30 nm (Figure 5). In this case, the gold nanoparticle acts as antenna (and not as quencher) increasing the excitation rates of the QDs and therefore the fluorescence intensity (Maye et al., 2010). Another effect leading to further example for optical signal enhancement is that QDs can interact with propagating surface plasmons on gold surfaces. These surface plasmons can excite QDs leading to light emission and vice versa - the light induced excited state of QDs are converted to propagating surface plasmons (Wei et al., 2009). Malic et al. demonstrated the beneficial combination of near infrared QDs and gold surfaces for drastic signal enhancement in a SPR imaging biosensor setup (Malic et al., 2011).

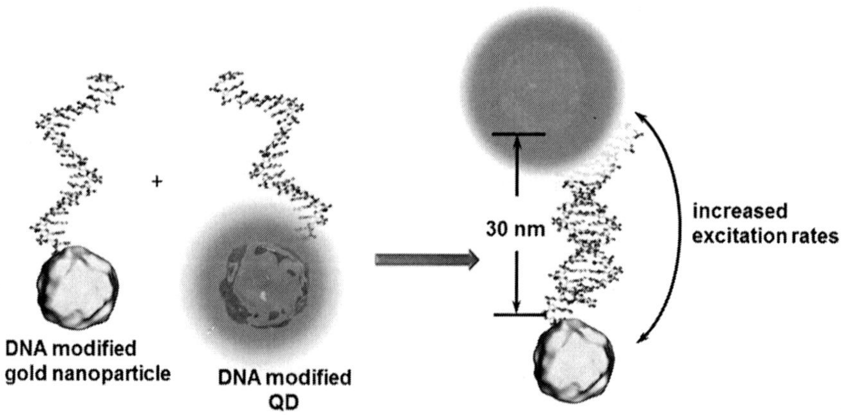

Figure 5. Principle of the antenna effect of gold nanoparticles enhancing the excitation rates of QDs when positioned at ~30 nm. This transduction principle might be limited to the identification of analyte DNAs composed of around 15 nucleotides.

QDs have demonstrated their appropriateness in bioanalytics either as transducer unit or as optical labels. A tremendous evolution of QD based

biosensors is therefore predictable. Furthermore, the described principles and examples about the use of QDs in bioanalysis let suggest that the combination of different nano-objects and their specific particularity is a promising way for future biosensors with new and original highly efficient transduction techniques.

4. MAGNETIC NANOPARTICLES

Magnetic nanoparticles are promising alternatives to fluorescent labels in biosensor devices. Nanosized magnetic nanoparticles show different magnetic behaviors compared to its bulk material due to the reduced number of magnetic domains (regions of parallel oriented magnetic moments caused by interacting unpaired electrons of an atom) leading to so called superparamagnetic behavior. This means that magnetization can flip the direction randomly within very short time (Neel's relaxation time) and the magnetization appears in average zero in absence of an external magnetic field. This temperature dependent phenomenon disappears by applying an external magnetic field aligning the magnetic moments. Even when this effect seems to be similar to this of classic paramagnetic materials, the magnetic susceptibility of superparamagnets are much higher (Bishop et al., 2009). Such superparamagnetic behavior prevents therefore from attractive or repulsive forces between the magnetic nanoparticles as long as no external magnetic field is applied.

Beside a wide range of ferromagnetic materials, iron oxide is mostly used for bioanalytical applications (Haun et al., 2010). The outstanding advantage using magnetic nanoparticles is the possibility to concentrate the analyte before the detection event. Receptor unit modified magnetic nanoparticles can simply be mixed with the analyte solution and interacts specifically with the specific target. After applying an external magnetic field, the nanoparticles agglomerate and can be separated from the solution. Efficient isolation of DNA strands in complex media was achieved in a fast and efficient manner using silica or gold coated core/shell nanoparticles (He et al., 2007; Li et al., 2011; Min et al., 2014).

Besides optical (Bi et al., 2009) or electrochemical (Mejri et al., 2011) detection techniques in combination with other often nanosized labels, magnetic nanoparticles offer a further high sensitive transduction technique in the domain of diagnostic magnetic resonance (Figure 6). In this context, high performant giant magnetoresistance, spin valve, or magnetic tunnel junction biosensors could be developed (Wang and Li, 2008; Konry et al., 2012). Magnetic labels are particular interesting for biosensing applications since

biological entities do not show any magnetic behavior or susceptibility and therefore, no interferences or noise is to expect during signal capturing (Tamanaha et al., 2008). For instance, an ultra-high sensitive magnetoresistant biosensor was developed for *Escherichia coli* (Mujika et al., 2009) detection or *Salmonella* were identified in skimmed-milk samples with a limit of detection (LOD) of 1 colony forming-unit (cfu)/mL using a magneto-genosensing setup (Liébana et al., 2009).

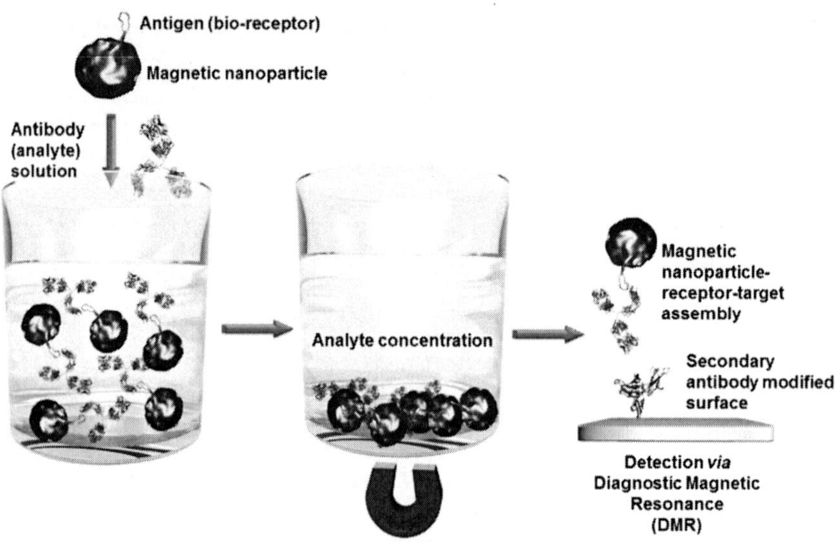

Figure 6. Scheme of analyte concentration in complex solutions via bioreceptor modified magnetic nanoparticles. After the recognition event, the captured analytes can be isolated by precipitation of the magnetic nanoparticles in an external magnetic field. When this magnetic field is removed, the nanoparticles can be redispersed. Only the analyte labeled nanoparticles are immobilized on surfaces which are modified with a secondary antibody and thus quantified using diagnostic magnetic resonance transduction.

A further advantage of magnetic nanoparticles is the possibility to carry the analytes to the transduction platform in microfluidic systems (Konry et al., 2012) enabling even complex multiplexed analyses for the simultaneous detection of different analytes. Nonetheless, these carrier properties are mainly applied for the transport of drugs and genes, or for magnetic resonance imaging (Sun et al., 2008).

5. CARBON NANOSTRUCTURES

The beneficial properties of nanostructured carbons such as carbon nanotubes or graphene made them a widely used material as electronic or electrochemical transducer in biosensor devices (Vamvakaki and Chaniotakis, 2007; Valentini et al., 2013). In particular, carbon nanotubes possess the outstanding combination of nanowire morphology, biocompatibility and electronic properties (Battigelli et al., 2013). Therefore, carbon nanotube interfaces present clearly enhanced capacities, e.g., to approach the active sites of a redox enzyme and to wire it to the bulk electrode. Furthermore, their ease and well-documented organic functionalization (Ménard-Moyon et al., 2010) brings new properties to nanostructured electrodes such as specific docking sites for biomolecules or redox mediation of bioelectrochemical reactions. Moreover, CNT films exhibit a high electroactive surface areas due to the natural formation of highly porous three-dimensional networks, suitable for the anchoring of a high amount of bioreceptor units, leading consequently to high sensitivities (Wang, 2005; Le Goff et al., 2011).

Nonetheless, since purely the increase of the specific surface is not discussed here, the presented examples of carbon nanomaterials focus the use of their electronic or electrochemical properties for biosensing applications.

As already mentioned, the nanowire morphology of CNTs enables the approach to the active centers of redox enzyme leading to fast and efficient electron transfers. In few cases, this enzyme wiring spontaneously occurs simply by adsorption of the enzyme on the nanocarbon. The driving force is mostly based on hydrophobic interactions. An important condition is that the hydrophobic domain of the enzyme has to be situated close to its active center for efficient electron transfer after immobilization and orientation.

The nitrite reductase *from Desulfovibrio desulfuricans* is a famous example of such an enzyme having the hydrophobic and catalytic domains at favorable localizations. The multihemic catalytic site for the reduction of nitrite ions is closely placed to a hydrophobic domain of the protein structure as illustrated in Figure 7. After catalytic reduction of nitrite to ammonia, the enzyme, in oxidized state, is directly regenerated by the carbon electrode. This concept was validated for pyrolitic graphite (Silveira et al., 2010a), and CNTs (Silveira et al., 2010b).

Since nature does not always provide such excellent conditions for electron transfers, appropriate functionalization of the carbon material with anchor molecules and/or with redox active species with corresponding redox potentials is necessary to establish so-called mediated electron transfer (via electron shuttle). One particular case of mediated electron transfer is the use

of the enzyme specific co-factor. As already demonstrated by Xiao et al. (2003) for glucose sensors using gold nanoparticles as electron carriers, FAD was also attached to carbon nanotubes enabling the wiring glucose oxidase that led to a performant glucose biosensor (Patolsky et al., 2004).

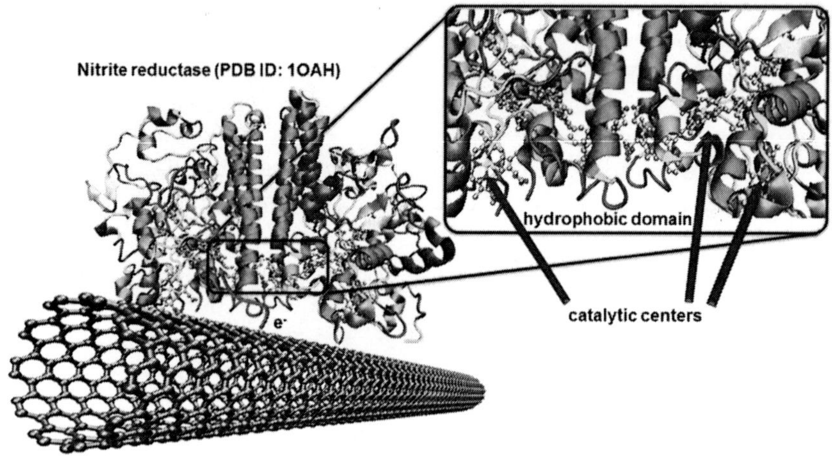

Figure 7. Oriented immobilization of nitrite reductase from *Desulfovibrio desulfuricans***on CNTs** *via* **hydrophobic interactions between the enzyme and the carbon nanotube surface**. This naturally occurring adsorption leads to an efficient electron transfer between the enzyme and the conducting surface where the redox center, after catalytic reduction of nitrite ions to ammonia, can directly be regenerated by the electrode.

However, the main purpose for efficient enzyme wiring on nanostructured carbon clearly targets bioenergy conversion (Holzinger et al., 2012).

The particular electric properties of CNTs were used in field effect transistor (FET) biosensor setups where changes of the conductivity of the CNT channel or the modulation of the Schottky barrier after the bio-recognition event (Gruner, 2006) led to high sensitivities and low detection limits down to single molecules (Besteman et al., 2003).

An original approach for CNT-FET based DNA sensor was proposed by taking advantage of the high affinity of CNTs and DNA strands. The nucleic bases of a ssDNA attach to the CNTs via π −π stacking leading to a wrapping of the DNA around the CNT (Gigliotti et al., 2006) thus eliminating the recognition capacity of this ssDNA toward its counterpart. Tang et al. reported a clear electrical

conductance change due to the modulation of energy level alignment between CNT and gold contacts of the CNT-FET device. The receptor unit was attached to the gold electrodes via thioethers where the wrapping of the ssDNA around the conductive CNT channel was accepted. The recognition event was therefore localized only on source and drain, modulating the Schottky barrier and finally the conduction of the CNT channel as sketched in Figure 8 (Tang et al., 2006). This is one of very little examples of CNT based DNA-sensors where the CNTs were not blended in composites to avoid this wrapping effect (Zhou et al., 2009). Else CNT-FET biosensors were mainly developed for enzyme and immunosensing applications (Chao et al., 2005; Gruner, 2006). Some other spectroscopic characteristics of CNTs like NIR photoluminescence and Raman scattering features used for biosensing applications are barely described which is most likely due to the low photoluminescence quantum efficiency (Biju, 2014).

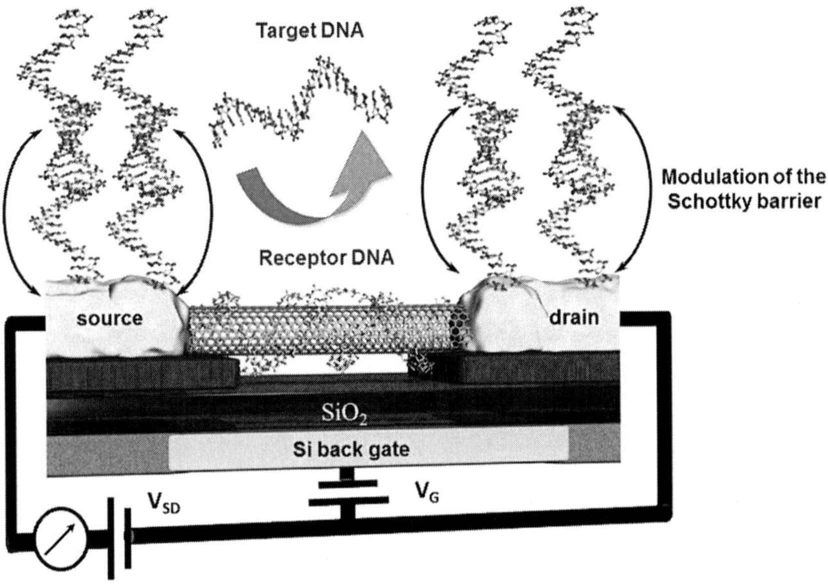

Figure 8. Principle of the modulation of the Schottky barrier at source and drain electrodes of CNT field effect transistor devices after hybridization of the receptor DNA with the target DNA. Here, the wrapping properties of DNA around nanotubes inhibit the capability of this receptor DNA to hybridize with its counterpart and thus deactivate indirectly the CNT channel which remains unaffected toward the recognition event.

CNT-based biosensors are clearly more developed than graphene-based biosensors (Yang et al., 2010) but there is a steady increasing interest in this two-dimensional material for bioanalytical applications. Besides the fabrication of isolation of real monolayer graphene sheets via mechanical cleavage using a scotch tape to exfoliate one sheet from highly oriented pyrolytic graphite (HOPG), by CVD on metal foils, or using epitaxy techniques where the graphene layer is formed out of silicon carbide, a wide series of graphite based bulk materials are also called graphene (Bonaccorso et al., 2012). These materials are mostly obtained after mechanic exfoliation (Coleman, 2012; Paton et al., 2014) or chemical oxidation of graphite based on the Hummers and Offeman method (Hummers and Offeman, 1958). This initially called graphitic oxide is now generally known as graphene oxide and allows obtaining soluble carbon oxide sheets of undefined layer composition and sizes. The electric conductivity of this isolating material can be reestablished by chemical, thermal, or electrochemical reduction (Kuila et al., 2013). Even when the exceptional conductivity of real monolayer graphene (Novoselov et al., 2004) cannot be obtained, reduced graphene oxide has other beneficial properties in high performance biosensor devices. As for CNTs, graphene-based materials are mostly used in electrochemical biosensors (Shao et al., 2010; Kuila et al., 2011; Ratinac et al., 2011) or field effect transistor setups (Liu and Guo, 2012) where graphene is principally used as electrode material with enhanced specific surface. Graphene materials can also act as transduction element itself in optical or colorimetric biosensor setups. High efficient graphene-based FRET biosensors were developed in combination with organic dyes (Li et al., 2010b) or quantum dots (Dong et al., 2010) modified receptor units for DNA, aptamer, immuno, or protein sensors (Ma et al., 2012). Contrary to CNTs, ssDNA or oligonucleotide receptors adsorb reversibly to graphene oxide and are released after the recognition event. This allows recovering the fluorescence of the dyes as shown in Figure9. An interesting example was the adsorption of several ssDNA receptors, each modified with differently colored dyes, leading to a fast multiplex colorimetric DNA sensor (He et al., 2010).

Graphene and CNTs are the most promising nanostructured carbon materials for biosensing applications where each allotrope has its particular advantage as transducer element.

6. CONCLUSION

Nanomaterials became important components in bioanalytical devices since they clearly enhance the performances in terms of sensitivity and detection limits down to single molecules detection. The specific properties of such nano objects also offer alternatives to classic transduction methods. Furthermore,

the combination of different nanomaterials, each with its characteristics, to increase even more the performances of biosensors is a well-accepted strategy. Due to the vast number of different nanomaterials all with its own specific properties, only few examples could be mentioned here by emphasizing the principal advantages of such materials. More detailed examples of the use of nanomaterials in biosensor applications are described in the references (Ju et al., 2011; Leí and Ju, 2012), and in the cited review articles.

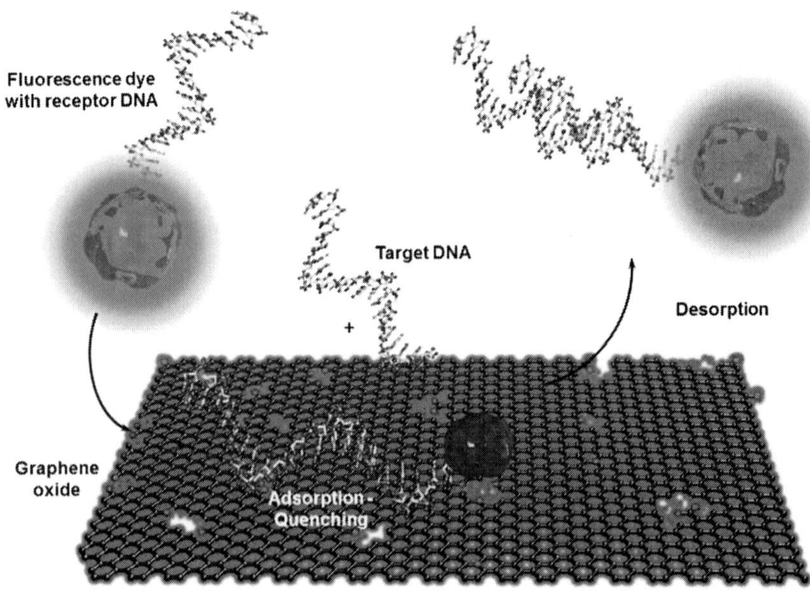

Figure 9. Adsorption and desorption of a fluorescence dye modified receptor DNA on graphene oxide before and after hybridization with the corresponding target DNA. Graphene oxide quenches the fluorescence of the dye when the receptor unit is adsorbed. By hybridization with the analyte DNA, the receptor DNA and therefor the dye is released which leads to the reestablishment of the fluorescence.

CONFLICT OF INTEREST STATEMENT

The authors declare that the research was conducted in the absence of any commercial or financial relationships that could be construed as a potential conflict of interest.

REFERENCES

1. Algar, W. R., Tavares, A. J., and Krull, U. J (2010). Beyond labels: a review of the application of quantum dots as integrated components of assays, bioprobes, and biosensors utilizing optical transduction. *Anal. Chim. Acta* 673, 1–25.

2. Battigelli, A., Ménard-Moyon, C., Da Ros, T., Prato, M., and Bianco, A. (2013). Endowing carbon nanotubes with biological and biomedical properties by chemical modifications. *Adv. Drug Deliv. Rev.* 65, 1899–1920.

3. Besteman, K., Lee, J. O., Wiertz, F. G. M., Heering, H. A., and Dekker, C. (2003). Enzyme-coated carbon nanotubes as single-molecule biosensors. *Nano Lett.* 3, 727–730.

4. Bi, S., Yan, Y., Yang, X., and Zhang, S. (2009). Gold nanolabels for new enhanced chemiluminescence immunoassay of alpha-fetoprotein based on magnetic beads. *Chemistry* 15, 4704–4709.

5. Biju, V. (2014). Chemical modifications and bioconjugate reactions of nanomaterials for sensing, imaging, drug delivery and therapy.*Chem. Soc. Rev.* 43, 744–764.

6. Biju, V., Itoh, T., and Ishikawa, M. (2010). Delivering quantum dots to cells: bioconjugated quantum dots for targeted and nonspecific extracellular and intracellular imaging. *Chem. Soc. Rev.* 39, 3031–3056.

7. Bishop, K. J. M., Wilmer, C. E., Soh, S., and Grzybowski, B. A. (2009). Nanoscale forces and their uses in self-assembly. *Small* 5, 1600–1630.

8. Bonaccorso, F., Lombardo, A., Hasan, T., Sun, Z., Colombo, L., and Ferrari, A. C (2012). Production and processing of graphene and 2d crystals. *Mater. Today* 15, 564–589.

9. Chao, L. I., Curreli, M., Lin, H., Lei, B., Ishikawa, F. N., Datar, R., et al. (2005). Complementary detection of prostate-specific antigen using In2O3 nanowires and carbon nanotubes. *J. Am. Chem. Soc.* 127, 12484–12485.

10. Chen, S., Svedendahl, M., Duyne, R. P. V., and Kaĺll, M. (2011). Plasmon-enhanced colorimetric ELISA with single molecule sensitivity.*Nano Lett.* 11, 1826–1830.

11. Clapp, A. R., Medintz, I. L., and Mattoussi, H. (2006). Förster resonance energy transfer investigations using quantum-dot fluorophores.*Chemphyschem* 7, 47–57.

12. Coleman, J. N. (2012). Liquid exfoliation of defect-free graphene. *Acc. Chem. Res.* 46, 14–22.

13. Cosnier, S., Le Goff, A., and Holzinger, M. (2014). "Enzymatic fuel cells: from design to implantation in mammals," in *Implantable Bioelectronics*, ed E. Katz (Weinheim: Wiley-VCH Verlag GmbH and Co. KGaA), 347–362.

14. Dabbousi, B. O., Rodriguez-Viejo, J., Mikulec, F. V., Heine, J. R., Mattoussi, H., Ober, R., et al. (1997). (CdSe)ZnS core-shell quantum dots: synthesis and characterization of a size series of highly luminescent nanocrystallites. *J. Phys. Chem. B* 101, 9463–9475.

15. Dai, Z., Zhang, J., Dong, Q., Guo, N., Xu, S., Sun, B., et al. (2007). Adaption of Au nanoparticles and CdTe quantum dots in DNA detection. *Chin. J. Chem. Eng.* 15, 791–794.

16. Dong, H., Gao, W., Yan, F., Ji, H., and Ju, H. (2010). Fluorescence resonance energy transfer between quantum dots and graphene oxide for sensing biomolecules. *Anal. Chem.* 82, 5511–5517.

17. Dyadyusha, L., Yin, H., Jaiswal, S., Brown, T., Baumberg, J. J., Booy, F. P., et al. (2005). Quenching of CdSe quantum dot emission, a new approach for biosensing. *Chem. Commun.* 3201–3203.

18. Eustis, S., and El-Sayed, M. A. (2006). Why gold nanoparticles are more precious than pretty gold: noble metal surface plasmon resonance and its enhancement of the radiative and nonradiative properties of nanocrystals of different shapes. *Chem. Soc. Rev.* 35, 209–217.

19. Frasco, M., and Chaniotakis, N. (2009). Semiconductor quantum dots in chemical sensors and biosensors. *Sensors* 9, 7266–7286.

20. Freeman, R., Girsh, J., and Willner, I. (2013). Nucleic acid/Quantum dots (QDs) hybrid systems for optical and photoelectrochemical sensing. *ACS Appl. Mater. Interfaces* 5, 2815–2834.

21. Geißler, D., Charbonnière, L. J., Ziessel, R. F., Butlin, N. G., Löhmannsröben, H.-G., and Hildebrandt, N. (2010). Quantum dot biosensors for ultrasensitive multiplexed diagnostics. *Angew. Chem. Int. Ed.* 49, 1396–1401.

22. Gigliotti, B., Sakizzie, B., Bethune, D. S., Shelby, R. M., and Cha, J. N (2006). Sequence-independent helical wrapping of single-walled carbon nanotubes by long genomic DNA. *Nano Lett.* 6, 159–164.

23. Gruner, G. (2006). Carbon nanotube transistors for biosensing applications. *Anal. Bioanal. Chem*. 384, 322–335.

24. Guo, X. (2012). Surface plasmon resonance based biosensor technique: a review. *J. Biophotonics* 5, 483–501.

25. Haddour, N., Cosnier, S., and Gondran, C. (2005). Electrogeneration of a poly(pyrrole)-NTA chelator film for a reversible oriented immobilization of histidine-tagged proteins. *J. Am. Chem. Soc*. 127, 5752–5753.

26. Hao, E., Schatz, G. C., and Hupp, J. T. (2004). Synthesis and optical properties of anisotropic metal nanoparticles. *J. Fluoresc*. 14, 331–341.

27. Haun, J. B., Yoon, T.-J., Lee, H., and Weissleder, R. (2010). Magnetic nanoparticle biosensors. *Wiley Interdiscip. Rev*. 2, 291–304.

28. He, S., Song, B., Li, D., Zhu, C., Qi, W., Wen, Y., et al. (2010). A graphene nanoprobe for rapid, sensitive, and multicolor fluorescent DNA analysis. *Adv. Funct. Mater*. 20, 453–459.

29. He, X., Huo, H., Wang, K., Tan, W., Gong, P., and Ge, J. (2007). Plasmid DNA isolation using amino-silica coated magnetic nanoparticles (ASMNPs). *Talanta* 73, 764–769.

30. Holzinger, M., Bouffier, L., Villalonga, R., and Cosnier, S. (2009). Adamantane/β-cyclodextrin affinity biosensors based on single-walled carbon nanotubes. *Biosens. Bioelectron*. 24, 1128–1134.

31. Holzinger, M., Le Goff, A., and Cosnier, S. (2012). Carbon nanotube/enzyme biofuel cells. *Electrochim. Acta* 82, 179–190.

32. Hossain, M. K., Huang, G. G., Kaneko, T., and Ozaki, Y (2009). Characteristics of surface-enhanced Raman scattering and surface-enhanced fluorescence using a single and a double layer gold nanostructure. *Phys. Chem. Chem. Phys*. 11, 7484–7490.

33. Huang, X., Li, L., Qian, H., Dong, C., and Ren, J. (2006). A resonance energy transfer between chemiluminescent donors and luminescent quantum-dots as acceptors (CRET). *Angew. Chem*. 118, 5264–5267.

34. Hummers, W. S, and Offeman, R. E (1958). Preparation of graphitic oxide. *J. Am. Chem. Soc*. 80, 1339–1339.

35. Jaiswal, J. K., Mattoussi, H., Mauro, J. M., and Simon, S. M. (2003). Long-term multiple color imaging of live cells using quantum dot bioconjugates. *Nat. Biotechnol.* 21, 47–51.

36. Ju, H., Zhang, X., and Wang, J. (eds.). (2011). “Nanomaterials for immunosensors and immunoassays,” in *NanoBiosensing, Biological and Medical Physics, Biomedical Engineering* (New York, NY: Springer), 425–452.

37. Kelly, K. L., Coronado, E., Zhao, L. L., and Schatz, G. C (2002). The optical properties of metal nanoparticles:? the influence of size, shape, and dielectric environment. *J. Phys. Chem. B* 107, 668–677.

38. Konry, T., Bale, S., Bhushan, A., Shen, K., Seker, E., Polyak, B., et al. (2012). Particles and microfluidics merged: perspectives of highly sensitive diagnostic detection. *Microchim. Acta* 176, 251–269.

39. Kuila, T., Bose, S., Khanra, P., Mishra, A. K., Kim, N. H., and Lee, J. H (2011). Recent advances in graphene-based biosensors. *Biosens. Bioelectron.* 26, 4637–4648.

40. Kuila, T., Mishra, A. K., Khanra, P., Kim, N. H., and Lee, J. H (2013). Recent advances in the efficient reduction of graphene oxide and its application as energy storage electrode materials. *Nanoscale* 5, 52–71.

41. Le Goff, A., Holzinger, M., and Cosnier, S. (2011). Enzymatic biosensors based on SWCNT-conducting polymer electrodes. *Analyst* 136, 1279–1287.

42. Lei, J., and Ju, H. (2012). Signal amplification using functional nanomaterials for biosensing. *Chem. Soc. Rev.* 41, 2122–2134.

43. Li, F., Huang, Y., Yang, Q., Zhong, Z., Li, D., Wang, L., et al. (2010b). A graphene-enhanced molecular beacon for homogeneous DNA detection. *Nanoscale* 2, 1021–1026.

44. Li, K., Lai, Y., Zhang, W., and Jin, L. (2011). Fe_2O_3@Au core/shell nanoparticle-based electrochemical DNA biosensor for Escherichia coli detection. *Talanta* 84, 607–613.

45. Li, Y., Schluesener, H., and Xu, S. (2010a). Gold nanoparticle-based biosensors. *Gold Bull.* 43, 29–41.

46. Liébana, S., Lermo, A., Campoy, S., Barbé, J, Alegret, S., and Pividori, M. I (2009). Magneto immunoseparation of pathogenic bacteria and

electrochemical magneto genosensing of the double-tagged amplicon. *Anal. Chem.* 81, 5812–5820.

47. Lim, D.-K., Jeon, K.-S., Hwang, J.-H., Kim, H., Kwon, S., Suh, Y. D., et al. (2011). Highly uniform and reproducible surface-enhanced Raman scattering from DNA-tailorable nanoparticles with 1-nm interior gap. *Nat. Nano* 6, 452–460.

48. Liu, J., and Lu, Y. (2004). Colorimetric biosensors based on DNAzyme-assembled gold nanoparticles. *J. Fluoresc.* 14, 343–354.

49. Liu, S., and Guo, X. (2012). Carbon nanomaterials field-effect-transistor-based biosensors. *NPG Asia Mater.* 4, e23.

50. Ma, H., Wu, D., Cui, Z., Li, Y., Zhang, Y., Du, B., et al. (2012). Graphene-based optical and electrochemical biosensors: a review. *Anal. Lett.* 46, 1–17.

51. Malic, L., Sandros, M. G., and Tabrizian, M. (2011). Designed biointerface using near-infrared quantum dots for ultrasensitive surface plasmon resonance imaging biosensors. *Anal. Chem.* 83, 5222–5229.

52. Maye, M. M., Gang, O., and Cotlet, M. (2010). Photoluminescence enhancement in CdSe/ZnS-DNA linked-Au nanoparticle heterodimers probed by single molecule spectroscopy. *Chem. Commun.* 46, 6111–6113.

53. Mejri, M. B., Tlili, A., and Abdelghani, A. (2011). Magnetic nanoparticles immobilization and functionalization for biosensor applications.*Int. J. Electrochem.* 2011:421387.

54. Mena, M. L., Yáñez-Sedeño, P., and Pingarrón, J. M. (2005). A comparison of different strategies for the construction of amperometric enzyme biosensors using gold nanoparticle-modified electrodes. *Anal. Biochem.* 336, 20–27.

55. Ménard-Moyon, C., Kostarelos, K., Prato, M., and Bianco, A. (2010). Functionalized carbon nanotubes for probing and modulating molecular functions. *Chem. Biol.* 17, 107–115.

56. Min, J. H., Woo, M-K, Yoon, H. Y., Jang, J. W., Wu, J. H., Lim, C.-S., et al. (2014). Isolation of DNA using magnetic nanoparticles coated with dimercaptosuccinic acid. *Anal. Biochem.* 447, 114–118.

57. Moskovits, M. (1978). Surface roughness and the enhanced intensity of Raman scattering by molecules adsorbed on metals. *J. Chem. Phys.* 69, 4159–4161.

58. Mujika, M., Arana, S., Castaño, E., Tijero, M., Vilares, R., Ruano-López, J. M., et al. (2009). Magnetoresistive immunosensor for the detection of Escherichia coli O157:H7 including a microfluidic network. *Biosens. Bioelectron.* 24, 1253–1258.

59. Mulvaney, P. (1996). Surface plasmon spectroscopy of nanosized metal particles. *Langmuir* 12, 788–800.

60. Murphy, C. J. (2002). Peer reviewed: optical sensing with quantum dots. *Anal. Chem.* 74, 520A–526A.

61. Murray, C. B., Norris, D. J., and Bawendi, M. G (1993). Synthesis and characterization of nearly monodisperse CdE (E = sulfur, selenium, tellurium) semiconductor nanocrystallites. *J. Am. Chem. Soc.* 115, 8706–8715.

62. Nie, S., and Emory, S. R (1997). Probing single molecules and single nanoparticles by surface-enhanced raman scattering. *Science* 275, 1102–1106.

63. Novoselov, K. S., Geim, A. K., Morozov, S. V., Jiang, D., Zhang, Y., Dubonos, S. V., et al. (2004). Electric field effect in atomically thin carbon films. *Science* 306, 666–669.

64. Oldenburg, S. J., Genick, C. C., Clark, K. A., and Schultz, D. A (2002). Base pair mismatch recognition using plasmon resonant particle labels. *Anal. Biochem.* 309, 109–116.

65. Park, J., Joo, J., Kwon, S. G., Jang, Y., and Hyeon, T. (2007). Synthesis of monodisperse spherical nanocrystals. *Angew. Chem. Int. Ed.* 46, 4630–4660.

66. Patolsky, F., Weizmann, Y., and Willner, I. (2004). Long-range electrical contacting of redox enzymes by SWCNT connectors. *Angew. Chem. Int. Ed.* 43, 2113–2117.

67. Paton, K. R., Varrla, E., Backes, C., Smith, R. J., Khan, U., O'Neill, A., et al. (2014). Scalable production of large quantities of defect-free few-layer graphene by shear exfoliation in liquids. *Nat. Mater.* 13, 624–630.

68. Pedersen, D. B, and Duncan, E. J. S. (2005). "Surface plasmon resonance spectroscopy of gold nanoparticle-coated substrates," in*Technical Report, DRDC Suffield TR 2005-109* (Suffield, CT Defence RandD Canada).

69. Petryayeva, E., and Algar, W. R. (2014). Multiplexed homogeneous assays of proteolytic activity using a smartphone and quantum dots.*Anal. Chem.* 86, 3195–3202.

70. Petryayeva, E., Algar, W. R., and Medintz, I. L (2013). Quantum dots in bioanalysis: a review of applications across various platforms for fluorescence spectroscopy and imaging. *Appl. Spectrosc.* 67, 215–252.

71. Pingarrón, J. M., Yáñez-Sedeño, P., and González-Cortés, A. (2008). Gold nanoparticle-based electrochemical biosensors *Electrochim.Acta.* 53, 5848–5866.

72. Putzbach, W., and Ronkainen, N. (2013). Immobilization techniques in the fabrication of nanomaterial-based electrochemical biosensors: a review. *Sensors* 13, 4811–4840.

73. Ratinac, K. R., Yang, W., Gooding, J. J., Thordarson, P., and Braet, F. (2011). Graphene and related materials in electrochemical sensing.*Electroanalysis* 23, 803–826.

74. Reiss, P., Protière, M., and Li, L. (2009). Core/Shell semiconductor nanocrystals. *Small* 5, 154–168.

75. Resch-Genger, U., Grabolle, M., Cavaliere-Jaricot, S., Nitschke, R., and Nann, T. (2008). Quantum dots versus organic dyes as fluorescent labels. *Nat. Meth.* 5, 763–775.

76. Reynolds, R. A., Mirkin, C. A., and Letsinger, R. L (2000). Homogeneous, nanoparticle-based quantitative colorimetric detection of oligonucleotides. *J. Am. Chem. Soc.* 122, 3795–3796.

77. Saha, K., Agasti, S. S., Kim, C., Li, X., and Rotello, V. M. (2012). Gold nanoparticles in chemical and biological sensing. *Chem. Rev.* 112, 2739–2779.

78. Shao, Y., Wang, J., Wu, H., Liu, J., Aksay, I. A., and Lin, Y. (2010). Graphene based electrochemical sensors and biosensors: a review.*Electroanalysis* 22, 1027–1036.

79. Silveira, C. M., Baur, J., Holzinger, M., Moura, J. J. G., Cosnier, S., and Almeida, M. G. (2010b). Enhanced direct electron transfer of a multihemic

nitrite reductase on single-walled carbon nanotube modified electrodes. *Electroanalysis* 22, 2973–2978.

80. Silveira, C. M., Gomes, S. P., Araújo, A. N., Montenegro, M. C. B. S. M., Todorovic, S., Viana, A. S., et al. (2010a). An efficient non-mediated amperometric biosensor for nitrite determination. *Biosens. Bioelectron*. 25, 2026–2032.

81. So, M.-K., Xu, C., Loening, A. M., Gambhir, S. S., and Rao, J. (2006). Self-illuminating quantum dot conjugates for *in vivo* imaging. *Nat. Biotechnol.* 24, 339–343.

82. Sun, C., Lee, J. S. H., and Zhang, M. (2008). Magnetic nanoparticles in MR imaging and drug delivery. *Adv. Drug Deliv. Rev*. 60, 1252–1265.

83. Tamanaha, C. R., Mulvaney, S. P., Rife, J. C., and Whitman, L. J (2008). Magnetic labeling, detection, and system integration. *Biosens. Bioelectron*. 24, 1–13.

84. Tang, X., Bansaruntip, S., Nakayama, N., Yenilmez, E., Chang, Y. I., and Wang, Q. (2006). Carbon nanotube DNA sensor and sensing mechanism. *Nano Lett*. 6, 1632–1636.

85. Valentini, F., Carbone, M., and Palleschi, G. (2013). Carbon nanostructured materials for applications in nano-medicine, cultural heritage, and electrochemical biosensors. *Anal. Bioanal. Chem*. 405, 451–465.

86. Vamvakaki, V., and Chaniotakis, N. A (2007). Carbon nanostructures as transducers in biosensors. *Sens. Actuators B Chem*. 126, 193–197.

87. Veitch, N. C (2004). Horseradish peroxidase: a modern view of a classic enzyme. *Phytochemistry* 65, 249–259.

88. Wang, J. (2005). Carbon-nanotube based electrochemical biosensors: a review. *Electroanalysis* 17, 7–14.

89. Wang, S. X., and Li, G. (2008). Advances in giant magnetoresistance biosensors with magnetic nanoparticle tags: review and outlook.*Magn. IEEE Trans*. 44, 1687–1702.

90. Wei, H., Ratchford, D., Li, X., Xu, H., and Shih, C.-K. (2009). Propagating surface plasmon induced photon emission from quantum dots.*Nano Lett*. 9, 4168–4171.

91. Weller, H. (1993). Colloidal semiconductor q-particles: chemistry in the transition region between solid state and molecules. *Angew. Chem. Int. Ed. Engl.* 32, 41–53.

92. Wijaya, E., Lenaerts, C., Maricot, S., Hastanin, J., Habraken, S., Vilcot, J.-P., et al. (2011). Surface plasmon resonance-based biosensors: from the development of different SPR structures to novel surface functionalization strategies. *Curr. Opin. Solid State Mater. Sci.* 15, 208–224.

93. Wilchek, M., and Bayer, E. (1988). The avidin-biotin complex in bioanalytical applications. *Anal. Biochem*. 171, 1–32.

94. Willner, B., Katz, E., and Willner, I. (2006). Electrical contacting of redox proteins by nanotechnological means. *Curr. Opin. Biotechnol*. 17, 589–596.

95. Xiao, Y., Patolsky, F., Katz, E., Hainfeld, J. F., and Willner, I. (2003). "Plugging into Enzymes": nanowiring of redox enzymes by a gold nanoparticle. *Science* 299, 1877–1881.

96. Xu, Q., Mao, C., Liu, N.-N., Zhu, J.-J., and Sheng, J. (2006). Direct electrochemistry of horseradish peroxidase based on biocompatible carboxymethyl chitosan–gold nanoparticle nanocomposite. *Biosens. Bioelectron*. 22, 768–773.

97. Xu, W., Xue, X., Li, T., Zeng, H., and Liu, X. (2009). Ultrasensitive and selective colorimetric DNA detection by nicking endonuclease assisted nanoparticle amplification. *Angew. Chem. Int. Ed*. 48, 6849–6852.

98. Yang, W., Ratinac, K. R., Ringer, S. P., Thordarson, P., Gooding, J. J., and Braet, F. (2010). Carbon nanomaterials in biosensors: should you use nanotubes or graphene? *Angew. Chem. Int. Ed*. 49, 2114–2138.

99. Zeng, S., Yu, X., Law, W.-C., Zhang, Y., Hu, R., Dinh, X.-Q., et al. (2013). Size dependence of Au NP-enhanced surface plasmon resonance based on differential phase measurement. *Sens. Actuators B Chem*. 176, 1128–1133.

100. Zhang, C.-Y., Yeh, H.-C., Kuroki, M. T., and Wang, T.-H. (2005). Single-quantum-dot-based DNA nanosensor. *Nat. Mater*. 4, 826–831.

101. Zhou, M., Zhai, Y., and Dong, S. (2009). Electrochemical sensing and biosensing platform based on chemically reduced graphene oxide.*Anal. Chem*. 81, 5603–5613.

Chapter 8

Templated Synthesis of Magnetic Nanoparticles through the Self-Assembly of Polymers and Surfactants

Vo Thu An Nguyen [1,2,3], Mario Gauthier [3] and Olivier Sandre [1,2,*]

[1]University of Bordeaux, Laboratoire de Chimie des Polymères Organiques (LCPO), UMR 5629, F-33600 Pessac, France

[2]National Center for Scientific Research (CNRS), LCPO, UMR 5629, F-33600 Pessac, France

[3]Department of Chemistry, University of Waterloo, Waterloo, ON N2L 3G1, Canada

ABSTRACT

The synthesis of superparamagnetic nanoparticles (NPs) for various technological applications continues to be an interesting research topic. The successful application of superparamagnetic NPs to each specific area typically depends on the achievement of high magnetization for the nanocrystals obtained, which is determined by their average size and size distribution. The size dispersity of magnetic NPs (MNPs) is markedly improved when, during the synthesis, the nucleation and growth steps of the reaction are well-separated. Tuning the nucleation process with the assistance of a hosting medium that encapsulates the precursors (such as self-assembled micelles), dispersing them in discrete compartments, improves control over particle formation. These

inorganic-organic hybrids inherit properties from both the organic and the inorganic materials, while the organic component can also bring a specific functionality to the particles or prevent their aggregation in water. The general concept of interest in this review is that the shape and size of the synthesized MNPs can be controlled to some extent by the geometry and the size of the organic templates used, which thus can be considered as molds at the nanometer scale, for both porous continuous matrices and suspensions.

KEYWORDS

superparamagnetic nanoparticles (NPs); templated synthesis; size and shape control; *in situ* synthesis

1. INTRODUCTION

The synthesis of superparamagnetic iron oxide nanoparticles (SPIONs) and other transition metal (e.g., cobalt) ferrite nanoparticles (NPs) targeting various technological applications such as magnetic storage media, magnetic actuators or biosensors is a topic of ongoing interest for many researchers. In the biological and medical fields, magnetic NPs (MNPs) with a large specific surface area provide innovative potential in targeted drug delivery, in magnetic resonance imaging (MRI) as both positive and negative contrast agents, and in the magnetic hyperthermia treatment of tumors. The successful application of SPIONs in each specific area depends on the attainment of high magnetization values for the nanocrystals obtained, which is necessarily determined by their average size, size distribution, and crystallinity (absence of defects in the magnetic order). Additionally, to facilitate their biomedical applications, the nontoxicity, biocompatibility and stability of the coated magnetic particles in biological media have also been under the topics of numerous studies.

MNPs can be synthesized by numerous chemical techniques such as the alkaline coprecipitation of aqueous salts, hydrothermal reactions, microemulsion and sol-gel syntheses, sonochemical reactions, the hydrolysis and thermolysis of precursors, flow injection and electrospray syntheses, *etc.* Irrespectively of the method used, the formation of superparamagnetic NPs always involves nucleation and growth steps. Their separation is extremely difficult in the aqueous coprecipitation route, involving complicated pathways through oxo-hydroxide phases and eventually to magnetic iron oxide nanocrystals [1]. The size dispersity of the NPs prepared is markedly improved when these two steps are well-separated, which is

unfortunately hard to achieve without relying upon kinetic parameters (for example, the temperature or mixing conditions) [2]. These barriers have been investigated and might be overcome by tuning the nucleation process with the assistance of a hosting medium, that encapsulates the precursors and disperses them in discrete compartments (such as small molecule or polymeric micelles), thereby providing control over particle formation. These inorganic-organic hybrids are particularly interesting since they inherit the properties of both the polymer and the inorganic material. In addition, coating by polymer chains can efficiently hinder aggregation of the particles [3]. The focus of this review is on methods providing some degree of control over the shape and the size of the synthesized NPs through variations in the geometry and size of the organic templates. Two classes of molds at the nanometer scale can be considered, consisting of either pores contained within a continuous matrix, or of individual objects suspended in a reaction medium. The organic component can also bring other advantages in addition to colloidal stability in aqueous media, such as the presence of specific functionalities.

2. *IN SITU* SYNTHESIS IN NON-POLYMERIC TEMPLATES

Complexing groups such as the carboxylate, phosphate, sulfonate, and sulfate functionalities are known to bind strongly to the surface of magnetite. Furthermore, surfactants containing such head-groups can be tailored for stabilization in oil or hydrocarbon carrier fluids, as well as with charged ligands (citrate, oxalate, *etc.*) for aqueous media. The same type of moieties can used to complex the divalent or trivalent iron precursors selected for the reactions leading to the MNPs, for example by the coprecipitation of Fe^{2+} and Fe^{3+} salts in water or in a polyol medium, or the oxidation, reduction, or thermal decomposition of a Fe(II/III)-ligand complexes in a high boiling point solvent.

2.1. Carboxylates

2.1.1 Oleic Acid Coating in Combination with Reduction

The templating ability of organic compounds bearing carboxylic functional groups, for example oleic acid and L-glutamic acid, was investigated in the aqueous phase reduction/hydrolysis of iron (III) acetylacetonate (acac) using $NaBH_4$ [4]. The ratio of reducing agent to iron precursor required to produce the largest diameter bare NPs (8.08 ± 0.93 nm by transmission electron microscopy (TEM) and 8.95 nm by X-ray diffraction (XRD) was first determined

and then applied to the *in situ*coating method. Both the uncoated and the coated NPs displayed clear lattice fringes in high resolution transmission electron microscopy (HRTEM) images (Figure 1a) with spacings of 2.9 Å and 2.5 Å, consistent with a cubic spinel iron oxide structure and confirming the highly crystalline character of the NPs. It is noteworthy that some of the NPs coated with hydrophobic oleic acid had a facetted particle shape on the HRTEM images (insert on Figure 1b) rather than the spherical morphologies observed in most other cases, indicating that monolayer formation by the hydrophobic tail of oleic acid molecules brings a different templating effect as compared to several hydrophilic carboxylic compounds tested in the literature [5], most often leading to a mere reduction in the final size of the MNPs due to capping of the crystallites.

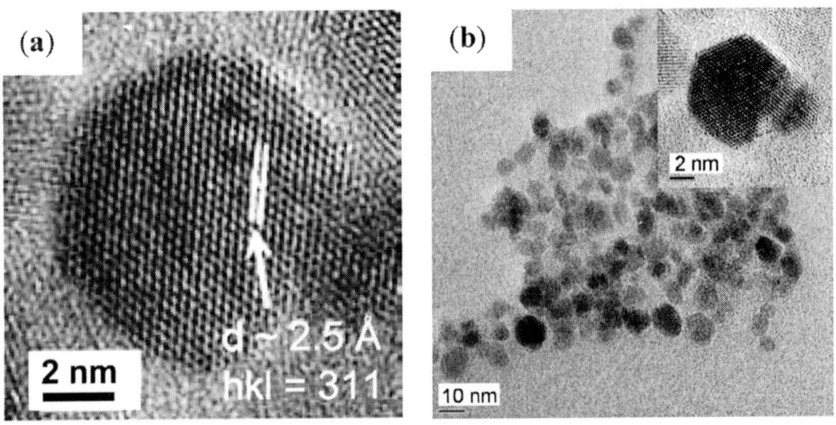

Figure 1. Transmission electron microscopy (TEM) images clearly showing: (**a**) lattice fringes; and (**b**) facetted shapes for nanoparticles (NPs) coated with oleic acid, using 10 equivalents of reducing agent ($NaBH_4$) at room temperature. Reprinted with permission from [4]. Copyright 2011 WILEY-VCH Verlag GmbH and Co. KGaA, Weinheim.

2.2. Sulfonates and Sulfates

2.2.1. Sodium Dodecyl Sulfate (SDS) Microemulsions

The use of non-polymeric templates in MNP synthesis was actually investigated earlier by the group of Liu [6], who performed $CoFe_2O_4$ NPs synthesis by a microemulsion process with SDS as a surfactant. The size of the $CoFe_2O_4$ NPs

was successfully controlled in a range from less than 4 nm to approximately 10 nm, by varying the concentration of metallic precursors and the methylamine base added [6]. Interestingly, the influences of NP size and temperature on the magnetic properties were systematically studied. For example, a decline in the remnant magnetization of 8.5 nm $CoFe_2O_4$ NPs with increasing temperature and its disappearance leading to the demagnetization of the NPs at 290 K were noted. Using field cooled (FC)/zero field cooled (ZFC) superconducting quantum interference device (SQUID) magnetometry, the magnetic susceptibility of NPs of different sizes was shown to exhibit the usual temperature dependence: initially increasing as the temperature was lowered, it reached a maximum at the point referred to as the blocking temperature (T_B), and eventually decreased. A correlation between higher T_B and increasing NPs diameters was convincingly pointed out. The superparamagnetism of the NPs above their T_B was confirmed by neutron diffraction measurements, while the increases in the T_B, coercivity (H_c) and saturation magnetization (M_s) with the size of the particles were in agreement with the Stoner-Wohlfarth theory on single domain particles. Such a study could not have been completed without the exceptionally good control achieved over the size and size dispersity by the microemulsion templating process.

2.2.2. Sodium Dodecylbenzenesulfonate (NaDBS) Microemulsions

The same technique to control the size of MNPs by varying the relative concentrations of iron salts, surfactant, and solvent in a water-in-oil microemulsion system was demonstrated by Lee *et al.* [7]. This effective approach using NaDBS microemulsions as nanoreactors and hydrazine as precipitating agent allowed the size of the NPs to be tuned in the range of 2–10 nm with narrow size distributions and high levels of crystallinity. The lattice fringes appearing in HRTEM imaging again hinted at the highly crystalline structure of the MNPs obtained. The matching XRD patterns obtained for the MNPS and for a magnetite standard further confirmed the structure of the Fe_3O_4 phase, while the peak widths calculated from the Debye-Scherrer equation were consistent with the observed TEM sizes. The magnetization *versus* temperature curves followed a pattern similar to the one previously described, in which the T_B and M_s for larger particles had higher values. These results were in good agreement with the Stoner-Wohlfarth theory in terms of the increase in activation energy for larger particles, although determining the origin of the smaller H_c values observed as compared to related studies would require further investigation. More importantly, the procedure was taken a step further by producing several mixed metal ferrites

in a size range of 4–8 nm, with uniform particle size distributions and an inverse spinel structure.

3. *IN SITU* SYNTHESIS WITH POLYMERS IN SOLUTION OR IN THE BULK STATE

The preparation of aqueous magnetic fluids was been achieved more recently using polymers as steric stabilizers in solution and polymer templates in the bulk state. These studies will be presented according to the different types of polymers used for that purpose.

3.1. Dextran (DEX) and Polysaccharide Derivatives

The term DEX refers to a class of branched polysaccharides composed from a majority of α-1,6- and a minority of α-1,3- or α-1,4-glycosidic linked glucose units, which are non-toxic, biocompatible and biodegradable hydrophilic polymers extensively used in biomedical applications and in NP preparation procedures. The general role of DEX in enhancing the biocompatibility of metal oxides, using green coprecipitation and gel template strategies, was previously discussed in a review by Patrinoiu *et al.* [8]. For the purposes of this discussion, we are primarily interested in the efficiency of DEX in templating and controlling the shape and size of MNPs.

3.1.1. Influence of DEX End-Group Reduction

In an attempt to optimize the nucleation of ultra-small superparamagnetic iron oxide (USPIO) particles in the presence of DEX, the effects of chemical functionalities and the molecular weight (MW) on the formation, stability, size, and magnetic properties of the hybrid particles obtained were investigated by Paul *et al.* [9]. Reduced DEX, generated by sodium borohydride treatment of native DEXs of different MW, has many hydroxyl groups existing in the native structure but only one additional hydroxyl group converted from the aldehyde (hemi-acetal) terminus. Despite this minor change in structure, reduced DEX was reported to successfully produce USPIOs with a volume-average diameter of less than 30 nm, while the size of native DEX-coated particles synthesized in the same conditions was larger than 700 nm in most cases and even reached 2.7 μm for 1 kDa DEX. Samples prepared with low MW reduced DEX (1 kDa and

5 kDa) possessed a magnetic susceptibility 5–25 times stronger than their native counterparts. However, the influences of the reduced terminal unit on the size and the magnetic susceptibility of the USPIOs were diminished for DEXs of higher MW.

The minor change at the terminal sugar unit of 10 kDa MW DEX resulting from the reduction reaction also led to more efficient utilization of the templating material, as more stable autoclaved USPIOs with much smaller diameters (23 nm *versus*587 nm) were formed, with a higher percentage of metal-bound DEX (0.96 g DEX/g Fe *versus* 0.59 g DEX/g Fe).

The size effect of reduced 10 kDa DEX was emphasized through similar experiments performed with 2.5–50 wt% addition of the simple sugar glucose *versus* the amount of DEX used. The samples derived from reduced DEX with added glucose had radii similar to those prepared without sugar added, thus convincingly demonstrating binding enhancement arising from the additional OH groups in the linear polyol termini as compared to the aldehyde functionality present in the monosaccharide. It is worth mentioning that the size of the USPIO particles increased with the MW of DEX (20 nm for 1 kDa*versus* 41 nm for 70 kDa DEX). However, stable colloids could only be created using reduced polysaccharides having a MW larger than 10 kDa as compared to 70 kDa for native DEX. This shows the dominant role of the MW of the chains in the growth of larger colloids from an initially unstable magnetic sol.

3.1.2. DEX Sulfate

Beyond the encapsulation enhancement brought by reduced DEX as compared to the native material, further modifications of this type of material may provide additional options leading to improved physicochemical properties. In addition to being considered a ligand for macrophage scavenger receptors, DEX sulfate was also proposed as an advanced coating material as compared to reduced DEX: a mixture of only 5 wt% DEX sulfate with reduced DEX yielded particles with a hydrodynamic diameter of over 30 nm and insignificantly larger core sizes (7.9 ± 2.0 nm *versus* 7.0 ± 1.9 nm for reduced DEX) [10]. Experimental parameters including the reaction time and the pH were varied to create DEX sulfate-coated NPs with good magnetic resonance properties. The delayed addition of Fe^{2+} was mentioned as a technique to avoid overgrowth and crosslinking of the particles caused by interactions between excess Fe^{3+} and the polysaccharide coating. While an increase in the polysaccharide/Fe^{3+} ratio decreased the core size and thus the relaxivity of the particles, failure of

particle formation was also reported at very low polysaccharide concentrations.

3.1.3. DEX-*b*-poly(methacrylic acid)(Fe_3O_4)-*b*-Poly(*N*-Isopropylacrylamide) with Thiol End Groups

The synthesis of quadruple-responsive nanocomposites was achieved in solutions of the copolymer DEX-*b*-poly(methacrylic acid)-*b*-poly(*N*-isopropylacrylamide) (DEX-poly(methacrylic acid)-PNiPAAm) where complexation was achieved by the poly(methacrylic acid) (PMAA) block, the PNiPAAm units provided thermo-sensitivity, the iron oxide NPs (IONPs) offered the magnetic response, and nucleated gold nanorods (AuNR) triggered near infrared radiation (NIR) response [11]. The diameter of the SPIONs before the AuNR encapsulation was 10 ± 3.5 nm (based on TEM analysis) with a narrow size distribution, while the aqueous hydrodynamic radius (R_h) measured by dynamic light scattering (DLS) was 66–69 nm, suggesting that a thick copolymer shell covered the SPIONs. The dependence of the lower critical solution temperature (LCST) on pH (29 °C at pH 4.5 *versus* 35 °C at pH 12) was explained by changes in the hydrophobic/hydrophilic balance caused by protonation of the PMMA blocks. Vibrating sample magnetometer (VSM) measurements at room temperature revealed superparamagnetic properties with neither remanence nor H_c. At room temperature, SPIONs 10 ± 3.5 nm in size had limited stability in a 300 mT magnetic field, displaying slow aggregation under these conditions. The magnetic interactions were strongly enhanced above the LCST, however, (e.g., at 50 °C), due to the phase transition of the hybrids that induced aggregation via hydrophobic as well as magnetic dipolar interactions. Owing to the presence of the PNiPAAm blocks the phase transition was reversible, providing water solubility to the recovered SPIONs when the temperature was brought back below the LCST, due to the hydrophilic DEX units.

3.1.4. Alginate (Alg) Beads

As a naturally occurring, biocompatible and inexpensive polysaccharide widely used in encapsulation processes, alginate is constituted of 1,4-linked α-L-guluronic acid (G) and 1,4-linked β-D-mannuronic acid (M) units (Figure 2), either segregated into homopolymer blocks or combined in copolymer structures. Commercially available alginates extracted from the cell wall of either brown algae or bacteria have different M/G compositions, MW,

sequences, and physicochemical properties which strongly affect their efficiency in encapsulation applications [12].

The loading capacity of alginates with different M/G ratios was investigated in the *in situ* alkaline oxidation of ferrous ions to produce maghemite MNPs [9]. More precisely, this synthesis relied upon the alkaline coprecipitation of Fe^{2+} and Fe^{3+}, but with the particularity of the *in situ* production of Fe^{3+} ions by the oxidation of Fe^{2+} in air. After a first oxidation cycle, a higher iron content (*ca.* 31%) was recorded in a sample for which the M/G ratio was 0.6, as compared to *ca.* 14% for the alginate with M/G = 1.6. This result highlights the superior binding ability of guluronic *versus* mannuronic acid units resulting from the "egg box" conformation of the guluronic acid segments that trap the Fe^{2+} ions more efficiently. Larger $M_s$values were recorded for the M/G = 0.6 samples, obviously due to their higher total iron content. As the total iron content in two samples reached the same maximum value after three oxidation cycles (approximately 47.09% *versus* 48.29%), these two samples displayed similar magnetization curves. The superparamagnetic character of these fully oxidized samples at room temperature was confirmed by their large values of M_s, zero hysteresis, and from the paramagnetic Mössbauer spectra obtained. Although no further data were provided about the size and the morphology of the maghemite NPs obtained, the hosting ability of alginate compounds was clearly demonstrated.

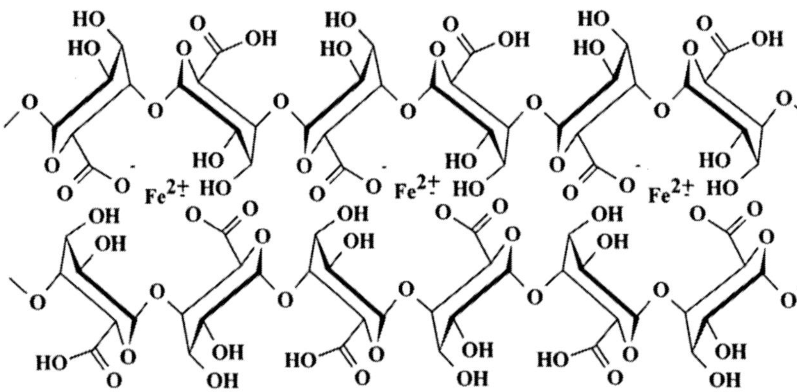

Figure 2. Representation of two 1,4-linked α-L-guluronic acid (G)-blocks forming an "egg box" structure with ferrous ions. Reprinted with permission from [9]. Copyright 2000 Elsevier Science B.V.

3.1.5. Precipitation in Aqueous Alginate Networks

Recently, questions related to the size and morphology of MNPs encapsulated in alginate networks were convincingly answered by the team of Srivastava *et al.* [12] who took the above approach one step further with the use of sodium alginate from *Laminaria hyperborea* (MW = 60,000–80,000 g·mol^{-1}, M/G = 32/68) to synthesize Fe_3O_4/Alg nanocomposites with a core-shell structure in a controlled low-temperature oxidation/precipitation process at pH 4, 10 and 14. The Fe loading, the size and, more interestingly, the magnetic properties of the core-shell composites were efficiently controlled through the final pH used in the precipitation process. It was indeed reported that higher Fe loading could be achieved at low pH (4–5) than at high pH (14). The Fe_3O_4 content of the alginate composites synthesized at pH 4–5 was confirmed by XRD analysis. Calculation of the crystallite size from the XRD patterns gave a mean value of *ca.* 14 nm, while the values determined using the Scherrer equation and in TEM micrographs were *ca.* 10 nm and 10.5 nm, respectively. As compared to bare Fe_3O_4 NPs, the alginate composites exhibited better dispersibility due to the alginate shell, essentially reducing the interactions between the MNPs. While neither H_c nor remnant magnetization were detected in the *M-H* curves for the magnetite alginate composites, confirming their superparamagnetic behavior, the saturated magnetization values exhibited by the hybrids were much lower (below 6 emu/g) than for pure Fe_3O_4 NPs prepared by the same process (above 31 emu/g). This was assigned to differences in morphology among these two sample series: the Fe_3O_4 synthesized at pH 4 was better dispersed than the sample produced at pH 10, which was agglomerated to some extent. The M_s values for the composites were larger for the NPs synthesized at a higher pH, which is obviously in agreement with the Fe loading efficiency reported above.

3.2. Synthetic Linear Homopolymers

To maximize the *in situ* templating effect as well as to enhance the stability of the ferrofluids obtained, numerous attempts have been made in terms of: (i) changing the chemical nature of the template material; and (ii) varying the method employed to produce the NPs. The first approach is more convenient experimentally for synthetic linear homopolymer templates than for natural polysaccharide derivatives, due to the fact that the former materials can possess controlled chemical structures, MWs, and MW distributions, which facilitates magnetic property improvement and particle surface tailoring in terms of enhanced colloidal stability and biocompatibility.

3.2.1. Thermal Decomposition of Poly(Vinyl Alcohol)-$Fe(OH)_3$

One of the first synthetic homopolymers investigated as template for the preparation of MNPs was poly(vinyl alcohol) (PVA), containing a large number of hydroxyl functional groups available to interact with Fe^{3+} ions and leading to water-soluble complexes. In their pioneering work, Yokoi and Kantoh [13] successfully combined different amounts of PVA to $Fe(OH)_3$ and obtained Fe_3O_4 precipitates by thermal decomposition. Although agglomerates of sintered magnetite particles were obtained instead of uniform products, the diameter of the Fe_3O_4 crystals in the composites evaluated from the line width of the (311) diffraction peak linearly increased with the heating temperature, indicating that the growth of Fe_3O_4crystals at higher temperatures, under the reducing effect of PVA, converted one third of the Fe^{3+} into Fe^{2+}, as required by the stoichiometry of the coprecipitation reaction.

3.2.2. $Fe^{2+/3+}$ Coprecipitation in PVA Membranes

In continuation of the earlier work using PVA solutions as *in situ* templates, Sairam *et al.* [14] performed the alkaline coprecipitation of various amounts of Fe^{2+} and Fe^{3+} previously impregnated in PVA membranes. The free-standing composite membranes prepared were expected to be useful as pervaporation separation systems to dehydrate organic solvents. The encapsulation of IONPs in PVA increased the mechanical strength of the membranes, but also enhanced their water affinity as shown by the improvement in solvent drying performance with increasing iron oxide content.

3.2.3. Conversion of Fe^{3+} in PVA into Hematite in Supercritical H_2O

Another study dealing with antiferromagnetic iron oxide (α-Fe_2O_3 or hematite)-PVA composites was conducted by Xu and Teja [15] using continuous hydrothermal processing (CHP) in supercritical H_2O, a process allegedly providing size and morphology control for metal oxide NPs. Although the influence of the experimental variables on the particle characteristics was not convincingly determined, a higher ferric nitrate concentration, longer residence time, or elevated temperature did result in increased average particle radii and crystallinity, as confirmed by narrowing of the peaks in XRD patterns. A larger

average particle size (27.4 nm *versus* 15.6 nm) and morphological transformations from spherical to rhombic (elongated) shapes were also reported as a consequence of a higher precursor concentration (0.06 M *versus* 0.03 M). More importantly, the particle size and size distribution (standard deviation) both decreased as the PVA concentration increased, as demonstrated in three experiments with a reaction time of 2.6 s (controlled by the residence time in the high pressure conducts). As the molar ratio of vinyl alcohol units to Fe^{3+} in the starting mixture was increased from 0:1 to 1:1 and 2:1, the average particle size of the samples decreased from 15.6 nm to 10.1 nm and 7.2 nm, respectively. This led the authors to conclude that a minimum amount of PVA was needed to coat all the particles, resulting in a polymeric layer with a thickness of approximately 10 nm that hindered further growth of the particles.

3.2.4. $Fe(acac)_3$ Decomposition/Diol Reduction in Octyl Ether (OE) with Polyvinylpyrrolidone (PVP)

It is widely accepted that the main limitation of the thermal decomposition of $Fe(CO)_5$ to prepare monodispersed Fe_3O_4NPs is the high toxicity of the carbonyl precursor. In an attempt to solve this issue, Fe^{3+} acac was used in an one-pot polyol reduction process in OE and in the presence of a water-soluble polymer, PVP, playing the role of polymeric stabilizer [16]. The PVP-coated, monodispersed Fe_3O_4 NPs obtained had diameters <5 nm and displayed well-defined superparamagnetic behavior at room temperature with zero H_c.

3.2.5. Controlled Precipitation in the Presence of Poly(4-vinylpyridine) (P4VP)

The role of polymers containing nitrogen-based functional groups in the synthesis of MNPs was furthered investigated in the *in situ* controlled precipitation of Fe^{3+} ions in Fe^{3+}-P4VP coordination complexes prepared beforehand in water/acetone mixtures and dried at 60 °C [17]. In this approach, the iron-P4VP solid matrix played the role of a growth-restricted medium, where ion diffusion was efficiently slowed down and the growing iron NPs were effectively "trapped" due to Fe–N interactions with the pyridine groups. More importantly, the hydrophilicity of protonated P4VP before precipitation, that facilitated N-Fe coordination, was described as an important factor controlling particle growth. At the onset of precipitation, the protonated pyridine groups became hydrophobic, inducing the collapse of the

homogeneous iron-polymer gel and the creation of a non-uniform microstructure consisting of hydrophilic and hydrophobic regions. The subsequent growth process was restricted by the amount of Fe ions available only in the hydrophilic regions, which led to small particles. The hybrid polymer nanocomposites obtained after washing and heating to 150 °C yielded rod-like NPs in TEM images. These nanorods, with lengths ranging from 15 nm to 75 nm and antiferromagnetic properties affected by the Cl/Fe atomic ratio, consisted of the antiferromagnetic chlorinated iron oxy-hydroxide "akaganeite" ($Fe_8O_8OH_8Cl_{1.35}$, also noted β-FeOOH). It was speculated that incomplete filling of the Cl^- sites affected the perfect compensation of antiferromagnetic sublattices, resulting in the appearance of a magnetic moment. Perfect sublattice alignment is indeed only achieved in a Cl^--saturated crystal lattice (*i.e.*, in a highly acidic P4VP environment). In that particular case, the templating effect of the polymer on the NPs can be partially ascribed to the chlorine counterions of this weak polyelectrolyte.

3.2.6. Poly(ethylene oxide) PEO in the "One-Pot" Synthesis of Pegylated Oxide NPs

Being known for its excellent solvating characteristics, good complexing ability for several transition metal cations, exceptional resistance to oxidation, reduction, and decomposition (by acids and bases at moderately high temperatures, hydrogen peroxide, and sodium borohydride), and the possibility of end-functionalization, poly(ethylene glycol) (PEG) or PEO has naturally become a good candidate to orient the synthesis of magnetic particles. Its stealth characteristics, that is, the ability to avoid non-specific interactions with proteins and uptake by the reticulo-endothelial system, provides to NPs pegylated on their surface a simple route for further medical applications [18].

3.2.7. PEO in Alkaline Hydrothermal Treatment

While conventional templating methods often result in the formation of polycrystalline nanowires or nanorods, monocrystalline 25 nm × 1.5 μm magnetite nanowires (Figure 3) were successfully produced in a PEG-assisted hydrothermal process [19]. There are numerous challenges in controlling the size, morphology, and crystallinity of Fe_3O_4 nanowires due to their complex inverse spinel structure. Full control over the iron oxide nanostructure phase through careful variations in the PEG and H_2O concentrations and the reaction

temperature was claimed (see Table 1 and Table 2). A maximum M_s value of ~91 emu/g, as obtained from vibrating magnetometer measurements, was achieved under the optimal conditions, confirming the crystalline perfection and extremely high purity of the sample (without γ-Fe_2O_3, having an M_s value in the bulk of 80 emu/g). The nanowire formation mechanism proposed involved PEG in a critical role as a stabilizer, mild reducing agent, and growth-orienting agent for the newborn Fe_3O_4 NPs into nanowires.

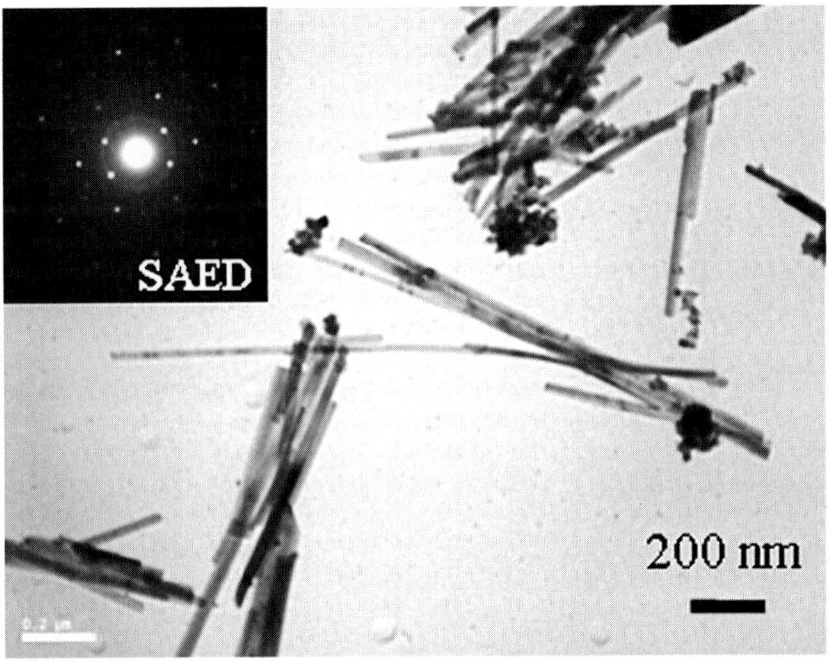

Figure 3. TEM image for magnetite nanowires synthesized with poly(ethylene glycol) (PEG)/H_2O (1:2 by volume) at 150 °C for 24 h in an autoclave. The selected area electron diffraction (SAED) pattern is shown in the inset. Reprinted with permission from [19]. Copyright 2008 Elsevier B.V.

3.2.8. Poly(acrylic acid) (PAA) Chains for Fe_3O_4 Growth Inhibition

Up to this point, the majority of synthetic polymeric stabilizers discussed did not contain functional groups able not only to bind efficiently to the surface of

Fe_3O_4 particles, but also to add strong enough repulsions (electrostatic and/or steric) between the MNPs and lead to an enhanced colloidal stability desirable for true "ferrofluid". In a search for different homopolymers with suitable functional groups, Lin *et al.* [20] investigated PAA MW~2000 g·mol^{-1} whose carboxylic acid groups can theoretically form strong ionic bonds with Fe_3O_4, and thus act as template for iron oxide nucleation. Furthermore, the presence of excess COOH groups at the surface of the coated particles should provide both electrostatic and electrosteric stabilization enhancing the stability of aqueous magnetite suspensions.

Table 1. Influence of the PEG:H_2O volume ratio on the identified phases and crystallite size collected from X-ray diffraction (XRD) data. All samples are prepared at 150 °C for 24 h. Reprinted with permission from [19]. Copyright 2008 Elsevier B.V.

PEG:H_2O (in volume)	Phases	Crystallite size (nm)
1:3	Fe_3O_4, α-Fe_2O_3	21.8
1:1	Fe_3O_4, α-Fe_2O_3, γ-Fe_2O_3	24.7
3:1	Fe_3O_4, α-Fe_2O_3, γ-Fe_2O_3	31.6
0:4	Fe_3O_4, α-Fe_2O_3, γ-Fe_2O_3	42.2
4:0	$NaFeO_2$	64.5

Table 2. Influence of reaction temperature and time on the phases of nanostructured iron oxides and crystallite sizes. All samples are prepared using PEG:H_2O = 1:2 in volume. Reprinted with permission from [19]. Copyright 2008 Elsevier B.V.

Conditions: temperature (°C)/time (h)	Phases	Crystallite size (nm)
100/24	Fe_3O_4 + $NaFeS_2(H_2O)_2$	10.6
125/24	Fe_3O_4 + α-Fe_2O_3 + $NaFeS_2(H_2O)_2$	13.8
150/24	Fe_3O_4	30.3
150/48	Fe_3O_4 + α-Fe_2O_3	24.1
150/72	Fe_3O_4 + α-Fe_2O_3 + γ-Fe_2O_3	28.4

The negative zeta potential obtained from laser light scattering velocimetry measurements demonstrated that the polymer molecules could bind to the surface of Fe_3O_4 NPs (particularly at alkaline pH), providing both electrostatic and steric stabilization against particle aggregation. This repulsion (especially at high concentrations of stabilizing polymer) sterically hindered the growth of

the particles, producing much smaller products with narrow size distribution, as compared to traditional methods.

Pseudo-aggregates of NPs, with a size of about 150–450 nm, were also observed several weeks after the ferrofluids were synthesized. The explanations suggested for this were hydrogen bonding between the COOH groups, or the entanglement of the PAA polymers when present at a high concentration in the synthesis. The fact that these large, broadly distributed aggregates were destroyed by sonication before light scattering measurements was confirmed by the appearance of 10–40 nm ferrofluid species with much narrower size dispersity. The M_s for the PAA-coated ferrofluid was about 35 emu/g, to compare with 50 emu/g for pure Fe_3O_4 NPs. The reduced value of M_s was explained by the presence in the particles of γ-Fe_2O_3 (which is ~20% less magnetized than pure Fe_3O_4) and by the significant reduction in M_s expected for particle sizes below 10 nm.

3.2.9. $FeCl_3$ Thermal Decomposition with PEG in Triethylene Glycol (Polyol)

In another study, PEG with carboxylic acid functionalities at both chain ends (PEG diacid, M_n ~ 600) was employed in a one-pot synthesis in a polar organic solvent, triethylene glycol (triEG), to yield surface-modified ultra-small SPIONs (USPIOs) with an average particle diameter (from TEM) of 1.7 nm [21]. These particles were nearly monodispersed and were highly water-dispersible, yielding a hydrodynamic diameter of about 5.4 nm, significantly larger than the diameter determined by TEM analysis (1.7 nm) due to the thickness of the hydrated PEG coating. The composition of USPIOs in the PEG-modified USPIOs, estimated from thermogravimetric analysis (TGA), was 30.65% and was used to correct the net magnetization values to yield 23.5 emu/g and 6.6 emu/g at T = 5 K and 300 K, respectively. Additional data obtained from the *M-H* curve indicated that the USPIOs were ferromagnetic at 5 K and superparamagnetic at 300 K. Fourier transform infrared spectroscopy (FTIR) spectra for PEG and the PEG-modified USPIOs exhibited a C–H stretch at 2890 cm^{-1}, and a C–O stretch at 1110 cm^{-1} in both cases, confirming the presence of PEG in the composites. More importantly, in the spectrum for the PEG-USPIO composites, the C=O stretch at 1745 cm^{-1} in PEG was not prominent and was shifted to 1645 cm^{-1} region, which suggests bonding between COO^- (rather than –O– of the polyether) and Fe ions on the surface of the particles, as shown on Figure 4. Relaxivity measurements and magnetic resonance images obtained highlighted the potential usefulness of these particles as both T_1 and T_2 MRI contrast agents, due to their ultra-small size, and the possibility of achieving target-specificity and a long blood half-life.

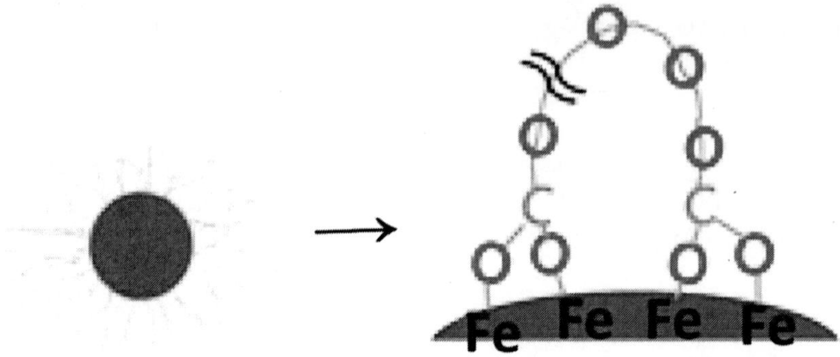

Figure 4. Suggested bond formation between COO^- and the Fe^{3+} ions at the ultra-small superparamagnetic iron oxide (USPIO) surface. Reprinted with permission from [21]. Copyright 2008 IOP Publishing Ltd.

3.2.10. Coprecipitation in the Presence of Trithiol-Terminated Poly(methacrylic Acid) (PMAA-PTTM)

Another homopolymer containing multiple carboxylic acid groups is PMAA. A dual-functionality PMAA with trithiol chain ends (PMAA-PTTM) was used in a single-step aqueous coprecipitation procedure leading to narrowly distributed Fe_3O_4 nanocrystals [22]. Interestingly, an increase in the molar ratio of COOH to the Fe (II) and Fe (III) ions (from 0.7 to 3.4) clearly led to a decrease in size (from 6.1 nm to 4.5 nm) as well as narrowing of the size distribution of the particles (standard deviation decreasing from 1.1 nm to 0.4 nm), thus indicating the dominant role of the COOH groups *versus* trithiol termini of the polymer in controlling the nucleating process. The less influential role of the trithiol chain ends in the impregnation step offers a facile post-synthesis functionalization path to the introduction of more reactive groups with further conjugation capacity, without causing large changes in size, size distribution or water-solubility of the NPs. The coated Fe_3O_4 nanocrystals had a size of 4.5 nm, clearly exhibited superparamagnetic characteristics on the *M-H* isotherms measured at 300 K, and possessed calculated M_s of 48 emu/g Fe_3O_4, which is surprisingly high as compared with conventional magnetite particles of the same size ($M_s <$ 15 emu/g). This low magnetization is usually attributed to several factors such as crystallinity defects including non-stoichiometry of the spinel structure, surface spin-canting, surface disorder caused by interactions with surface ligands, cation site distribution, *etc.* In a study by another team [23], PMAA was also used to produce coated agglomerates by an aqueous coprecipitation method, having magnetic core sizes around 8 nm by TEM analysis and hydrodynamic mean sizes ranging from 1.5 μm to 3 μm. The PMAA-

encapsulated magnetite particles did not exhibit superparamagnetic behavior however, but rather a small remnant magnetization and H_c, probably due to the aggregation of the magnetite NPs, although the individual 8 nm magnetite NPs displayed superparamagnetic properties. A low magnetization of 40 emu/g determined at 7 T, as compared to the magnetization values for bulk magnetic iron oxides (92 emu/g for Fe_3O_4, magnetite, and 80 emu/g for γ-Fe_2O_3, maghemite), was assigned to the particle surface spin disorder effects listed above [23].

3.2.11. Poly(Methyl Glutarimide) (PMGI) Templated Iron NPs by Spin-Coating

While investigating a novel method to generate high-density and uniformly distributed carbon nanotube (CNT) mats, Lu*et al.* [24] employed spin-coating to prepare a film of PMGI loaded with various iron precursors (Fe^{3+} chloride, nitrate or acac, and the organometallic complex ferrocene) for the templated formation of IONPs, subsequently exploited as catalysts for CNT growth. Following the trend of using polymeric templates with specific characteristics, PMGI, a polymer widely used in the bilayer lift-off process, was expected to form complexes with Fe^{3+} ions by multidentate coordination.

The particle size, distribution, and overall yield in the spin-coating process were strongly influenced by the complexation tendency of the iron precursors used with the PMGI templates. The particles derived from strongly coordinated compounds like ferrocene and iron acac had a low density of iron oxide and broad size distributions, since the Fe^{3+} ions in these materials were unable to complex with the PMGI chains due to steric hindrance caused by the bulky counterions. Iron nitrate and iron chloride produced the highest and second highest density of IONPs, respectively, with small mean diameters, since both monodentate anions were less bulky and more loosely bound to iron.

3.3. Synthetic Linear Copolymers

Block copolymers exhibiting microphase separation into ordered morphologies such as lamellar, cylindrical and spherical microdomains, provide a convenient self-assembled template for the synthesis of nanocomposites. These microdomains represent nanoreactors within which a variety of NP clusters can be synthesized.

3.3.1. Aqueous Coprecipitation with Double-Hydrophilic Block Copolymers (DHBCs)

To investigate the templating effects of DHBCs as well as tailored functional groups on the surfaces, Wan *et al.* [25] prepared stable aqueous ferrofluids using copolymers with poly(glycerol monoacrylate) (PGA) or poly(glycerol methacrylate) (PGMA) as one block, and either PAA, poly[(*N*,*N*-dimethylamino)ethyl methacrylate] (PDMAEMA) or poly(ethylene glycol) monomethyl (mPEG) ether as the other block.

When coated with PAA-*b*-PGA the surface of SPIONs carries negative charges, while the surface of particles obtained from PDMAEMA-*b*-PGMA, of which the PDMAEMA chains are protonated at neutral pH, possesses positive charges.

Multidentate coordination arising from the formation of five-membered rings between the 1,2-diol groups by the PGA or PGMA blocks and Fe^{2+} or Fe^{3+} cations at the surface of the magnetic particles was proposed to be more stable than the four-membered rings formed by the COOH groups in PAA. The disappearance of the COO^- adsorption band at 1592 cm^{-1} in the FTIR spectra of PAA-*b*-PGA-coated IONPs indicated that the carboxylic groups of PAA-*b*-PGA were not coordinated with the $Fe^{2+/3+}$ cations. Additionally, magnetic fluids obtained with PGA or mPEG-*b*-PGA were stable over a wider pH range as compared to those prepared from PAA or mPEG-*b*-PAA. PGA was considered as a better stabilizer than PAA for the dispersions, such that the PGA block of the copolymer was coordinated to the surface of the magnetite NPs, while the PAA block was extended into water.

The same conclusion was drawn in the cases of PDMAEMA-*b*-PGMA or mPEG-*b*-PGA block copolymers, of which the PGMA or PGA blocks were chemisorbed onto the surface of the IONPs, thus playing decisive role in orienting particle synthesis while the PDMAEMA or mPEG blocks, with a swollen conformation in water, contributed to post-synthesis stability.

3.3.2. Aqueous Coprecipitation with Triple-Hydrophilic Block Copolymers (THBCs)

Following this research trend, triblock copolymers constituted from the monomers listed above, but particularly mPEG-*b*-PMAA-*b*-PGMA, were used in the alkaline coprecipitation of Fe^{2+}/Fe^{3+} [26]. PGMA influenced the nucleation step by tightly binding to the magnetite cores, while PMMA created an intermediate layer with a double function of bearing the mPEG corona and

encapsulating the antibiotic and antitumor drug adriamycin (ADR), also called doxorubicin. The drug encapsulation was driven by both electrostatic interaction at pH 7.4 between anionic COO^- of PMMA and the protonated glycosidic amine of ADR (pK_a = 8.2) and hydrophobic interaction between the PMAA main chain and the hydrophobic anthracycline ring of ADR (Figure 5). The size of the Fe_3O_4 core was estimated to be about 7–8 nm from TEM micrographs, whereas a narrow size distribution and an average R_h of 23 nm were determined for the coated particles by DLS analysis. Superparamagnetic behavior at 300 K without magnetic hysteresis was observed for the MNPs obtained.

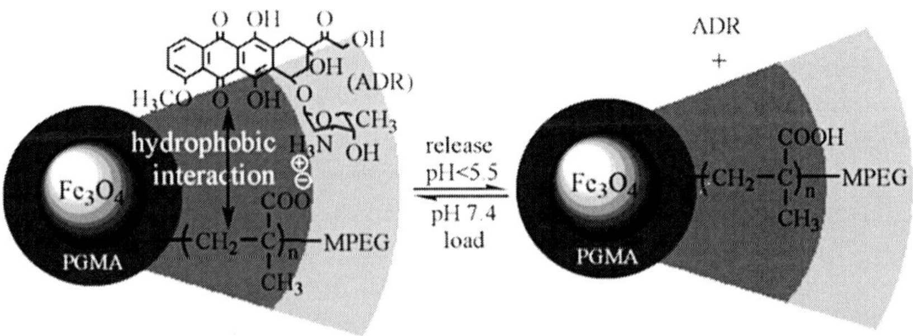

Figure 5. Schematic structure of mPEG-*b*-PMAA-*b*-PGMA-Fe_3O_4 NPs loaded with adriamycin at neutral pH, designed to release the anticancer drug in the acidic environment of a tumor. mPEG: poly(ethylene glycol) monomethyl; PMAA: poly(methacrylic acid); PGMA: poly(glycerol methacrylate); and ADR: antitumor drug adriamycin. Reprinted with permission from [26]. Copyright 2008 Royal Society of Chemistry.

3.3.3. Fe^{2+}/Fe^{3+} Coprecipitation with Silicon-Containing Random Copolymers

The effectiveness of an amphiphilic random block copolymer of 3-(trimethoxysilyl)propyl methacrylate and ethylene glycol methyl ether methacrylate, denoted as poly(TMSMA-*r*-PEGMA), as stabilizer for MNPs was assessed by Lee *et al.*[27]. In particular, the *in situ* synthesis of Fe_3O_4 was compared to the stepwise addition of the same copolymer to preformed iron oxide colloids. In both cases, covalent binding was achieved between the magnetite surface and the siloxane functionalities of the polymer hydrolyzing to silanol groups, while PEG more likely remained on the outside of the SPIONs.

The cross-linking reaction of the silanol groups by self-condensation, promoted by heating, resulted in more stable coating layers. The hydrodynamic radii of SPIONs prepared with and without copolymer in the NP synthesis step were 16.0 ± 2.2 nm and 12.3 ± 1.2 nm, respectively, with narrow size distributions and core sizes in the range of 4–8 nm in both cases, consistent with a thicker layer of coating by the *in situ* preparation as compared to sequential addition. However, the SPIONs synthesized *in situ* had an M_s value of 65 emu/g Fe (equivalent to 45.5 emu/g Fe_3O_4), significantly lower than the 80 emu/g Fe (56 emu/g Fe_3O_4) value obtained for the post-synthesis coated SPIONs. This effect can be interpreted as reflecting the influence on crystallinity or magnetic order of the NPs of surface complexation by the silanol groups of the copolymer during the synthesis. However, unlike ligands such as carboxylic acids and polyethers, that can enter the "inner coordination sphere" of Fe^{2+} and Fe^{3+} cations, silanols mainly interact with the Fe-OH hydroxyls at the surface of the iron oxide nuclei already formed in the sol-gel process, *i.e.*, the "outer sphere" of the metal centers. The risk of magnetic properties decreases trough surface ligand or spin-canting effects is therefore lower than for strong ligands introduced during the synthesis. Furthermore, the SPIONs synthesized were found to allow the detection of tumors in T_2-weighted MR images within 1 h, as a result of the accumulation of the nanomagnets at the tumor site. Protein uptake by the surface of these coated SPIONs was reported to be significantly lower than by Feridex™ I.V., a popular contrast agent, thereby confirming the stealth effect in blood plasma of the PEG component of the copolymer.

3.3.4. Coprecipitation in Water in the Presence of Poly(ethylene oxide)-*Block*-Poly(Methacrylic Acid) (PEO-*b*-PMAA) DHBCs

In an attempt to confirm the roles of PEO-*b*-PMAA DHBC in directing the nucleation, controlling the growth, and sterically stabilizing the dispersions, a comparison was made among three cases: precipitation without polymer present, in the presence of "non-interacting" PEO, and in the presence of PEO-*b*-PMAA [28]. TEM images (Figure 6c) revealed that the particles coated with PEO-*b*-PMAA were much smaller (5 ± 4 nm) and more uniformly dispersed than those created using either PEO homopolymer or without polymer, highlighting the decisive role of carboxylic acid groups in the PMAA block in promoting nucleation and inhibiting growth of the magnetic iron oxide particles. Although crystallinity was confirmed by powder XRD (PXRD) analysis, the magnetite and maghemite phases could not be distinguished by that technique because these two phases have the same spinel structure and their XRD peaks positions are quite close to each other. The PEO-*b*-PMAA-coated oxide exhibited substantial

magnetism, as seen on Figure 6d, with a magnetization of 50 emu/g at a field strength H = 275 kA/m or B = 0.34 T), from which we can infer a saturation level around 53 emu/g (so the value of 60 emu/g given by the authors appears a little over-estimated). The hydrophilic layer covering the particles facilitated their re-dispersion in hydrophilic mixtures of hydroxyethyl methacrylate and methacrylic acid monomers, before evaporation of the water. The waxy solid obtained was re-dispersed in an oil phase (decane) via sonication with poly(ethylene-*co*-butylene)-*block*-poly(ethylene oxide) (PBO-*b*-PEO) as emulsifier. Inverse emulsion polymerization of this sol yielded magnetic poly(hydroxyethyl methacrylate-*co*-methacrylic acid) P(HEMA-*co*-MMA) latexes with sizes broadly distributed from 35 nm to 250 nm, with the same superparamagnetic properties as the initial magnetic core-shell NPs, but also a lower overall magnetization (10 emu/g latex) corresponding to 18 wt% of iron oxide in the polymer matrix.

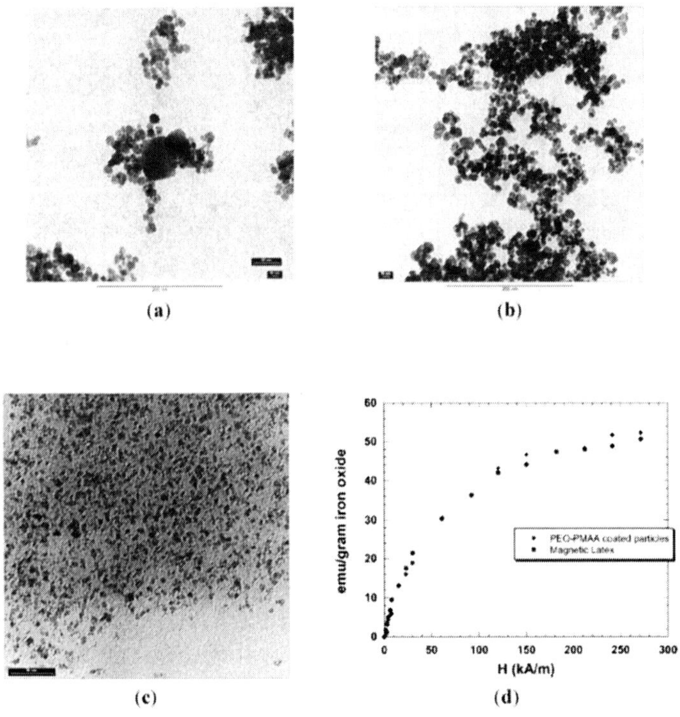

Figure 6. Transmission electron micrographs for magnetic iron oxide precipitated in: (**a**) water alone; (**b**) the presence of poly(ethylene oxide) (PEO) homopolymer; and (**c**) the presence of PEO-*b*-poly(methacrylic acid) (PMAA) block copolymer. Note thatFigure 6c is at somewhat higher magnification than the others; (**d**) magnetization curves for the PEO-*b*-PMAA coated NPs and for P(HEMA-*co*-MMA) magnetic latexes obtained by inversion emulsion polymerization. Reprinted with permission from [28]. Copyright 2001 Elsevier.

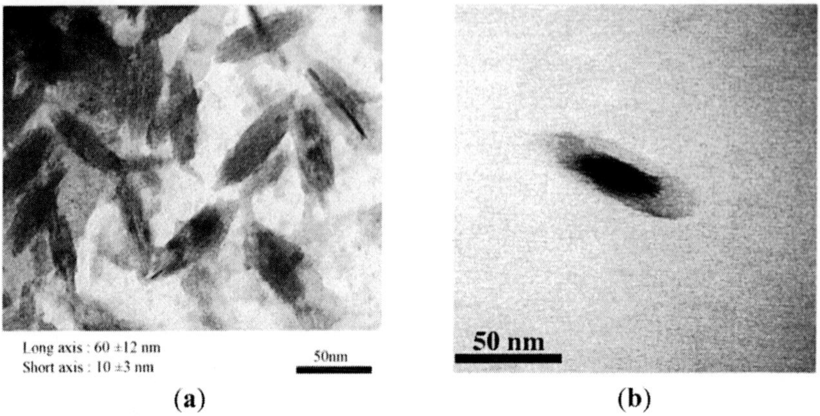

Figure 7. Energy filtering transmission electron microscopy (EFTEM) zero-loss images of: (**a**) β-FeOOH NPs; and (**b**) poly(ethylene glycol)-*block*-poly(aspartic acid) (PEG-*b*-PAsp)-coated β-FeOOH NPs. Reprinted with permission from [29]. Copyright 2007 Elsevier B.V.

3.3.6. Coprecipitation in Water with Poly(Ethylene Oxide)-*Block*-Poly(Acrylic Acid) (PEO-*b*-PAA)

In a study by Hatton *et al.* [30], the use of PEO-*b*-PAA in an aqueous coprecipitation route followed a size sorting step using a high-gradient magnetic separator (HGMS) yielded MNPs clusters with sizes larger than 100 nm, containing several particles with average core sizes of 9.5 nm. Both the bleed-off and captured clusters in the magnetic separator were superparamagnetic (M_s = 62.9 emu/g for the latter sample), thus indicating a single domain structure for the particles. The iron-to-polymer loading ratio (expressed as the molar ratio of the $Fe^{3+/2+}$ cations to the polymer carboxylates) was varied, thus changing the polymer coverage of the MNPs. These results suggested that the formation of large clusters was induced by the limited amount of polymer added. Indeed, when a large excess of copolymer was used, the MNPs clusters obtained became small enough to pass though the magnetic separator without being arrested. This HGMS method is thus very interesting to separate constituents such as free DHBC and individually dispersed MNPs from clusters with a relatively well-controlled spherical morphology and a finite size, as shown in Figure 8.

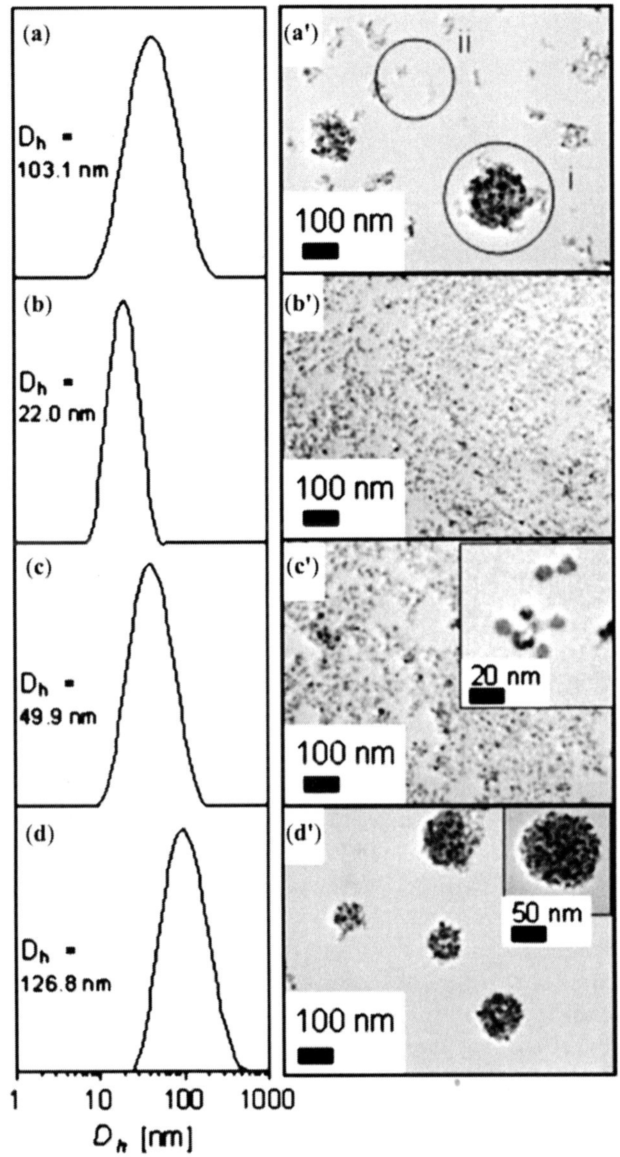

Figure 8. Size distribution: (**a**–**d**) by dynamic light scattering (DLS); and (**a'**–**d'**) from TEM images. (**a**,**a'**) the initial suspension after coprecipitation in presence of poly(ethylene oxide)-*block*-poly(acrylic acid) (PEO-*b*-PAA); (**b**,**b'**) the non-captured particles; (**c**,**c'**) the bleed-off samples; and (**d**,**d'**) the captured clusters. Reprinted with permission from [30]. Copyright 2009 Elsevier B.V.

3.3.7. Deuterated Poly(Norbornene) (PNOR) Block Copolymers

PNOR is a rubbery material with interesting properties, such as the ability to absorb up to 10 times its own weight of hydrocarbons while still retaining its high tear strength and vibration damping ability. Amphiphilic block copolymers of this material, suitable as templates for loading and co-precipitating iron salts, were obtained by the sequential ring-opening metathesis polymerization (ROMP) of norbornene and deuterated norbornene dicarboxylic acid (to provide better contrast between the two blocks in small-angle neutron scattering (SANS) experiments). The sodium salt form of poly(norbornene)-*block*-poly(norbornene dicarboxylic acid) (PNOR-*b*-PNORCOOH) copolymer was dissolved in a THF solution, to achieve ion exchange between Fe^{3+} and the Na^{+} ions of the carboxylate block. Thin films prepared from this solution were then soaked in NaOH solutions to produce γ-Fe_2O_3 particles [31]. Variations in the constituent block volume fraction ratio ($\Phi_{PNOR/PNORCOOH}$), while ensuring a constant 1:1 ratio of COOH/Fe^{3+}, led to changes in the morphology and characteristic distances for the microphase-separated block copolymer that were correlated with the size and the mean inter-particle distance between the MNPs (as assessed by both SANS and the T_B measured by magnetometry). Starting from 10.4 nm diameter disordered spherical particles at $\Phi_{PNOR/PNORCOOH}$ = 0.64/0.36 (Figure 9a), the samples changed to interconnected particles with 16 nm diameter at $\Phi_{PNOR/PNORCOOH}$ = 0.50/0.50 and 0.40/0.60 (Figure 9b), thus suggesting the possibility of controlling the nucleating process through the separation of the iron-doped hydrophilic blocks by the hydrophobic walls between the domains. The *d*-spacing between the metal oxide NPs measured by TEM (e.g., 62 nm for 0.64/0.36) was consistent with the correlation distance obtained from SANS analysis (e.g., 53 nm for 0.64/0.36), which proved the success of using microphase-separated domains as template for the synthesis of IONPs. The ratio $d/(2R)$, where d is the average distance between the particles given by SANS and $2R$ is the individual particle diameter from TEM, was found to be inversely related to the T_B of interconnected NPs. The sample at $\Phi_{PNOR/PNORCOOH}$ = 0.40/0.60, with the largest hydrophilic block, lead to the lowest $d/(2R)$ ratio, and thus the highest T_B (115 K) and the highest M_s (76 emu/g at 300 K), in agreement with the largest magnetic moments and magnetic dipolar interactions between them.

The same *in situ* synthesis strategy was applied to a DHBC template of poly(norbornene methanol)-*block*-poly(norbornene dicarboxylic acid) (PNORMEOH-*b*-PNORCOOH) [32], resulting in a lamellar geometry with IONPs confined within the lamellar domains. Electron micrographs (with iodine staining) revealed that 6 nm average diameter γ-Fe_2O_3 particles decorated the lamellar structure of the copolymer in thin film samples, appearing in the

images as thin lines of NPs, the spacing between the lines being 80 ± 10 nm (Figure 10).

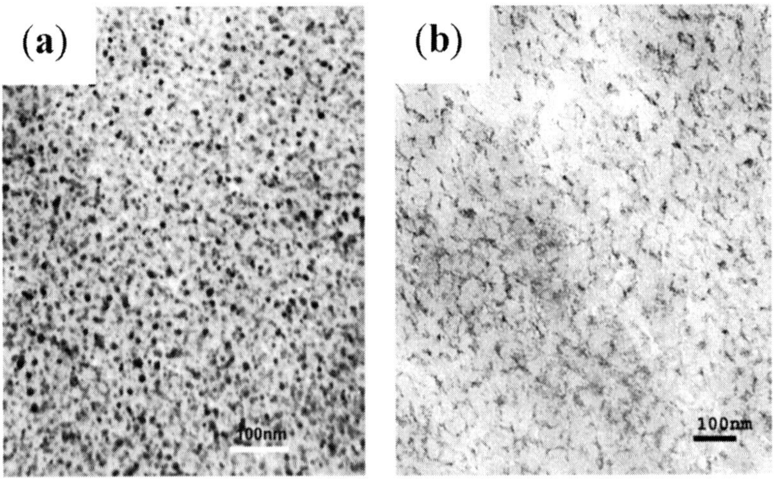

Figure 9. Morphology of poly(norbornene-*block*-deuterated norbornene dicarboxylic acid) loaded with iron oxide NPs (IONPs) at: (**a**) $\Phi_{PNOR/PNORCOOH}$ = 0.64/0.36 (disordered spheres); and (**b**) $\Phi_{PNOR/PNORCOOH}$ = 0.40/0.60 (interconnected spheres). PNOR: poly(norbornene); and PNORCOOH: poly(norbornene dicarboxylic acid). Reprinted with permission from [31]. Copyright 2005 Elsevier Ltd.

3.3.8. Poly(Styrene Sulfonate-*Alt*-maleic Acid) (PSS-*Alt*-MA) Shell with Subsequent Cross-Linking

The nucleating effects for *in situ* coprecipitation of amphiphilic PSS-*alt*-MA and of hydrophilic PAA were nicely demonstrated by Yoon *et al.* [33]. The coprecipitation was followed by cross-linking of the polymers with 1,6-hexanediamine, to prevent subsequent desorption of the polymer shell from the surface of the as synthesized IONPs and to reach a good colloidal state through electro-steric stabilization. The zeta potential remained highly negative after cross-linking (Table 3), even at high salt concentrations (e.g., 8%) added to reduce the thickness of the electric double layer thus weakening the electrostatic repulsions. The sub-100-nm IONPs clusters obtained were therefore stabilized by a balance of van der Waals and electrostatic interactions. The primary, highly crystalline IONPs exhibited

superparamagnetism without hysteresis and an excellent M_s *ca.* 90 emu/g_{Fe}, approaching the theoretical magnetization value for magnetite (92–100 emu/g_{Fe3O4}, equivalent to 127–138 emu/g_{Fe}), suggesting that the stabilizer did not lower the magnetization. These results suggest good stabilizing effects of PAA and PSS-*alt*-MA, while not demonstrating a true template effect since the MNPs were still polydisperse.

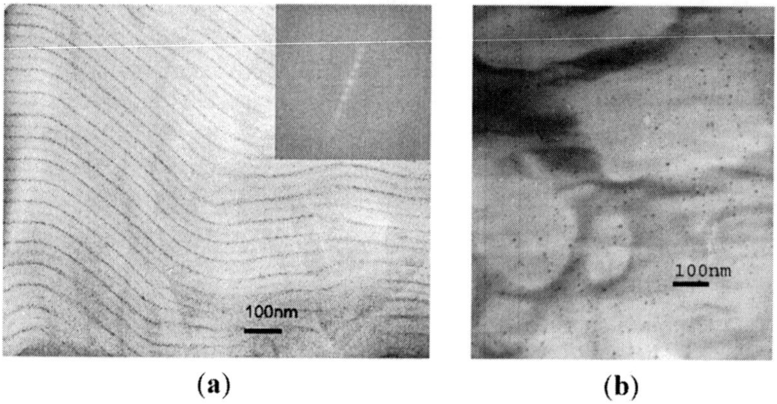

Figure 10. Electron micrographs for poly(norbornene methanol)/poly(norbornene dicarboxylic acid) (PNORMEOH/PNORCOOH) diblock copolymer: (**a**) stained with I_2vapor; and (**b**) doped with iron oxide by submerging a thin film in $FeCl_3$ solution. Reprinted with permission from [32]. Copyright 2006 Elsevier Ltd.

Table 3. Zeta potential and average hydrodynamic diameters of poly(acrylic acid) (PAA) or poly(styrene sulfonate-*alt*-maleic acid) (PSS-*alt*-MA)-coated IONPs with different cross-linking densities for an iron oxide concentration of 0.14 wt% at pH 8. Adapted with permission from [33]. Copyright 2011 American Chemical Society.

Cross-linking (%)	Hydrodynamic diameter (DLS) (nm)	Zeta potential (mV)
PSS-*alt*-MA before cross-linking	44 ± 9	−48.6 ± 4.7
PAA before cross-linking	83 ± 2	−39.7 ± 1.5
PAA, 12.5% cross-linking	69 ± 17	−48.6 ± 1.1
PAA, 50% cross-linking	126 ± 9	−44.3 ± 2.2
PAA, 12.5% cross-linking	77 ± 16	−47.1 ± 2.2
PAA, 100% cross-linking	91 ± 2	−38.3 ± 3.0

3.3.9. Brush Linear Poly(oligo(ethylene Glycol) Methacrylate-*co*-methacrylic Acid) P(OEGMA-*co*-MAA) as Nucleating and Stabilizing Ligand

Instead than relying upon the same copolymer block for both nucleation control and stabilization of the MNPs, Lutz *et al.*[34] developed back in 2006 the idea of using a double hydrophilic brush-linear block copolymer in which each block played a different, specific task. The coordinating ability of methacrylic acid units and the stabilizing capability of oligo(ethylene glycol) segments were examined in a coprecipitation process of iron ions, using hydrophilic P(OEGMA-*co*-MAA). The hydrodynamic diameter of the MNPs obtained was readily tuned in the range of 10–24 nm (Z-average values, polydispersity indices around 0.2) by variation of the initial polymer amount used as compared to the iron salts (from 1:1 to 0.42:1 wt:wt), thus also tuning their zeta potential from −14 mV to −7 mV. The 10 nm hydrodynamic diameter NPs were revealed as nearly mono-disperse 7 nm particles in the TEM micrographs. As evidenced by TEM the NPs had a quasi-spherical topology, in contrast to the "rock-like" shapes often obtained in coprecipitation methods. The stabilizing chain density was estimated as being relatively constantly in the range of 0.2–0.3 chains/nm^2 in all cases, which could explain the correlation found between the size of the NPs and the polymer amount used in the reaction. Moreover, the colloidal stability of the ferrofluids in physiological media, enhanced by POEGMA block, allowed their *in vivo* use as MR imaging contrast agents.

3.3.10. Coprecipitation with DHBCs of Poly(oligoethylene Glycol Acrylate) (POEGA) and Different Binding Blocks

The groups of Boyer and Davis described the synthesis by the xanthate reversible addition-fragmentation chain-transfer (RAFT/MADIX) technique of DHBCs composed of the same repelling block POEGA and anchoring blocks with either phosphonic acid ethyl acrylate (PAEA), carboxylic acid (PAA), or glycerol acrylate anchoring moieties in different copolymers [35]. These DHBCs were used for the *in situ* synthesis of IONPs by alkaline coprecipitation. The expected binding strength PAEA > PAA > PGA correlated nicely with the iron oxide loading, that also increased linearly with the feed-weight ratio of the iron salts (Figure 11b). The size of the MNPs measured both by TEM and XRD followed the reverse order (PAEA < PAA < PGA) for a given polymer-iron feed ratio, *i.e.*, the stronger the affinity for iron of the ligand, the smaller the magnetic core size. The size also decreased for increasing polymer contents (Table 4). All

these observations confirmed a high degree of control over the nucleation and growth of the IONPs achieved with the copolymer template. The final products consisted in hybrid magnetic multi-core clusters (as seen on the TEM images), exhibiting good colloidal stability in biological buffers and high T_2 relaxivity, which had the authors envision their use for *in vivo* MRI.

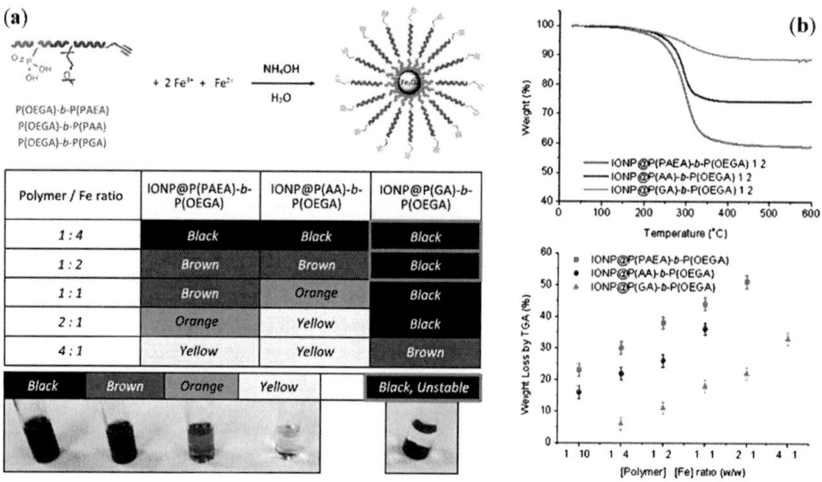

Figure 11. (**a**) *In situ* coprecipitation of iron oxide NPs (IONPs) at different polymer-to-Fe ratios; and (**b**) polymer content measured by thermogravimetric analysis (TGA) for IONPs coated with different anchoring groups at a 1:2 [Polymer]:[Fe] weight concentration ratio. Reprinted with permission from [35]. Copyright 2014 Royal Society of Chemistry.

3.3.11. Spherical Micelles Loaded with IONPs

In the same context, amphiphilic block copolymers were also used as templates by introducing bidentate moieties, such as in poly[2-(acetoacetoxy) ethyl] methacrylate (PAEMA), to promote the nucleation of iron oxide. Papaphilippou *et al.* [36] investigated the use of this hydrophobic polymer in a copolymer with PEGMA as a hydrophilic and biocompatible second block, but also having a LCST near 60 °C. Highly stable hybrid micelles, visualized by atomic force microscopy (AFM) as spherical micelles loaded with IONPs (Figure 12), were obtained at various precursor salt concentrations. The maximum amount of iron that could be encapsulated without precipitation was unfortunately not determined. A plateau magnetization M_s of 300 A/m was

reached for the PEGMA$_{70}$-*b*-AEMA$_{16}$-coated MNPs at a 3:1 [Fe^{3+}]/[AEMA] molar ratio, but the specific magnetization cannot be calculated for this system since the corresponding iron concentration used was not specified. *In vitro* biocompatibility and macrophage uptake tests were also conducted and were encouraging in terms of toxicity and the minimization of recognition and phagocytosis.

Table 4. Summary of weight loss TGA results and particle size (d_{TEM} and d_{XRD}) of polymer-coated IONPs at different polymer-to-iron ratios [35]. PAEA: phosphonic acid ethyl acrylate. Copyright 2014 Royal Society of Chemistry.

Hybrid magnetic core-shells	[Polymer]:[Fe] weight ratio	Wt loss TGA	d_{TEM} (nm)	d_{XRD} (nm)	Grafting density (nm^{-2})
	1:10	23%	10.45	9.83	0.083 ± 0.005
	1:4	30%	9.80	9.10	0.124 ±0.003
IONP@P(PAEA)-*b*-P(OEGA)	1:2	38%	9.40	8.52	0.168 ± 0.008
	1:1	44%	8.64	7.32	0.192 ± 0.01
	2:1	51%	-	-	-
	1:10	16%	10.25	10.09	0.065 ± 0.001
IONP@P(AA)-*b*-P(OEGA)	1:4	22%	8.91	8.75	0.084 ± 0.003
	1:2	26%	7.83	7.44	0.091 ± 0.003
	1:1	36%	-	-	-
	1:4	6%	12.39	11.47	0.023 ± 0.001
	1:2	11%	10.24	10.62	0.039 ± 0.001
IONP@P(GA)-*b*-P(OEGA)	1:1	18%	9.14	9.78	0.063 ± 0.002
	2:1	22%	8.65	8.41	0.073 ± 0.001
	4:1	33%	-	-	-

As an intermediate case between (fully hydrophilic) DHBCs and amphiphilic block copolymers, graft polymers of composition PAA-*g*-PEO/PPO (PPO: poly(propylene oxide)) were studied by the group of Hatton [37]. In analogy with the commercial Pluronics® systems (PEO-*b*-PPO-*b*-PEO), the idea was to vary the hydration level of the micelles by tuning their PEO/PPO content while using the PAA backbone to complex with Fe^{2+}/Fe^{3+} ions. After alkaline coprecipitation, the water-based magnetic fluids obtained consisted of *ca.* 7.3 nm (TEM) magnetite NPs decorating the amphiphilic graft copolymer micelles. The magnetic core radii obtained by TEM and neutron scattering analyses exhibited broad size distributions and irregular particle shapes, however. Therefore the self-assembly of such copolymers with iron oxide precursors does not appear to provide a sufficient template effect for the synthesis of well-calibrated MNPs, and better controlled nanostructures need to be explored as reactors.

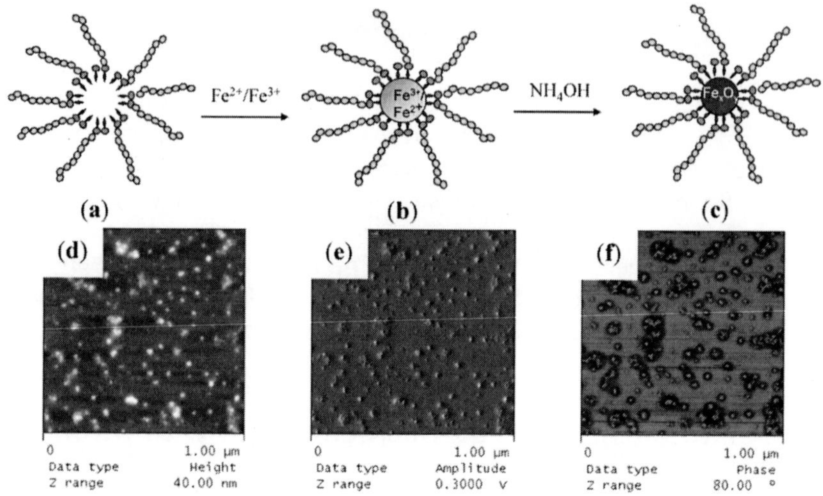

Figure 12. Strategy for the preparation of stabilized magnetic IONPs in aqueous media (**top**): (**a**) micelle formation by PEGMA$_x$-*b*-AEMA$_y$ diblock copolymers in water; (**b**) addition of the Fe^{3+}/Fe^{2+} mixture to the micellar solution, leading to complexation of the iron salts by the β-ketoester units inside the micellar core; and (**c**) transformation of the iron salt "precursor" into IONPs inside the micellar core upon addition of NH_4OH solution; atomic force microscopy (AFM) images of the block copolymer micelles loaded with IONPs (**bottom**): (**d**) height image; (**e**) amplitude image; and (**f**) phase image. Reprinted with permission from [36]. Copyright 2009 American Chemical Society.

4. SYNTHESIS TEMPLATED BY PREFORMED STRUCTURES

Rather than using individual polymer chains and inducing their self-assembly in the presence of metallic salts, loading can be achieved by exposing preformed self-assembled structures to these precursors. This strategy is expected to minimize changes in morphology or size dispersity, but the ability of the metallic precursors to reach the anchoring moieties might be affected. Consequently, several studies aimed at optimizing the conditions for the *in situ* synthesis of MNPs in individual nanosized, well-defined objects.

4.1. Microemulsions in an Organic Solvent

The size, shape, morphology and physical properties of *in situ* synthesized MNPs are not only strongly influenced by the nature of the templates, but also by the methods used to load the precursors and create the MNPs. In emulsion methods, the nanosized droplets of water-in-oil microemulsions appear ideal as reactors inside which NPs can be synthesized, with a monodispersed distribution of diameters below 20 nm. The results obtained in this approach depend on the type of amphiphilic species used to stabilize the microemulsion.

4.1.1. Chitosan Shells in Fe^{2+} Microemulsions with Triton®-X

Due to the presence of numerous amine and hydroxyl groups, chitosan, a derivative of chitin with an acetylation level below 60%, has been considered as a template for loading metallic ions. Combined with other useful properties such as being non-toxic, hydrophilic, biocompatible, biodegradable and anti-bacterial, chitosan can be useful to shorten the gap between the synthesis of the MNPs and their biomedical applications. In research combining the advantages of microemulsion systems with the benefits of chitosan templates, Zhi *et al.* [38] proposed a route in which NaOH solution was quickly added to an microemulsion system of cyclohexane, *n*-hexanol, HCl, chitosan, and a ferrous salt, stabilized by the Triton®-X 100 surfactant with controlled exposure to oxygen. The type of iron oxide obtained in the NPs was efficiently controlled by the amount of oxygen in the N_2 gas added during the synthesis, with 0.05% of oxygen yielding 60–80 nm spherical chitosan particles encapsulating cubic-shaped Fe_3O_4 cores (Figure 13). Glutaraldehyde (GDA) crosslinking of the magnetic chitosan NPs was suggested as a method to reduce the diameter of the obtained particles in the range of 10–50 nm without affecting the Fe_3O_4 cores. M_s values of 11.15 emu/g were reported for the composite NPs, in accordance with their iron oxide content.

4.1.2. Quaternary Microemulsions to Produce Aligned Spinel $CoFe_2O_4$ Nanorods

Quaternary microemulsions of cetyl trimethylammonium bromide (CTAB), water, cyclohexane and pentanol were used as microreactors for the coprecipitation of Co^{2+}, Fe^{2+} and oxalate anions, followed by the decomposition at high temperature of $CoFe_2(C_2O_4)_3$ (as "sacrificial template") and oxidation of Fe^{2+} into Fe^{3+} to produce spinel $CoFe_2O_4$ nanorods [39]. The calcination

treatment was indispensable to convert the amorphous nanorods into crystalline nanorods while maintaining control over their dimensions (50–100 nm in diameter, several micrometers in length), by stimulating internal nucleation and crystallization, in analogy to the natural petrification process which requires centuries to complete. scanning electron microscope (SEM) images highlighted the overall rod-like shape of the particles (Figure 14), and TEM images revealed the morphological transformation and the ripening of the rod-like particles upon heating, leading to the final state with individual $CoFe_2O_4$ crystals linearly arranged along the annealed nanorods (Figure 15a). The crystallographic alignment of the $FeCo_2O_4$ nanocrystals became visible after calcination in the high resolution TEM images as atomic planes (Figure 15c) and grain boundaries between the interconnected nanocrystals (Figure 15b). After annealing, the nanorods also exhibited magnetic H_c with a Curie temperature well above room temperature. Regarding their formation mechanism, the authors noted that the direct calcination of a co-precipitated $CoFe_2(C_2O_4)_3$ suspension in the absence of the microemulsion template did not lead to long crystalline anisotropic magnetic structures. Therefore an "oriented attachment" mechanism between the inorganic nuclei growing through the organic template was proposed. The ionic nature of the CTAB surfactant may also have played a role at the beginning of the calcination process, by forming a molten salt also favoring oriented nucleation.

4.2. Spherical Micelles in Water

Templates with predetermined sizes and structures are good candidates to orient the nucleation process, as they can be designed with an appropriate anchoring block, carry highly stabilizing units, and optionally a cross-linkable segment providing additional versatility.

4.2.1. Triblock Polyisoprene-*Block*-poly(2-cinnamoylethyl Methacrylate)-*Block*-Poly(*Tert*-Butyl Acrylate) (PI-*b*-PCEMA-*b*-PtBA) Copolymer Hollow Nanospheres

At first, spherical micelles with a structure locked by photo-cross-linking were prepared using a triblock copolymer PI-*b*-PCEMA-*b*-P*t*BA, in which PI forms the corona, PCEMA a solvent-insoluble shell, and P*t*BA the hydrolysable core [40]. Then, hydroxylation of the PI corona and cleavage of the *tert*-butyl groups to convert PtBA to PAA enabled the coordination of Fe^{2+} ions in the template. Finally, alkaline oxidation with NaOH and H_2O_2 yielded water-soluble Fe_2O_3-

impregnated nanospheres. This multi-step pathway is represented schematically on Figure 16. The diameter of the polymeric template (90 nm) did not change after the *tert*-butyl removal step (as seen by comparison of Figure 17b with Figure 17a), indicating the benefits of using templates with a predetermined size, shape and morphology. Inorganic cores with diameters ranging from 4 nm to 16 nm were visible on TEM images (Figure 17c). The Bragg peaks detected in selected area electron diffraction (SAED) experiments were matched with the Miller indices of the maghemite (γ-Fe_2O_3) phase. Therefore, such templates with a cross-linked polymer shell surrounding a porous polymer core coordinating ferrous precursors were efficient as templates for the preparation of IONPs, although further studies were suggested to control the size distribution of the magnetic cores.

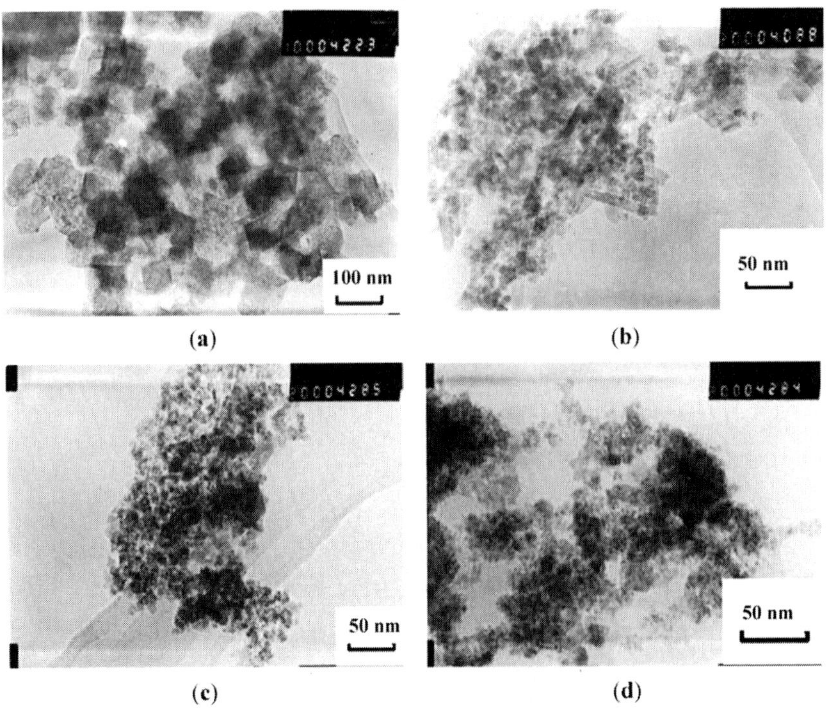

Figure 13. TEM of magnetic NPs (MNPs) obtained with: (**a**) molecular weight (MW) = 5.0 × 10^5 g·mol^{-1} non cross-linked chitosan; (**b**) cross-linked; (**c**) MW = 1.0 × 10^5 g·mol^{-1}non cross-linked chitosan; (**d**) cross-linked. Reprinted with permission from [38]. Copyright 2006 Elsevier B.V.

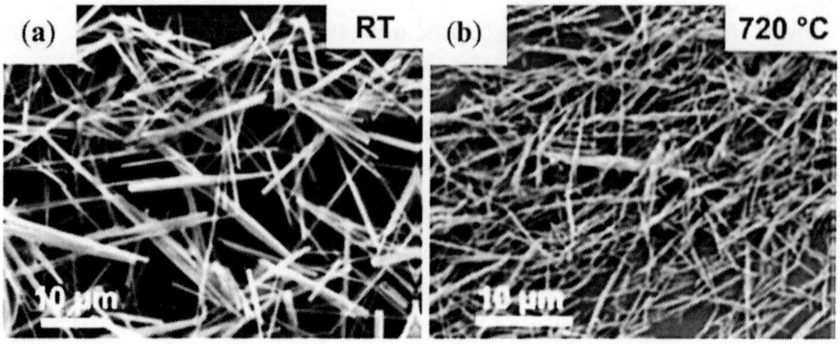

Figure 14. Scanning electron microscope (SEM) images for: (**a**) as-synthesized $CoFe_2(C_2O_4)_3$ suspension after coprecipitation in a microemulsion; and (**b**) $CoFe_2O_4$rods annealed at 720 °C. Reprinted with permission from [39]. Copyright 2005 Wiley-VCH.

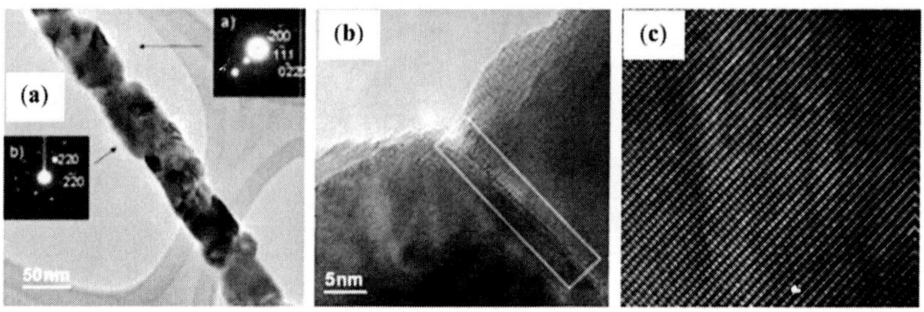

Figure 15. (**a**) TEM image of an individual $CoFe_2O_4$ nanorod annealed at 720 °C; the insets illustrate SAED patterns acquired from two individual nanocrystals of size about 80 nm. Insert (**a**) shows the indexed diffraction pattern for the *fcc* crystals in the [011] beam direction, and insert (**b**) in the [001] beam direction; (**b**) high resolution transmission electron microscopy (HRTEM) image showing a grain boundary between two $CoFe_2O_4$ nanocrystals; and (**c**) HRTEM image for a $CoFe_2O_4$ nanocrystal. Reprinted with permission from [39]. Copyright 2005 Wiley-VCH.

4.2.2. PEG-*block*-Poly(4-Vinylbenzylphosphonate) (PEG-*b*-PVBP) Micelles

Among the papers collected for this review, a small number concerned the density of surface coatings of pegylated IONPs (PIONs) for *in vivo* applications such as MRI. A recent study by Ujiie *et al.* [41] utilized PEG-*b*-PVBP block

copolymers for *in situ* alkaline coprecipitation (Figure 18). A correlation between the (PEG-*b*-PVBP)/(iron salt) feed ratio and the surface density of PEG chains in PEG-PIONs was revealed, suggesting an optimal ratio to obtain sufficiently small particles with a high PEG chains density. Interestingly, while the core sizes measured by TEM (≈7.7 nm) were typical for the aqueous coprecipitation route, they were unrelated to the feed ratios used. Thus although a "template effect with micelle-like aggregates" was invoked, the main interest of this method is for the preparation of individually dispersed MNPs stable in biological buffers rather than achieving precise control over the size of the magnetic cores. It is worth noting that the phosphonate groups significantly contributed to the incorporation efficiency of the block copolymer, due to its well-known binding ability for various metal oxides. The $R_h \approx 30$ nm observed for the micelle-like aggregates before the ammonia addition was considered as reflecting a "templating effect" for metallic nucleation. This is the main reason why this study is reported in the preformed structures section instead of the block-copolymer section.

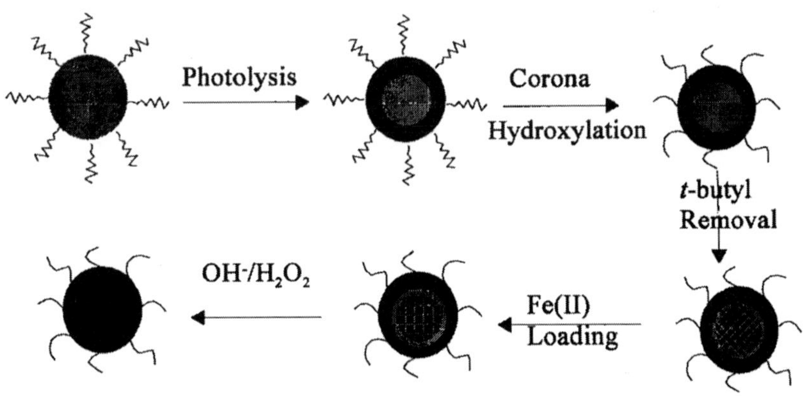

Figure 16. Polyisoprene-*block*-poly(2-cinnamoylethyl methacrylate)-*block*-poly(*tert*-butyl acrylate) (PI-*b*-PCEMA-*b*-P*t*BA) as template: Photolysis cross-links the PCEMA shell (gray to black); the PI corona chains are made water-soluble by hydroxylating the double bonds (wavy lines to free-hand lines); the core is made inorganic-compatible by removing the *tert*-butyl groups (light gray to gridded pattern). Soaking the nanospheres in aqueous $FeCl_2$ leads to proton exchange (slanted to vertical grids) and the Fe^{2+} ions are precipitated (NaOH) and oxidized (H_2O_2) to yield cubic γ-Fe_2O_3magnetic particles (last step). Adapted with permission from [40]. Copyright 2000 American Chemical Society.

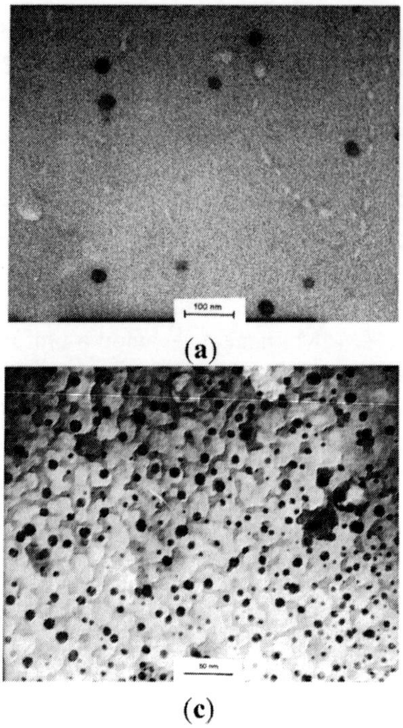

Figure 17. TEM images for PI-*b*-PCEMA-*b*-PtBA nanospheres at each step of the synthesis: (**a**) after PCEMA cross-linking and PI hydroxylation (stained with OsO_4); (**b**) after removal of the *tert*-butyl groups (stained with OsO_4); (**c**) after Fe_2O_3 loading (no staining); and (**d**) attraction by a magnet. Adapted with permission from [40]. Copyright 2000 American Chemical Society.

4.2.3. Multiarm Star-Like Amphiphilic or DHBCs

For linear block copolymer templates, the size, shape, and the colloidal stability of the nanocrystals are greatly influenced by experimental conditions used (e.g., concentration, solvent, temperature, and pH). Multi-arm star-like unimolecular micelles derived from either amphiphilic or DHBCs were successfully employed to achieve better control over the size and shape of the MNPs. Starting from a multi-arm ATRP initiator (suitably functionalized β-cyclodextrin (CD)), the polymeric blocks (P*t*BA or P4VP) serving to bind the inorganic precursor were synthesized first. These star-like homopolymers we then coupled by "click" chemistry with either hydrophobic (PS) or hydrophilic polymeric (PEO) segments which imparted solubility to the micelles in either organic or aqueous environments. The $FeCl_2$ and $FeCl_3$ precursors were loaded in the hosts in appropriate solvents. After ammonium hydroxide addition,

Fe_3O_4 NPs were produced with different sizes (6, 10 and 16 nm, e.g., Figure 19a and Figure 20a), depending on the of the PAA block derived from P*t*BA (4.5, 8.4 and 16.8 kDa). These exhibited superparamagnetic properties at 300 K without hysteresis, and M_s increasing with the NP size. The advantages of triblock structures (β-CD-P4VP-P*t*BA-PS) and (β-CD-P4VP-P*t*BA-PEO) were also highlighted by producing NPs with Fe_3O_4 core and $PbTiO_3$ (Figure 19b,c) or Au shell (Figure 20b). More accurately, the magnetite NPs were encapsulated within the P4VP core, and the $PbTiO_3$ or Au layers were grown within the PAA shell (derived from the P*t*BA block after the first encapsulation step), thus retaining the magnetic properties of the NPs. TEM imaging revealed a uniform size and narrow size distribution for the NPs, while the crystalline lattices were characterized by HRTEM, XRD and energy-dispersive X-ray spectroscopy (EDS) measurements. Hollow gold NPs were also reported, aside from the magnetic-gold core-shell systems.

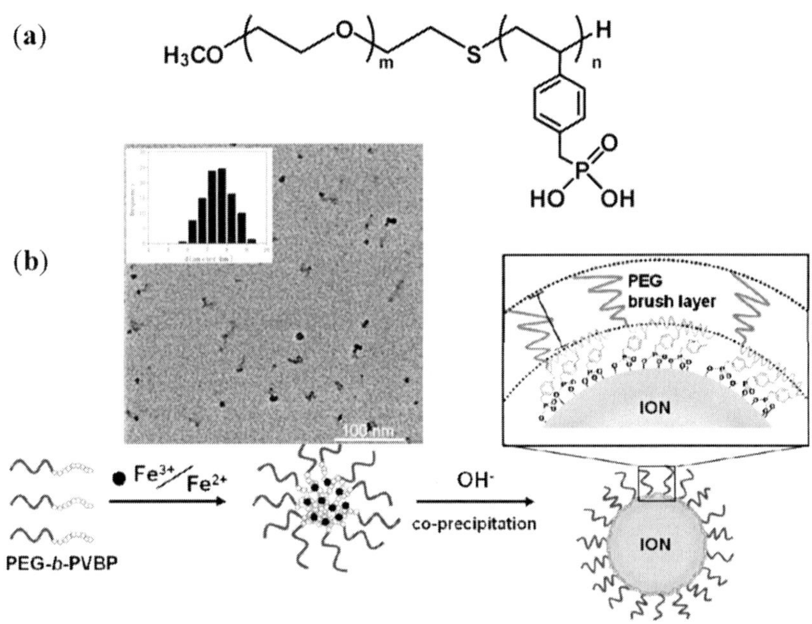

Figure 18. (**a**) Chemical structure of PEG-*b*-PVBP and schematic illustration of the proposed morphology of the pegylated IONPs (PIONs); and (**b**) TEM image of the PIONs and their size distribution as inset. PVBP: poly(4-vinylbenzylphosphonate). Adapted with permission from [41]. Copyright 2011 Elsevier B.V.

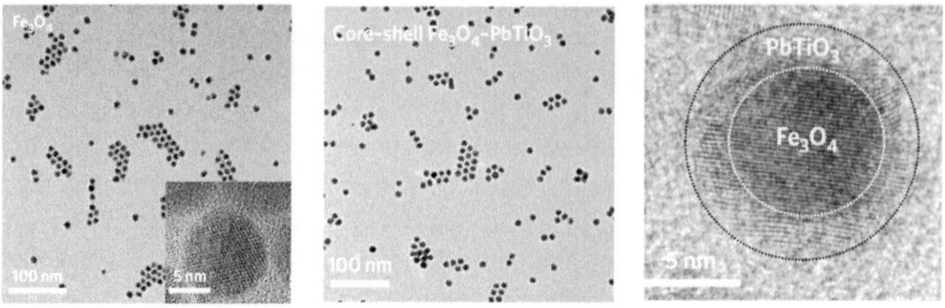

(a) (b) (c)

Figure 19. (**a**) Representative TEM images for Fe_3O_4NPs synthesized with star-like PAA-*b*-PS templates, *D*(Fe_3O_4) = 10.1 ± 0.5 nm; (**b**) TEM; and (**c**) HRTEM images of Fe_3O_4-$PbTiO_3$ core-shell NPs formed with the star-like triblock copolymers nanoreactors. Reprinted with permission from [42]. Copyright 2013 Macmillan Publishers Limited.

This work undoubtedly evidences a strong templating effect in the *in situ* synthesis of inorganic NPs, in particular magnetic ones, using unimolecular star-like PAA-PEO micelles.

4.3. Cylindrical Multimolecular Micelles

Different morphologies were obtained by self-assembly of commercial Pluronic® F127 PEO-*b*-PPO-*b*-PEO triblock copolymers in water/alcohol mixtures, to serve as templates for iron ion loading by coprecipitation [43]. Surprisingly, as the alcohol content in the solvent mixture increased, a gradual phase transformation from 15 nm Fe_3O_4 NPs (Figure 21a) to uniform (*ca*. 20 nm diameter, 200–300 nm length) α-FeOOH (goethite) nanorods (Figure 21b) was observed. This shows that Pluronics® can serve as structure-directing agents that control both the metal oxide mesoscale size and the crystalline structure. The decrease in M_s of the samples observed was in agreement with this phase transformation from the superparamagnetic (maghemite) to the anti-ferromagnetic state (goethite).

4.3.1. Fe^{3+} Loading in Poly(Acrylic Acid)-*Graft*-Poly(*n*-Butyl Acrylate) Brush

In an attempt to preparing hybrid magnetic unimolecular comb-like polymer micelles, Zhang and Muller reported the procedure of employing amphiphilic

core-shell cylindrical polymer brushes with poly(acrylic acid)-*block*-poly(*n*-butyl acrylate) (PAA-*b*-PnBA) side chains to impregnate Fe^{3+} [44]. It was worth noting that in order to increase the rate and extent of iron ion uptake, COOH groups in PAA blocks were prior deprotonated by NaOH before the encapsulating step performed with an excess amount of $FeCl_3$. The polychelate effect was proved by FTIR and ultraviolet-visible spectroscopy (UV-VIS) measurements, while the TEM and AFM images revealed the core of the unimolecular wormlike cylinder hybrids that have been mineralized by the ferric cations (Figure 22). However, the authors did not go further than the precursor loading, *i.e.*, they did not describe the alkaline coprecipitation *in situ*.

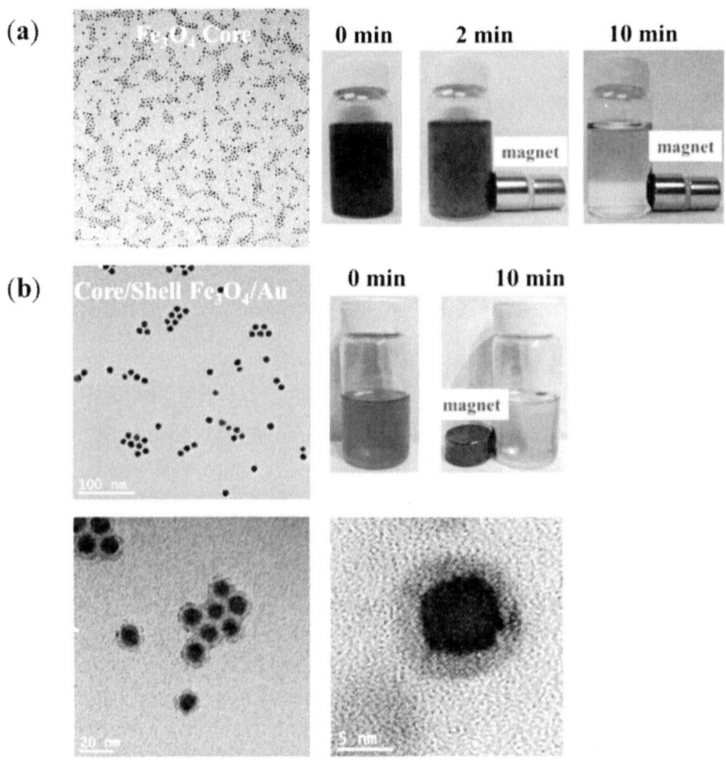

Figure 20. TEM, HRTEM and digital camera images for core/shell Fe_3O_4/Au NPs: (**a**) Fe_3O_4 core (D = 6.1 ± 0.3 nm); and (**b**) Fe_3O_4/Au core/shell NPs at different magnifications (Au shell thickness is 2.9 ± 0.2 nm). Fe_3O_4 appears dark in the center. The magnetic properties of Fe_3O_4 were retained, as evidenced by the response of the NP dispersion in toluene to a magnet (right panel in (**b**)). Reprinted with permission from [42]. Copyright 2013 Macmillan Publishers Limited.

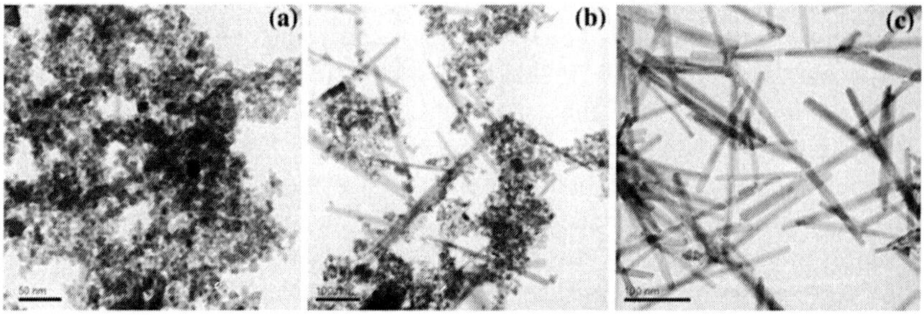

Figure 21. TEM images for samples prepared in alcohol/water mixtures with various volume ratios of alcohol to water: (**a**) 0:1; (**b**) 1:1; and (**c**) 5:1. Reprinted with permission from [43]. Copyright Springer Science Business Media, LLC 2008.

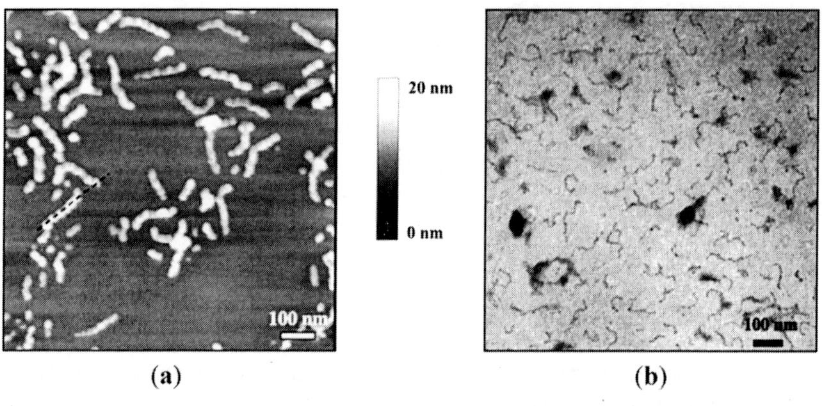

Figure 22. (**a**) AFM height image of the polychelate of Brush 1 and $FeCl_3$ after dialysis; and (**b**) TEM image of uni-molecular polymer brush polychelate of $FeCl_3$ (after dialysis). Adapted with permission from [44]. Copyright 2004 Springer-Verlag.

4.3.2. $Fe^{3+/2+}$ Loading and Precipitation in Poly[Poly(Ethylene Glycol) Methylether Acrylate]-*Graft*-Poly(Methacrylic Acid) (PPEGMEA-*g*-PMAA) Brush

Both nucleating and stabilizing abilities were introduced into a densely grafted bottle-brush double hydrophilic copolymer constituted of two different types of hydrophilic segments, respectively a PMAA backbone, providing COOH

groups for nucleation of Fe_3O_4 particles, and poly[poly(ethylene glycol) methylether acrylate] (PPEGMEA) side chains (Figure 23), introducing a shielding effect to prevent the unfavorable aggregation. Such template was investigated for an *in situ* coprecipitation process [3]. Narrow size distributed Fe_3O_4/polymer clusters with hydrodynamic sizes were 50–210 nm, carrying 7–15 nm diameter Fe_3O_4 crystals were obtained. The magnetic core sizes were efficiently tuned by varying the polymer/Fe_3O_4 feed ratios and the length of PMAA side chains. For example, the diameter of the MNPs increased from 8.5 nm to 9.5 nm and 10.4 nm when this ratio was varied from 1/1 to 1/2 and 1/3. Measurements on a VSM setup revealed the superparamagnetism of the hybrids at room temperature, with a concomitant increase of the M_s value from 48 emu/g, to 49 emu/g and 50 emu/g. Such dense polymeric combs were thus considered as influent templates to orient the distribution of sizes and to improve the magnetization properties of the Fe_3O_4 NPs.

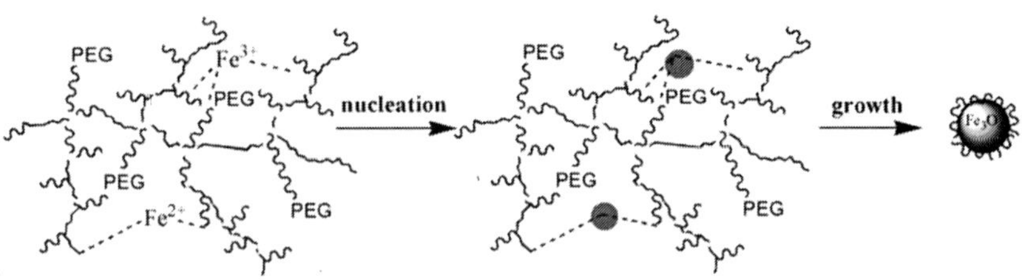

Figure 23. Synthesis of Fe_3O_4/polymer nano-composite in the presence of PPEGMEA-*g*-PMAA densely grafted double hydrophilic copolymer. PPEGMEA: poly(ethylene glycol) methylether acrylate. Reprinted with permission from [3]. Copyright 2008 Royal Society of Chemistry.

4.3.3. Poly(Glycerol Monoacrylate)-*Graft*-Poly(PEG Methyl Ether Acrylate) (PGA-*g*-PEG) Copolymers

In a detailed study [45] aimed at investigating the influence of the graft density *g* and MW of the graft polymer PGA-*g*-PEG on the average size and magnetic properties of MNPs, a possibility of changing the diameter of particles in 4–18 nm range by employing polymers with different grafting densities was suggested. It is understandable that as the hydroxyl groups in PGA block were the anchoring functionalities, the decreased feed ratio of the monomeric precursors of glycerol monoacrylate to the PEG methyl ether acrylate side chains in graft copolymerization resulted in the rise of the graft densityg, yet reducing the coordinating capability of each individual graft polymer micelle,

thus leading to the ascending particle sizes. This observation was confirmed by the data obtained from XRD measurement were the Bragg peaks widths were analyzed by the Scherrer equation: the average particle sizes were 4, 7, 9 and 18 nm, respectively, for g = 0.14, 0.23, 0.30 and 0.42. Interestingly, graft copolymers with varied MW (from 4.2 kD to 7.6 kD and 15 kD) bearing the same graft density g = 0.24 produced particles of similar size. The particles were superparamagnetic at room temperature with normalized $M_s$decreasing in the range of 30–42 emu/g with reducing particle sizes, due to the smaller domains crystallized.

4.3.4. Poly(Ethylene Oxide)-*Graft*-Poly(Acrylic Acid) (PEO-*g*-PAA) Graft Copolymer

The four-arm graft copolymers PEO-*co*-ethoxyethyl glycidyl ether-*graft*-PAA $[PEO_x$-*co*-Gly_y-*g*-$PAA_z]_4$ were employed in the same coprecipitation process as reported above leading to highly dispersed narrow size distributed particles encapsulating 10–20 nm iron oxide monocrystals inside the polymer chains [46]. The TEM results were in good agreement with those from XRD analysis with the exception of the peak broadening which was assigned for the matrix constrain of the nanosized particles. The superparamagnetic properties were confirmed, and the calculated M_s was 56 emu/g Fe_3O_4. The size and the size distribution of hybrid NPs were tuned by changing the mass ratio of polymer compared to Fe_3O_4 and by varying the ratio of PAA/PEO segments in the graft copolymer (better template effect with smaller particles and narrower size distribution was reported with larger proportion of PEO segments). More importantly, star-shaped graft polymers with more confined structure exhibited better template effect as compared to linear graft copolymer consisting of similar components.

From the three preceding examples, which originate from different teams, one can conclude that despite some little changes of the chemical nature of the polymer segments, the graft polymer structures always induce a reproducible trend in the variation of the average size of the magnetite NPs synthesized by *in situ* coprecipitation. However, these studies may have overestimated the template effect, in the sense that the size distributions observed on TEM images are always broad, and do not correspond to true monodisperse samples.

4.4. Lamellar Films

Polystyrene-*Block*-Poly(2-Vinylpyridine) (PS-*b*-P2VP) Lamellae Hosting Pristine Iron Nanoparticles

The complexation ability of vinylpyridine moieties was investigated by the groups of Cohen and Ross [47] not only in the form of linear homo- and copolymers as discussed above, but also in bulk films of PS-*b*-P2VP. Nucleation was induced by the thermal decomposition of $Fe(CO)_5$ or $Co_2(CO)_8$. The impregnation of the organometallic complexes $Fe(CO)_5$ and nickelocene in vacuum-dried films produced monodispersed iron, iron-cobalt, and cobalt-nickel metallic NPs residing within the P2VP domains, that were superparamagnetic at room temperature. The lamellar structure of these composites was shown in TEM micrographs as a "cluster of grapes" morphology of pristine metal particles that were monodispersed in size when the metallic feed ratio of Fe to Co was 100:0 (Figure 24a) or 80:20 (Figure 24b), and bimodal when increasing the Co content to 50:50 (Figure 24c) or 20:80. The magnetization values of the samples varied from 1.1 emu/g P2VP in the first two cases to 8.4 emu/g P2VP and 10.9 emu/g P2VP for the higher Co contents. The increase in M_s observed for higher Co contents was explained by the preference for cobalt atoms to remain at the surface of the particles to minimize their surface energy, which led to thinner layers and more walls (as shown by scanning transmission electron microscopy (STEM)) separating smaller particles with a larger surface-to-volume ratio. The smaller particles nucleated on the film were considered a dominant factor explaining the non-zero H_c observed at room temperature for some samples.

4.5. Hexagonal Ordered Films

4.5.1. Monolayer Films of Polystyrene-*block*-Poly(4-Vinylpyridine) (PS-*b*-P4VP) Copolymer Micelles

To create monolayer films where the particles were temporarily arranged in a hexagonal array, PS-*b*-P4VP micelles (formed by self-assembly in toluene) encapsulating $FeCl_3$ within their P4VP core were spread onto cleaned silicon wafers by spin coating [48]. Oxygen plasma treatment was needed to convert the Fe^{3+} precursor into iron oxide and to remove the polymer template. The γ-Fe_2O_3 crystals formed at 1.0, 0.5, and 0.2 of $FeCl_3$/P4VP molar ratios appeared on the TEM images (Figure 25a) as 25.0, 18.0, and 6.0 nm sized particles, respectively, arranged in a hexagonal self-assembled pattern. The wide range of IONP diameters obtained and the constant spacing of the IONP patterns in

all three cases demonstrated that PS-*b*-P4VP is a powerful template that effectively controls the nucleation and the growth of the particles. AFM (Figure 25b) and field emission-scanning electron microscopy (FE-SEM) imaging (Figure 25c) revealed the hemispherical topology of the IONPs. Contrary to what was expected, the annealing process aiming to improve the crystallinity of the particles did not cause irregular shapes but slightly reduced the average diameter and height of the particles for all the samples with different $FeCl_3$/P4VP ratios, as shown in the tilt-view image of Figure 25c. The 25.0 nm-sized particles exhibited ferrimagnetism with normalized magnetization M_s = 100 emu/g and H_c = 50 Oe, while the 6.0 nm crystals possessed 30 emu/g saturated magnetization and zero hysteresis, corresponding to superparamagnetic properties. The impressive variation in the magnetic properties of the particles obtained was achieved by only by changing the $FeCl_3$/P4VP feed ratio in the mixture.

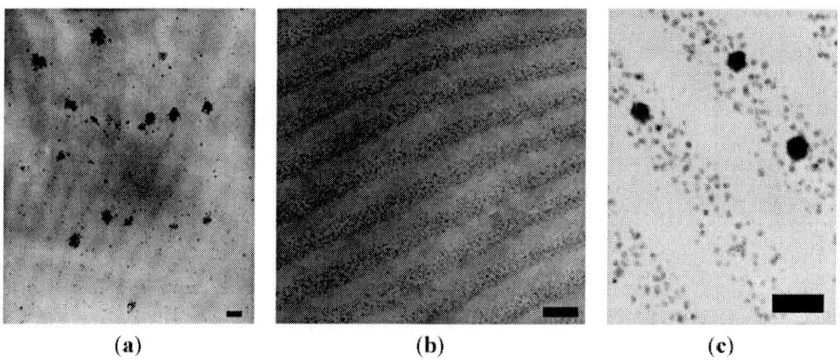

Figure 24. (**a**) TEM micrograph for a lamellar PS-*b*-P2VP copolymer containing 1.2 meq of iron/g P2VP; sample heated to 195 °C for 24 h, scale bar = 50 nm; (**b**) TEM micrograph for the same copolymer containing a total of 1.2 meq of Fe and Co/g P2VP, with an atomic ratio of Fe to Co of 80:20; sample heated to 161 °C for 24 h, scale bar = 100 nm; and (**c**) scanning transmission electron microscopy (STEM) micrograph of the copolymer containing a total of 1.2 meq of Fe and Co/g P2VP, atomic ratio of Fe to Co = 50:50; sample heated to 161 °C for 24 h; the atomic ratio of Fe:Co at the center of the large particles was 48.9:51.1 from energy dispersive X-ray spectroscopy (EDX) spectroscopy; scale bar = 50 nm. Reprinted with permission from [47] Copyright 2003 Elsevier B.V.

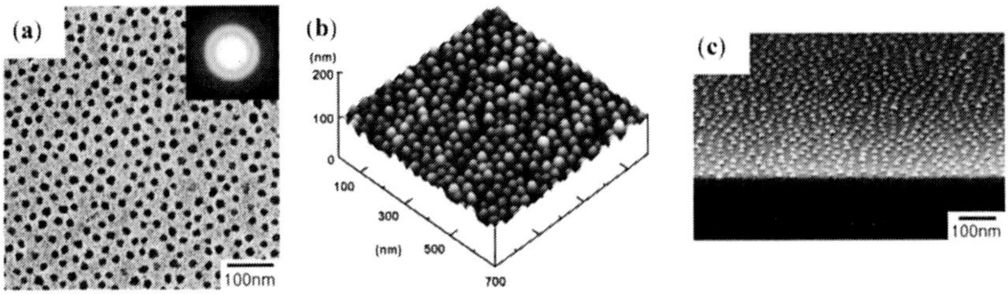

Figure 25. (**a**) TEM image for a self-assembled pattern of IONPs with hexagonal packing. The inset is a SAED pattern typical of γ-Fe_2O_3; (**b**) AFM image for the hexagonal pattern of IONPs on a silicon wafer; and (**c**) field emission-scanning electron microscopy (FE-SEM) image in tilt view. The molar ratio of $FeCl_3$ precursor to 4-vinylpyridine was 0.5 in all cases. Reprinted with permission from [48]. Copyright 2005 American Chemical Society.

4.5.2. Monolayer Films of Polystyrene-*block*-Poly(Ethylene Oxide) (PS-*b*-PEO) Copolymers

Also targeting the preparation of hexagonal arrays of NPs, the *in situ* synthesis of metal oxide NPs through a gas phase oxidation pathway in oriented cylinder structures was investigated by Ghoshal *et al.* [49]. The ordered thin films were prepared using a commercial diblock copolymer of poly styrene (PS) with PEO as the metal-anchoring block. The block weight ratio was selected to yield cylindrical PEO domains 20 nm in diameter and 28 nm in height, regularly distributed within the PS matrix. One of the strategic details differentiating this research from the methods reported earlier was the microphase separation induced by a solvent mixture after spin coating of a toluene solution of the PS-*b*-PEO block copolymer onto cleaned silicon wafers. Annealing with a water/toluene mixture efficiently weakened the preferential interactions between PEO and the substrate, thus allowing the vertical alignment of the PEO cylinders (Figure 26A). A subsequent treatment with an ethanol solution at 40 °C for 15 h (Figure 26B) appeared necessary as an "activation step" for the PEO cylinders to create a functional chemical pattern for nanodot development, which was interpreted by the authors as being due to the crystallization of the PEO blocks induced by ethanol. Then the affinity of PEO for $Fe(NO_3)_3$ as compared to PS allowed selective inclusion of the metallic salt by spin-coating of the 0.4 wt% precursor solution (Figure 26C). Finally, ultraviolet (UV)/ozone treatment (Figure 26D) produced highly ordered Fe_3O_4 particles (or CeO_2 or CuO, depending on the precursor). This treatment not only produced the oxides from the precursors but also removed the polymer template. The Fe_3O_4 particles appeared in HRTEM images as nanodots

24 nm in diameter and 9 nm in height, with a regular spacing of 42 nm between their centers. Control over the diameter and the height of the oxide nanodots was attained by varying the precursor concentration, which changed the height and the diameter of the nanodots without affecting their spacing. Precursor solution concentrations exceeding 1% led to saturation of the PEO cylinder template, however. This method nevertheless offers great potential to easily tune the NP size for iron oxide and several other inorganic materials. In addition, the nanodots could be sintered without losing their structural order, as seen on Figure 27. For example, at 800 °C in air, the IONPs shrunk in diameter by 3 nm and in height by 2 nm, as ascribed to oxide densification.

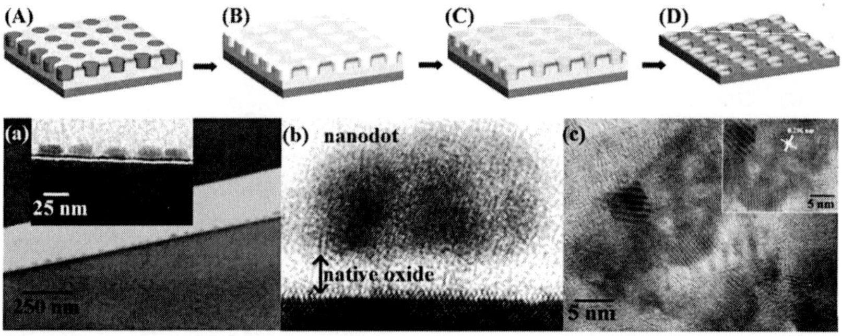

Figure 26. Schematic illustration of the fabrication of oxide nanodots: (**A**) highly ordered PS-*b*-PEO thin film prepared by a solvent annealing process; (**B**) nanoporous template produced by activation of the PEO cylinders upon exposure to ethanol at 40 °C for 15 h; (**C**) the metal oxide precursor diffuses into the cylinders after spin coating the metal nitrate solution; and (**D**) oxide dots remaining after the ultraviolet (UV)/ozone treatment. (**a**) TEM cross-sectional image of iron oxide nanodots; the inset shows a higher magnification image; (**b**) cross-sectional HRTEM image for a single nanodot; and (**c**) HRTEM image for the nanodots after the UV/ozone treatment; the inset shows crystalline fringes corresponding to Fe_3O_4. Reprinted with permission from [49]. Copyright 2012 Royal Society of Chemistry.

4.6. Holey Membranes

4.6.1. PVA-Fe^{3+}/Citric Acid (CA)/Ethylene Glycol (EG) Calcination

PVA, containing a high density of hydroxyl functional groups available to complex ferric ions, was employed in a so-called "polymer precursor" (PP)

method to produce porous iron oxide films [50]. The first step in the PVA-PP method involved the preparation of a viscous solution of Fe^{3+} ions complexed with PVA and the polyether produced *in situ* by the reaction of CA and EG. The solution was then spin-coated on a Pyrex® glass substrate before repeated calcination to decompose the PVA into gas and yield a uniform porous iron oxide film consisting of sintered 100 nm diameter ellipsoidal particles, with a micro-porosity volume (assessed by the N_2 adsorption isotherm method) increasing with the PVA/CA weight ratio. A change in crystal structure from α-Fe_2O_3 to γ-Fe_2O_3 was detected by PXRD and X-ray photoelectron spectroscopy (XPS) measurements when the PVA/CA ratio exceeded 1, which was interpreted as being due to the surface reduction of α-Fe_2O_3 into Fe_3O_4 crystals. The kinetic constant for this transformation was enhanced by a decrease in Gibbs energy for the reduction due to improved elimination of O_2 for the higher porosity films.

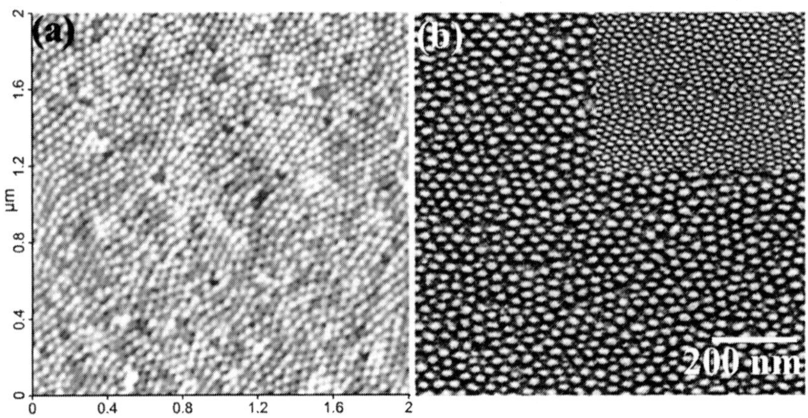

Figure 27. (**a**) AFM; and (**b**) SEM images for hexagonally ordered iron oxide nanodots after UV/ozone treatment. The inset of (**b**) shows the iron oxide nanodots after annealing at 800 °C for 1 h. Reprinted with permission from [49]. Copyright 2012 Royal Society of Chemistry.

4.6.2. Loading of the Pores of a Nafion® Membrane Followed by Reduction

In comparison to commonly used methods like coprecipitation or thermal decomposition, fewer studies relied upon a reduction reaction. This route was selected by Yoon *et al.* [51] using a polymeric ion-exchange membrane to

produce nanosized magnetic pristine iron particles. The method basically consisted in the diffusion of Fe^{3+} ions into a commercial porous perfluorinated sulfonate ionomer membrane, and the *in situ* reduction with $NaBH_4$ into magnetic Fe NPs. This led to the disappearance of the hysteresis loop at 120 K, corresponding to the T_B for very tiny Fe NPs, as well as ferromagnetism at 10 K and superparamagnetism at 200 K. The observation of 1.5–2.5 nm diameter spherical particles dispersed in a clear polymer background in the TEM images, combined with crystalline lattices revealed by EDS measurements, confirmed the single-domain structure of the iron NPs obtained.

4.7. Tridimensional Scaffolds (Macroscopic Samples)

In an encapsulation process, the strength of interactions between the anchoring moieties in a polymer matrix and the NPs is considered the most important criterion preventing the diffusion of the NPs out of the template and leading to successful directed nucleation and growth. One approach suggested realizing this condition was to rely on the large surface area and the complicated inner structure of polymer gel networks to perform the *in situ* synthesis of MNPs. Tridimensional polymeric scaffolds containing further structural hierarchical microstructures offer great potential for size, shape, crystalline structure, and orientation control, and have consequently motivated a large number of studies.

4.7.1. Porous Preformed Carbon Foams (CF)

In an innovative study by Yoon *et al.* [52], mesoporous CF produced by the resorcinol-formaldehyde method, with a very large pore volume (0.87 $cm^3 \cdot g^{-1}$) and specific surface area (834 $m^2 \cdot g^{-1}$) were loaded with a $Fe(NO_3)_3$ solution by repeated impregnation. After *in situ* thermal decomposition of the precursor at 450 °C under argon, Fe_3O_4 nanocrystals were obtained in the confined pores of the carbon template (Figure 28). HRTEM images and XRD patterns confirmed the Fe_3O_4composition of the crystals, which had a non-uniform size and shape distribution in the 5–50 nm range, *i.e.*, larger and broader than the 7–15 mm particles obtained by the conventional (non-templated) polyol synthesis (same precursor but at 180 °C in EG). The width of the XRD peaks also yielded a significantly larger average crystallite size of 21.7 nm (as compared to 10.1 nm for the polyol pathway). Although physical adsorption was invoked rather than chemical interactions, due to the absence of coordinating functional groups, the roles of the carbon foam in orientating the growth of the iron oxide nanocrystals and preventing their agglomeration were clearly shown. When

used as anode material for lithium-ion batteries, the as-prepared CF/Fe_3O_4 composites exhibited superior charge/discharge capacities as compared to the unsupported materials.

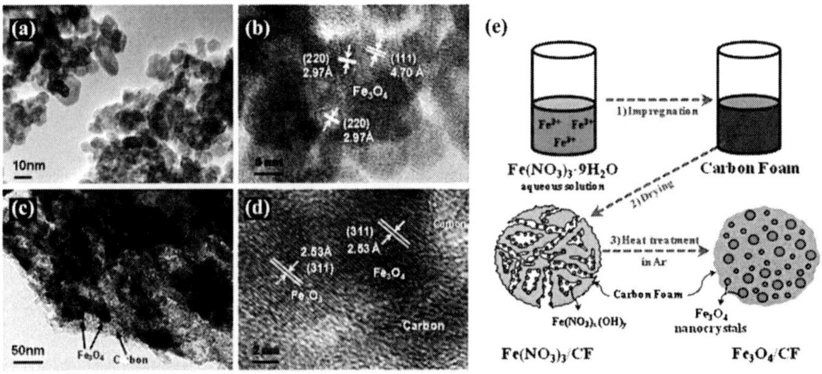

Figure 28. TEM images for unsupported Fe_3O_4 NPs synthesized by: (**a**,**b**) the hydrothermal method; (**c**,**d**) the *in situ* formation of Fe_3O_4 NPs on the CF composite; and (**e**) schematic representation of the *in situ* formation of Fe_3O_4 nanocrystals in the confined pores of carbon foam. Reprinted with permission from [52]. Copyright 2011 Royal Society of Chemistry.

4.7.2. Sponge-Like Polystyrene (PS)/Polyacrylate Copolymer Gel

In the early work, the team of Antonietti and Mann [53] successfully developed magnetic iron oxide nanocolloids using elastic sponge-like PS-polyacrylate copolymer gels, providing a mesoporous open-cell structure with carboxylic groups readily coordinating with metallic ions. The hierarchical structures of the micro-spherical pores, whose size could be further adjusted during synthesis, offered enormous surface areas that effectively facilitated the templating and nucleation processes (allowing up to 20 wt% of iron loading). It is worth noting that the gel structure was not changed after the loading step (Figure 29a). There were no crystalline structures in the air-dried composites, but rather amorphous hydrated Fe^{3+}oxide which gradually converted to 24 nm magnetic Fe_3O_4 crystals (based on wide-angle X-ray scattering (WAXS) line width analysis). These results were in agreement with TEM analysis of gel sections, showing homogeneous spherical and cubic particles with a mean size of 16.3 ± 9.4 nm trapped in the wall of the gels (Figure 29b). No hysteresis was

reported in room temperature measurements, indicating the superparamagnetism of the particles. This pioneering research using an easily tailored material was followed by many other studies on gel templates.

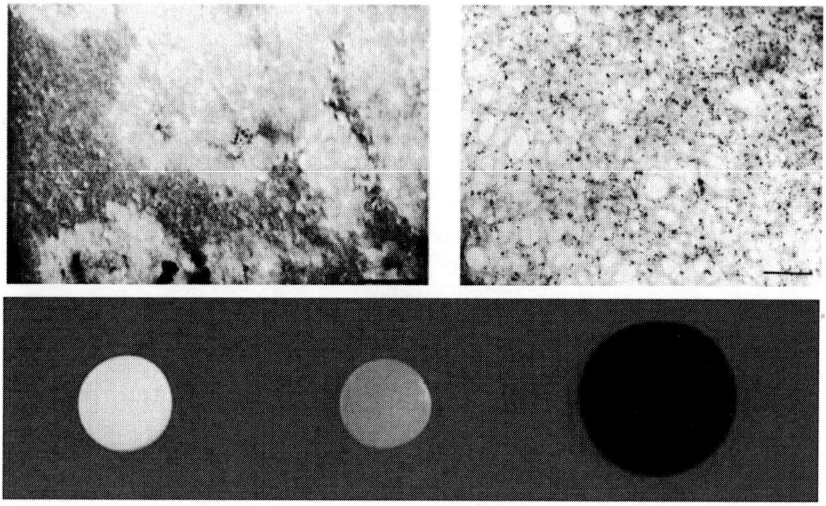

Figure 29. TEM images for sample sections: (**a**) 3.5 wt% Fe-loaded polystyrene (PS)-polyacrylate gel prior to reaction with NaOH (scale bar = 500 nm); (**b**) after reaction with NaOH, showing the distribution and uniformity of the magnetite nanocrystallites (scale bar = 500 nm); and (**c**) synthesis of the magnetic sponge-like copolymer gel. Comparison between: (**c**, **left**) an unloaded polymer gel; (**c**, **middle**) after exposure to a 0.2 M Fe^{2+} solution; and (**c**, **right**) after reaction with NaOH. The gels imaged with a digital camera on (**c**) are in the swollen state. Reprinted with permission from [53]. Copyright 1998 WILEY-VCH Verlag GmbH.

4.7.3. Cross-Linked Poly[Styrene-*co*-(*N*-4-Carboxybutylmaleimide)] Copolymer

Taking advantage of the facile polymerization of styrene and maleimide-based monomers, cross-linked networks were created from poly[styrene-*co*-(*N*-4-carboxybutylmaleimide)] copolymers poly(St-*co*-CBMi) in which CBMi acts as a metal-anchoring moiety. This copolymer was successfully prepared and used for the *in situ* synthesis of IONPs by deposition of Fe^{2+} and multi-step NaOH treatment [54]. The data obtained from XRD analysis after two particle

deposition cycles revealed the initial existence of maghemite (γ-Fe_2O_3) NPs which were gradually transformed to goethite (α-FeOOH). This was likely not due to oxidation as ascribed by the authors, but rather a result of the unavoidable pH increase in such a multi-step NaOH treatment. The constant value of mean particle size (14 nm) calculated from the Scherrer equation and XRD diffractograms after several deposition cycles indicate that the growth of the particles was inhibited by the constrained architecture of the polymer network. The maghemite content increased after each loading cycle, but the fourth cycle led to agglomeration. The magnetization of the hybrids also increased after each loading cycle, but was maximized after the third one (8.04 emu/g). There was no subsequent change in M_s due to the appearance of the goethite crystals, that are known to be antiferromagnetic. The non-zero H_c values observed for the hybrids, even at room temperature, were assigned to the same reason, *i.e.* mixing of the two phases giving rise to spontaneous ferrimagnetism.

4.7.4. Semi-Interpenetrating (Semi-IPN) Polymer Networks of Alginate and Poly(*N*-Isopropylacryl Amide) (PNiPAAm)

In an attempt to synthesize IONPs within networks of mixed polymer hydrogels, Hernández *et al.* [55] initially prepared semi-IPN polymer networks from alginate bringing chelating carboxylate groups and PNiPAAm providing thermo-sensitivity to these materials. Cylindrical samples cut from the semi-IPN were submerged into a 1:2 solution of Fe^{2+} and Fe^{3+}ions, and then into a NH_3 solution for the *in situ* preparation of semi-IPN ferrogels. For samples with different numbers of alkaline bath treatments, the phase type of the IONPs within the gels was determined to be a combination of magnetite and maghemite based on the diffraction patterns in wide angle X-ray analysis and characteristic peaks in Raman spectroscopy. The diameters calculated from the Debye-Scherrer equation for the (311) Bragg peak suggested larger sizes for the NPs seeded in the polymer networks (D_{311} = 10.1 ± 2.5 nm and D_{311} = 11.2 ± 2.8 nm for one and two alkaline treatments, respectively, as compared to D_{311} = 6.7 ± 1.6 nm for NPs created in an alginate solution.

4.7.5. Cross-Linked Polyacrylamide (PAAm) Hydrogels

To prepare water-dispersible magnetite NPs, Xiong and Sun applied an *in situ* synthesis approach involving cross-linked PAAm hydrogels and NaOH solution treatments. As schemed on Figure 30a, the PAAm hydrogel was photo-

polymerized with pentaerythritol triacrylate (PE-3A) as a hydrolysable cross-linker and BDMBP as photo-initiator [56]. The polymer network was expected to act as template possessing both sufficient capacity for ion impregnation and a restrained mesh structure limiting the growth of MNPs and preventing their aggregation. Magnetite NPs with narrow size distribution were obtained by this method, as seen on Figure 31, with a size inversely related to the cross-linker density (8.3 nm for 0.46 mol% PE-3A, 6.3 nm for 6.14 mol%). In a second step, the hydrolysis of the acrylic ester cross-linker PE-3A in alkaline water liberates well-dispersed hydrophilic MNPs (Figure 30c). Interestingly, XPS results showed the existence of a thin layer of polymer chains bearing carboxylate and acrylamide groups covering the surface of these NPs, which explains their long term stability in water. Hysteresis and H_c were observed at 5 K, indicating ferromagnetism, but the particles were superparamagnetic at 300 K with negligible H_c. The saturated magnetization was 44.6 emu/g at 300 K.

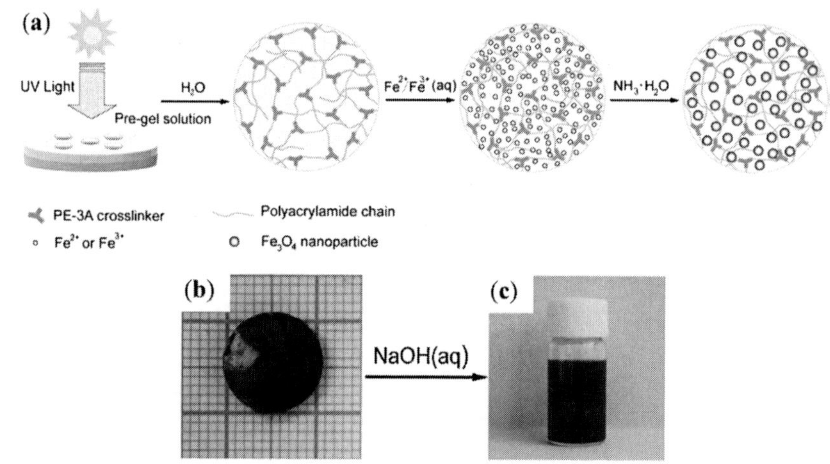

Figure 30. (**a**) Schematic representation; (**b**) appearance of a magnetic hybrid hydrogel obtained by photo-polymerization and *in situ* coprecipitation; and (**c**) in the second step, the ferrogel was exposed to NaOH to hydrolyze the cross-link points and induce the release of water-dispersible IONPs. Adapted with permission from [56]. Copyright 2011 Elsevier B.V.

4.7.6. Hydrogels of Poly(2-Acrylamido-2-Methyl-1-Propansulfonic Acid) (PAMPS) and P4VP

Hydrogels prepared by radical copolymerization, either with a photo-initiator for PAMPS [57], or a redox (persulfate) initiator for poly(4-vinyl pyridine)

(P4VP), poly(2-hydroxyethyl methacrylate) p(HEMA), and poly(4-vinyl pyridine-*co*-2-hydroxyethyl metacrylate) (P4VP-*co*-HEMA) [58], were impregnated with $FeSO_4$ and $FeCl_3$ and employed in a coprecipitation reaction to create magnetic composite hydrogels. These hydrogels, designed for water remediation, were able to absorb toxic heavy metal ions efficiently in aqueous media. Unfortunately, no information was provided on the size and the morphology of the magnetic particles obtained.

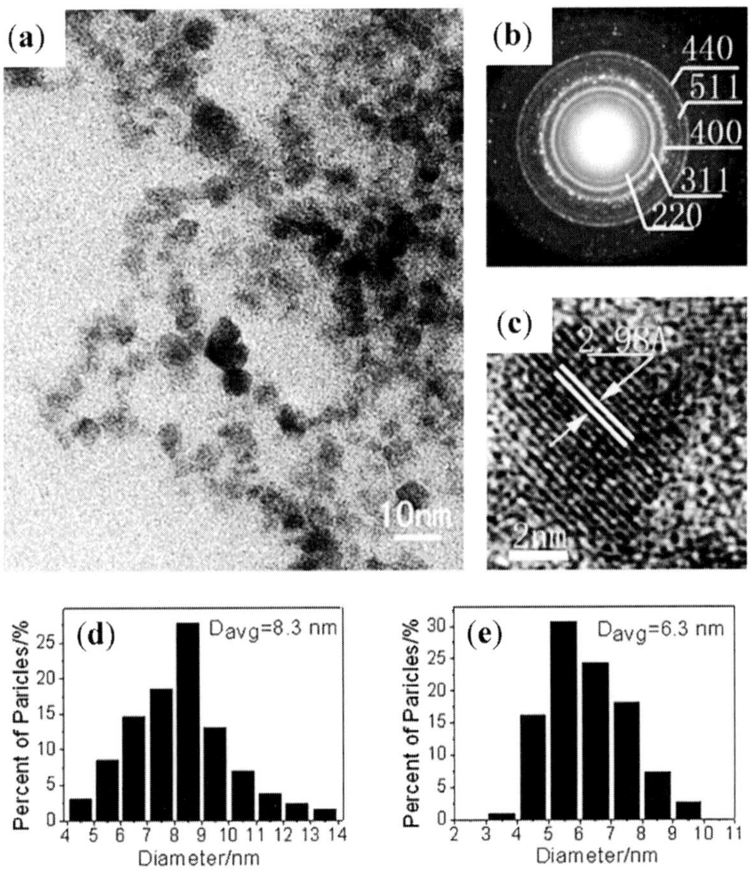

Figure 31. (**a**) TEM image; (**b**) SAED pattern for magnetic Fe_3O_4 NPs synthesized by coprecipitation of iron salts within polymer disk; (**c**) typical HRTEM image for a single Fe_3O_4 NP, and size distribution of the Fe_3O_4 NPs for samples containing the cross-linker at: (**d**) 0.46 mol% and (**e**) 6.14 mol%. Reprinted with permission from [56]. Copyright 2011 Elsevier B.V.

4.7.7. PAAm Hydrogels

PAAm hydrogels were employed as templates for the coprecipitation of Fe^{2+} and Fe^{3+} to create pH-responsive magnetic hydrogels [59]. The magnetite particles obtained were 3–5 nm in size and homogeneously distributed inside the hydrogel network (Figure 32). The population of Fe_3O_4increased for higher iron ion feeds (from 25 wt% to 100 wt% relatively to PAAm), indicating the good loading capacity of the template. Despite a low M_s value of 4 emu/g ascribed to the low size range of the MNPs and to the large weight proportion of the polymer matrix, the composite exhibited superparamagnetism with only a small H_c of 14.8 Oe.

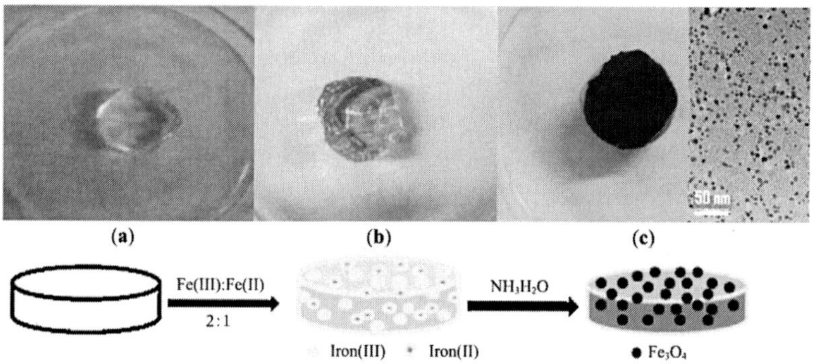

Figure 32. Schematic views and pictures of polyacrylamide (PAAm) hydrogel networks at three steps of loading by magnetite NPs: (**a**) swollen hydrogel; (**b**) iron ion-loaded hydrogel; and (**c**) magnetite NPs in the hydrogel matrix, inset: TEM image showing well-dispersed MNPs of uniform size. Reprinted with permission from [59]. Copyright 2009 Elsevier B.V.

4.8. Dispersed Colloids (Microscopic)

4.8.1. Poly(N-Isopropyl Acrylamide-co-Acrylic acid-co-2-Hydroxyethyl Acrylate) (Poly(NIPAM-co-AA-co-HEA)) Microgels Cross-Linked with N,N'-methylene Bisacrylamide (BIS)

Sub-micrometer gels of poly(NIPAM-*co*-AA-*co*-HEA), prepared by suspension polymerization and cross-linked with BIS, were loaded with ferrous iron ions in an ion exchange process [60]. A post-loading oxidation step with H_2O_2 was used

to produce the IONPs (Figure 33a). The size of these microgels was successfully controlled by the polymerization conditions: a rise of pH from 2.3 to 9.2 transformed the PAA blocks from a highly compact and weakly hydrophobic conformation to a highly hydrophilic and expanded coil one, and greatly increased the repulsive forces between COO^- groups, leading to larger particle sizes (from 200 nm to 600 nm). Not surprisingly, the concentration of COOH functionalities in the polymer composition was the dominant parameter affecting the size and the loading capacity of the microgels. However, the Fe^{2+}precursor concentration and the pH value range useable for the reactions were limited by the solubility product of $Fe(OH)_2$($K_s = 4.87 \times 10^{-17}$): for instance, the maximum allowable pH value was 7.1 for [Fe^{2+}] = 1.5 mM. Interestingly, the size distribution of the NPs was relatively narrow (e.g., 8.5 ± 1.0 nm diameter by TEM). These Fe_3O_4 NPs trapped in the 500 nm diameter microgels exhibited superparamagnetic properties at 300 K (both the remnant field and the H_c were zero) with a specific M_s of 32.4 emu/g (Figure 33b).

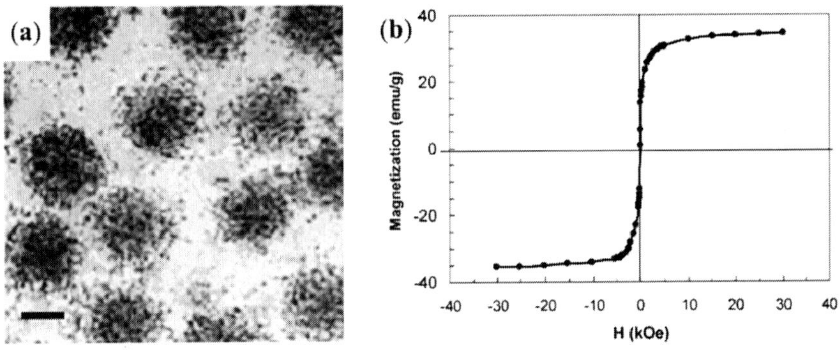

Figure 33. (**a**) TEM image for hybrid poly(*N*-isopropyl acrylamide-*co*-acrylic acid-*co*-2-hydroxyethyl acrylate) (poly(NIPAM-*co*-AA-*co*-HEA)) microgels loaded with 0.618 g Fe_3O_4/g polymer; the scale bar is 150 nm; and (**b**) magnetization curve at 300 K. Reprinted with permission from [60]. Copyright 2004 American Chemical Society.

4.8.2. Acetoacetoxyethyl Methacrylate (AAEM)-*N*-Vinylcaprolactam (VCL) Microgels

An interesting characteristic of these systems is that the chemical composition of the microgels can be adjusted to provide additional properties to the MNPs.

One example of this is the work of Pich *et al.* [61], who successfully employed microgels obtained by the copolymerization of AAEM and VCL to synthesize thermosensitive magnetic hybrids (Figure 34). It is common knowledge that thermoresponsive polymers such as PVCL can extensively and reversibly swell in appropriate solvents at temperatures below the LCST, but will shrink rapidly above the LCST. The PVCL-rich shell of the microgels, exhibiting an LCST, thus offered the possibility to control their size through temperature changes, while the AAEM-rich portions provided a high iron loading capacity via bidendate coordination. The observed non-monotonic increase of R_h of the microgels for increasing magnetite loadings was explained in terms of the chemical interactions formed. The slight decrease in R_h observed when increasing the Fe_3O_4 loading from 0 wt% to 7 wt% was attributed to the multivalent complexation between the β-diketone groups of AAEM and the Fe^{2+}/Fe^{3+} ions linking simultaneously several AAEM units, thus leading to additional cross-linking and shrinkage of the microgels. When the loading level surpassed 7 wt%, the repulsive forces among crowded Fe_3O_4 particles led to swelling of the microgels (increase in hydrodynamic size). An increase in magnetite content (e.g., from 4% to 9.4%), leading to larger magnetite-filled microgel cores, also affected the colloidal stability of the microgels in suspension as measured by their sedimentation velocity. Superparamagnetism was reported, with plateau values of the magnetization curve corresponding to 75 emu/g Fe_3O_4. The variation in the magnetic core size *versus* magnetite-to-polymer content in the hybrid microgels was not investigated in this study.

4.8.3. Sulfonated Copolymer Beads

Polymer beads produced by the suspension copolymerization of styrene and a small amount of divinylbenzene were sulfonated to add complexing moieties to the polymer chains and facilitate the encapsulation of Fe^{3+} and Co^{2+} cations for the synthesis of MNPs *in situ* [62]. The presence of a $CoFe_2O_4$ spinel phase was confirmed from its characteristic XRD peaks. TEM images obtained from ground powder displayed 7 nm, nearly spherical particles well-distributed within the polymer matrix. The absence of hysteresis, remanence and H_c demonstrated the superparamagnetic character of the hybrids, the saturated magnetization being also low (0.818 emu/g) due to a small amount of magnetic content in the hybrid beads.

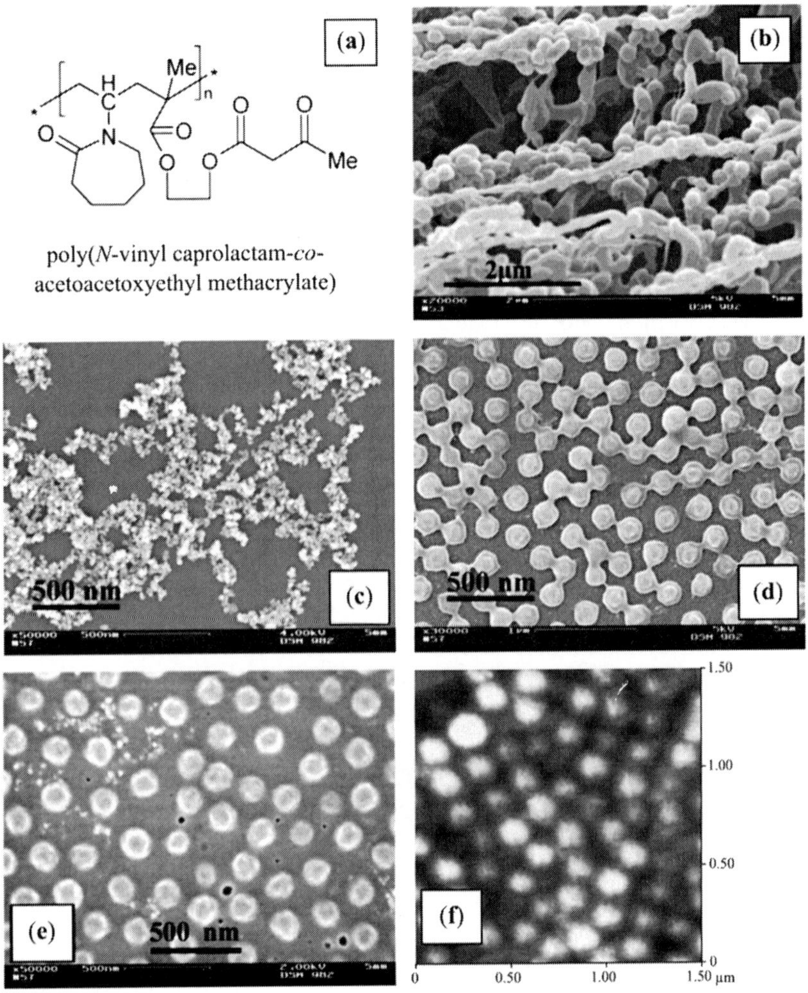

Figure 34. (**a**) Chemical structure of an *N*-vinylcaprolactam (VCL)/acetoacetoxyethyl methacrylate (AAEM) copolymer; and SEM images for: (**b**) empty cross-linked microgels; (**c**) magnetite NPs; (**d**) composite microgels with 4% magnetite; (**e**) SEM; and (**f**) AFM images for microgels with 9.4% magnetite (the height scale of the AFM image in the height mode is 0–100 nm). Reprinted with permission from [61]. Copyright 2004 American Chemical Society.

4.8.4. Poly[Poly(Ethylene Glycol Diacrylate-*co*-Acrylic acid) (PEGDA-*co*-PAA) Copolymer Hydrogels Prepared by Microfluidics

Janus particles, with their structure combining two distinct materials on the opposite sides of the NP, were designed in the groups of Doyle and Hatton [63] to work as selective nanoreactors for the *in situ* synthesis of magnetic species. Figure 35shows Janus-like and other non-spherical hydrogel magnetic particles prepared by stop-flow lithography (SFL), with one part of PEG diacrylate (PEGDA) non-interacting with iron salts, and another part constituted of a copolymer of PEGDA and acrylic acid (AA) to coordinate the iron ions. Since the polymerization was initiated by a photosensitive radical initiator, the shape of the hydrogels could be varied at will (e.g., disks, triangles, *etc.*) by an appropriate mask shaping the UV beam illuminating the reactants in a Y-shaped laminar flow chamber (Figure 36). Then the transfer of the minigels into an alkaline solution converted the COOH groups into COO^- and induced swelling by electrostatic repulsion between the anionic carboxylate groups. Interestingly, the Fe^{3+}/Fe^{2+} feed ratio used (1:75) was much lower than the required 2:1 stoichiometric ratio due to the different adsorption affinities of these two ions toward COO^-. The iron content of the particles was increased by repetition of the loading process, which allowed the adsorption of more ions at unoccupied nucleation sites and/or the surface of the magnetic particles, leading to larger, visibly darker and more magnetic particles. The hydrogel particles obtained after eight loading cycles reached an M_s value of 42 emu/g. It is worth noting that the maximum magnetic domain diameter reported was 5.7 nm (but with a high σ of 0.3–0.4), which was considered to be limited by the swollen mesh size of the polymer network. This suggests partial size control for the MNPs synthesized *in situ* in such hydrogel scaffold.

4.8.5. Multi-Responsive Microgels Made from VCL, AAEM and Vinylimidazole (VIm)

The wide range of polymeric materials available in terms of number and structural variety, but also in terms of physical and chemical characteristics, have inspired many researchers to look for novel materials with combining different properties. To prepare "multi-stimuli-responsive" magnetic hybrid microgel particles, Bhattacharya *et al.* [64] thus performed coprecipitation reactions of iron ions in the presence of microgels obtained by the batch copolymerization of VCL, AAEM, and VIm, cross-linked with BIS. The VCL-*co*-AAEM-*co*-VIm hybrid microgels were expected to exhibit thermosensitivity

brought by the VCL units, pH responsiveness ascribed to the AAEM-rich core, and magnetic switchable properties owning to encapsulated iron oxide particles. Unfortunately, the concentration of VIm units in the copolymer was found to strongly influence the swelling behavior and the transition temperature of the hydrogels. Shrinkage of the hybrid particles was also observed at higher magnetite contents (which led to partial collapse of the network in some cases), in agreement with a previous study [61]. The magnetite particles had a diameter of *ca.* 10 nm, as determined from TEM images (Figure 37). The highest M_s measured was 30 emu/g for the highest loading obtained of 15.4 wt% magnetite. Evidence for pH responsiveness and temperature-sensitive properties was shown for the hybrids, indicating the success of monomer blending to obtain copolymer microgels with combined characteristics and the creation of multifunctional materials.

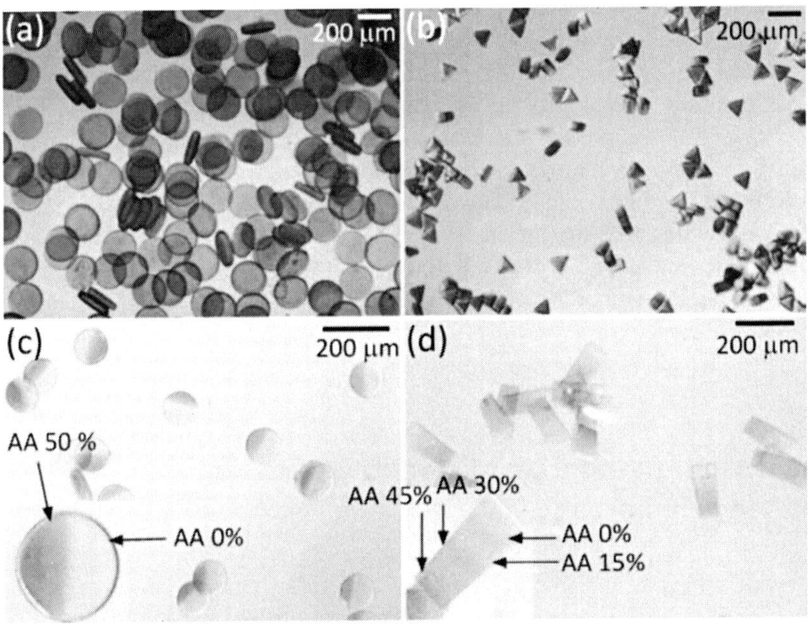

Figure 35. Optical images for various magnetic microparticles: (**a**) homogenous magnetic disks; (**b**) homogenous triangular particles; (**c**) Janus disks; and (**d**) gradient particles. Reprinted with permission from [63]. Copyright 2012 American Chemical Society.

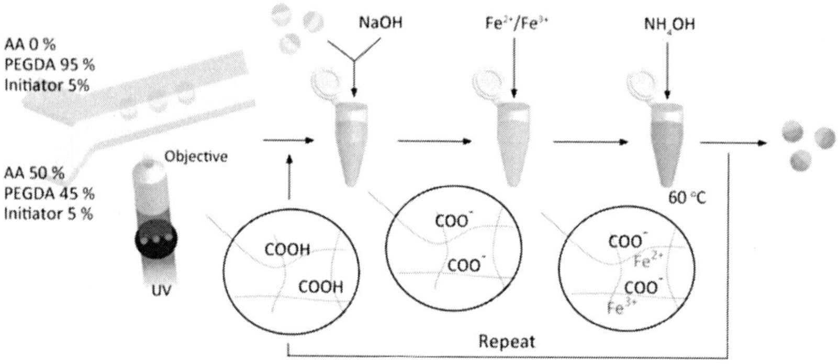

Figure 36. Schematic representation of the coprecipitation process in photo-polymerized microgels with ~20 μm diameter prepared in a microfluidic channel. After deprotonation in 0.5 M NaOH and rinsing with 0.5% Tween 20 to reach pH 7, 0.2 M $FeCl_3$ and 1 M $FeCl_2$ solutions were mixed with microgels at a Fe^{3+}:Fe^{2+} ratio of 1:75. After diffusion of the iron ions in the polymer particles, excess salts were removed. After the addition of NH_4OH at 60 °C, MNPs were nucleated and grown *in situ* before a final rinse with Tween 20. All these steps were repeated several times for successive growth cycles. Reprinted with permission from [63]. Copyright 2012 American Chemical Society.

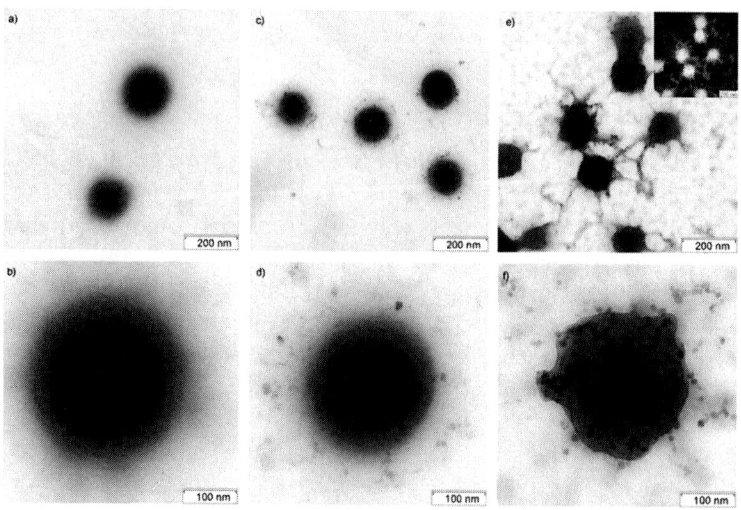

Figure 37. TEM images of microgels for different magnetite contents: (**a**,**b**) no magnetite; (**c**,**d**) 8.4 wt%; (**e**,**f**) 15.4 wt%. Inset: EDX iron-mapping image. Reprinted with permission from [64] Copyright 2007 Wiley-VCH Verlag GmbH & Co., Weinheim.

4.9. Preformed Microspheres in Organic Solvents

4.9.1 PS Microsphere Swollen in $CHCl_3$

Interestingly, even water-insoluble cross-linked PS microspheres were found to be capable of serving as hosts for the thermal decomposition of encapsulated ferric tri-oleate, after swelling of the microspheres in an organic solvent ($CHCl_3$) to facilitate the impregnation [65]. The magnetic character of the NPs obtained strongly depended on the decomposition temperature used, since the dissociation of the Fe(oleate)$_3$ complex only started around 200–240 °C but was completed at 300 °C, while crystal growth only took place above 310 °C and led to strong magnetism. TGA results showed that decomposition of the polymer matrix occurred between 380 °C and 450 °C, yielding an inorganic content of 15.9 wt% for the composite spheres. SEM and TEM images (Figure 38) revealed nearly spherical IONPs with a mean diameter of 11 nm and low size dispersity (21% standard deviation), distributed uniformly within the network serving as template, by allowing nucleation but inhibiting the overgrowth of NPs. XRD analysis confirmed the Fe_3O_4 composition of the iron oxide. The particles exhibited superparamagnetic characteristics at room temperature (M_s = 12.6 emu/g and T_B = 171 K), with excellent magnetic responsiveness but readily redispersing after removal of the magnetic field.

5. CONCLUSIONS

In recent years, multiple routes have been explored for the synthesis of IONPs. This includes the alkaline coprecipitation of ferric and ferrous salts, the oxidation of Fe_2SO_4, the thermal decomposition of iron complexes, the polyol route, and hydrothermal processes in various organic templates, as summarized in Table 5. One feature common to all the systems explored so far is the ability of one of the components used to interact with the iron oxide precursor, most often through coordination with the iron ions. A number of the *in situ* syntheses did not give results significantly different from bulk synthesis methods, as these were realized under similar physicochemical conditions (pH, temperature, *etc.*). For instance, syntheses by alkaline coprecipitation leading to IONPs with rock-like shapes and a broad distribution of diameters around 7 nm do not make a strong case for a true templating effect by the organic matrix. In contrast, several investigation listed in this review (in particular those using well-defined polymer structures) highlighted a significant influence of the polymer/precursor ratio on the outcome of the synthesis. This is the case, for instance, for tridimensional networks such as the hydrogels, for which the importance mesh size was recognized as a parameter

limiting the diameter of the MNPs grown*in situ*. It was also demonstrated on several occasions that strong ligands (such as phosphate groups) directed the synthesis towards lower particle sizes, which seems to be a general well-known trend in coordination chemistry that can be ascribed to hampering of ion diffusion through the ligand shell during the growth of nanocrystals. On the contrary, other studies reported MNP diameters that increased from 10 nm to 20 nm with the template. This is the case for chemical reactions performed at high temperatures, were an important property required from the organic template is to prevent the aggregation of the growing nanocrystals. This protecting effect was observed in particular in the last study reported, where a PS matrix could withstand temperatures as high as 310 °C for the thermal decomposition of an iron-oleate complex.

Regarding shape control, several studies claimed success of the organic templating method to orient the synthesis towards rod-like NPs. However, caution should be used in that the physicochemical conditions used may themselves favor the precipitation of antiferromagnetic ferrihydrite phases such as goethite or akaganeite. A more convincing case of needle-like cobalt ferrite microparticle formation was reported for a microemulsion process, based on the oriented aggregation of nanocrystals. One interesting aspect that can be brought by polymer-templated synthesis is the ability to prepare NPs with highly magnetic properties (useful for MRI contrast or magnetic hyperthermia) directly covered by a polymer shell, providing colloidal stability and stealth properties in biological media. This was achieved in several investigations by employing either double hydrophilic or amphiphilic copolymers, in particular those forming micelles with a dense core strongly binding to the precursor species and a shell with multiple hydrophilic arms providing steric repulsion.

In all these studies, both with macroscopic matrices and individually dispersed objects, loading of the inorganic precursor into the adsorption sites was critical and needed to be optimized (by iterative process or by careful removal of excess $Fe^{2+/3+}$ions) to achieve a true templating effect.

ACKNOWLEDGMENTS

The authors thank the International Doctoral Program in Functional Materials (IDS-FunMat) funded by the ERASMUS MUNDUS Program of the European Union, and the Precision Polymer Materials Program funded by the European Science Foundation.

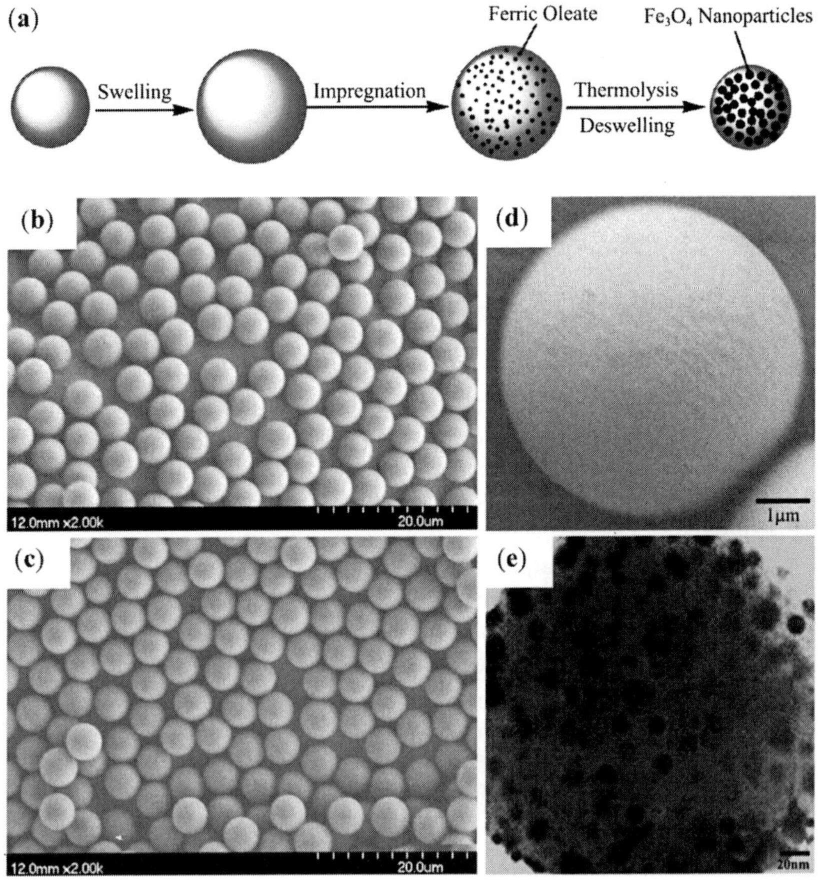

Figure 38. (**a**) Schematic representation of the preparation of Fe_3O_4 NPs embedded in PS microspheres by thermal decomposition of $Fe(oleate)_3$ at 300 °C; SEM images for: (**b**) the polymer seed microspheres prepared by dispersion polymerization; (**c**) the magnetic polymer microspheres prepared by swelling and thermolysis; (**d**) the outer surface of magnetic polymer microspheres; and (**e**) TEM image for the ultramicrotomed magnetic polymer microspheres, showing the location of the IONPs in the microspheres. Reprinted with permission from [65]. Copyright 2009 American Chemical Society.

Table 5. Summary of the different organic templates used as host matrices for the synthesis of MNPs. In this review, the examples gathered from the literature were sorted according to their chemical nature and structural properties (e.g., geometry, dimensionality, size). DEX: dextran; Alg: alginate; PNiPAAm: poly(*N*-isopropylacrylamide); PVA: poly(vinyl alcohol); PVP: polyvinylpyrrolidone; P4VP: poly(4-vinylpyridine); PMGI: poly(methyl glutarimide); DHBC: double-hydrophilic block copolymer; PGA: poly(glycerol acrylate); PAsp: poly(aspartic acid); PAEMA: poly[2-(acetoacetoxy) ethyl] methacrylate; PGMA: poly(glycerol methacrylate); PDMAEMA: poly[(*N*,*N*-dimethylamino)ethyl methacrylate]; and PAMPS: poly(2-acrylamido-2-methyl-1-propansulfonic acid).

Surfactants: $-COO^-$, $-SO_3^-$, $-SO_4^-$ **Natural polysaccharides**: DEX, DEX sulfate, Alg, DEX-*b*-PMAA-*b*-PNiPAAm	**Synthetic linear polymers**: PVA, PVP, P4VP, PEO, PAA, PMAA, PMGI **Linear DHBC or amphiphilic copolymers**	**Hydrophilic blocks**: iron-complexing (PGMA, PGA, P(norbornene-methanol or di-carboxy), PMAA, PAA, PAsp, PAEA, PAEMA) or repulsive block (PEO, PEOGMA, PDMAEMA)	**Hydrophobic blocks**: PMMA, PNOR, PVBP **Tri-blocks**: Pluronics®, PI-*b*-PCEMA-*b*-PtBA
Three-dimensional scaffold macroscopic matrices: Carbon foam, sponge-like PS-*co*-PAA, P(S-*co*-CBMi) networks, Alg/PNiPAAm semi-IPN, PAAm, PAMPS, P(4VP-*co*-HEMA) hydrogels	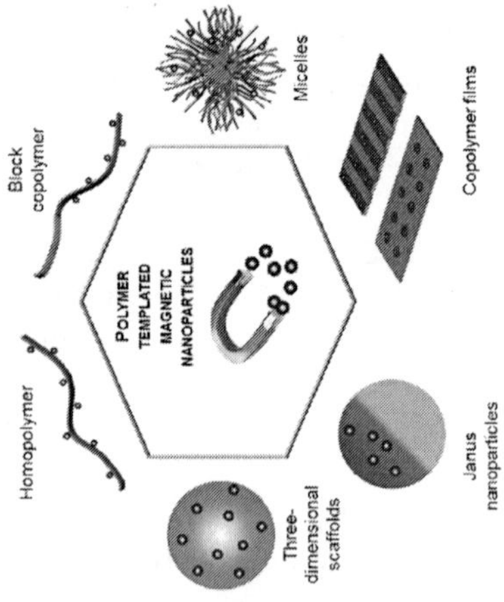		**Star-like** (PPO-*b*-PAA, PEO-*b*-PS) **Comb-like** (PPEGMEA-*g*-PMAA, PEO-*g*-PAA) **Spherical micelles** (PS-*b*-P2VP) **Cylindrical multi-molecular/ unimolecular micelles**
Dispersed colloids: microgels, hydrophobic PS microbeads	**Janus, gradient or triangle**: PEGDA, PVCL-*co*-AAEM core-VIm shell microgels	**Lamellar** or **hexagonally ordered cylinders** (PS-*b*-P2VP, PS-*b*-PEO) **Membranes** (PVA, Nafion®)	**Microemulsions**: hexane/Triton®-X/Chitosan, CTAB

AUTHOR CONTRIBUTIONS

Vo Thu An Nguyen wrote the first draft of this review article that was later corrected by Olivier Sandre and Mario Gauthier. Olivier Sandre made the final revision.

CONFLICTS OF INTEREST

The authors declare no conflict of interest.

REFERENCES

1. Ahn, T.; Kim, J.H.; Yang, H.-M.; Lee, J.W.; Kim, J.-D. Formation pathways of magnetite nanoparticles by coprecipitation method. J. Phys. Chem. C 2012, 116, 6069–6076.

2. Laurent, S.; Forge, D.; Port, M.; Roch, A.; Robic, C.; vander Elst, L.; Muller, R.N. Magnetic iron oxide nanoparticles: Synthesis, stabilization, vectorization, physicochemical characterizations, and biological applications. Chem. Rev. 2008,108, 2064–2110.

3. Gu, L.; Shen, Z.; Feng, C.; Li, Y.; Lu, G.; Huang, X.; Wang, G.; Huang, J. Synthesis of PPEGMEA-g-PMAA densely grafted double hydrophilic copolymer and its use as a template for the preparation of size-controlled superparamagnetic Fe3O4/polymer nano-composites. J. Mater. Chem. 2008, 18, 4332–4340.

4. Yathindranath, V.; Rebbouh, L.; Moore, D.F.; Miller, D.W.; van Lierop, J.; Hegmann, T. A versatile method for the reductive, one-pot synthesis of bare, hydrophilic and hydrophobic magnetite nanoparticles. Adv. Funct. Mater. 2011,21, 1457–1464.

5. Bee, A.; Massart, R.; Neveu, S. Synthesis of very fine maghemite particles. J. Magn. Magn. Mater. 1995, 149, 6–9.

6. Liu, C.; Rondinone, A.J.; Zhang, Z.J. Synthesis of magnetic spinel ferrite CoFe2O4 nanoparticles from ferric salt and characterization of the size-dependent superparamagnetic properties. Pure Appl. Chem. 2000, 72, 37–45.

7. Lee, Y.; Lee, J.; Bae, C.J.; Park, J.G.; Noh, H.J.; Park, J.H.; Hyeon, T. Large-scale synthesis of uniform and crystalline magnetite nanoparticles using reverse micelles as nanoreactors under reflux conditions. Adv. Funct. Mater. 2005, 15, 503–509.

8. Patrinoiu, G.; Visinescu, D.; Tirsoaga, A.; Carp, O. Green synthetic strategy for oxide materials: Polysaccharides-assisted synthesis. Part III. Dextran-assisted synthesis of nanosized metal-oxides. Rev. Roum. Chim. 2011, 56, 145–150.

9. Paul, K.G.; Frigo, T.B.; Groman, J.Y.; Groman, E.V. Synthesis of ultrasmall superparamagnetic iron oxides using reduced polysaccharides. Bioconjug. Chem. 2004, 15, 394–401.

10. Benjamin, R.J.; Michele, F.; Jacob, V.; Angelique, Y.L. Size-controlled synthesis of dextran sulfate coated iron oxide nanoparticles for magnetic resonance imaging. Nanotechnology 2007, 18.

11. Feng, W.; Lv, W.; Qi, J.; Zhang, G.; Zhang, F.; Fan, X. Quadruple-responsive nanocomposite based on dextran–PMAA–PNIPAM, iron oxide nanoparticles, and gold nanorods. Macromol. Rapid Commun. 2012, 33, 133–139.

12. Srivastava, M.; Singh, J.; Yashpal, M.; Gupta, D.K.; Mishra, R.K.; Tripathi, S.; Ojha, A.K. Synthesis of superparamagnetic bare Fe3O4 nanostructures and core/shell (Fe3O4/alginate) nanocomposites. Carbohydr. Polym.2012, 89, 821–829.

13. Yokoi, H.; Kantoh, T. Thermal decomposition of the iron(III) hydroxide and magnetite composites of poly(vinyl alcohol). Preparation of magnetite and metallic iron particles. Bull. Chem. Soc. Jpn. 1993, 66, 1536–1541.

14. Sairam, M.; Naidu, B.V.K.; Nataraj, S.K.; Sreedhar, B.; Aminabhavi, T.M. Poly(vinyl alcohol)-iron oxide nanocomposite membranes for pervaporation dehydration of isopropanol, 1,4-dioxane and tetrahydrofuran. J. Membr. Sci. 2006, 283, 65–73.

15. Xu, C.; Teja, A.S. Continuous hydrothermal synthesis of iron oxide and pva-protected iron oxide nanoparticles. J. Supercrit. Fluids 2008, 44, 85–91.

16. Liu, H.-L.; Ko, S.P.; Wu, J.-H.; Jung, M.-H.; Min, J.H.; Lee, J.H.; An, B.H.; Kim, Y.K. One-pot polyol synthesis of monosize PVP-coated sub-5 nm Fe3O4 nanoparticles for biomedical applications. J. Magn. Magn. Mater. 2007, 310, e815–e817.

17. Millan, A.; Urtizberea, A.; Natividad, E.; Luis, F.; Silva, N.J.O.; Palacio, F.; Mayoral, I.; Ruiz-González, M.L.; González-Calbet, J.M.; Lecante, P.; et al. Akaganeite polymer nanocomposites. Polymer 2009, 50, 1088–1094.

18. Karakoti, A.S.; Das, S.; Thevuthasan, S.; Seal, S. PEGylated inorganic nanoparticles. Angew. Chem. Int. Ed. 2011, 50, 1980–1994.

19. Harraz, F.A. Polyethylene glycol-assisted hydrothermal growth of magnetite nanowires: Synthesis and magnetic properties. Phys. E Low Dimens. Syst. Nanostruct. 2008, 40, 3131–3136.

20. Lin, C.-L.; Lee, C.-F.; Chiu, W.-Y. Preparation and properties of poly(acrylic acid) oligomer stabilized superparamagnetic ferrofluid. J. Colloid Interface Sci. 2005, 291, 411–420.

21. Ja Young, P.; Patel, D.; Gang Ho, L.; Seungtae, W.; Yongmin, C. Highly water-dispersible PEG surface modified ultra small superparamagnetic iron oxide nanoparticles useful for target-specific biomedical applications. Nanotechnology2008, 19.

22. Li, Z.; Tan, B.; Allix, M.; Cooper, A.I.; Rosseinsky, M.J. Direct coprecipitation route to monodisperse dual-functionalized magnetic iron oxide nanocrystals without size selection. Small 2008, 4, 231–239.

23. Tural, B.; Özkan, N.; Volkan, M. Preparation andcharacterization of polymer coated superparamagnetic magnetite nanoparticle agglomerates. J. Phys. Chem. Solids 2009, 70, 860–866.

24. Lu, J.Q.; Moll, N.; Fu, Q.; Liu, J. Iron nanoparticles derived from iron-complexed polymethylglutarimide to produce high-quality lithographically defined single-walled carbon nanotubes. Chem. Mater. 2005, 17, 2237–2240.

25. Wan, S.; Zheng, Y.; Liu, Y.; Yan, H.; Liu, K. Fe3O4 nanoparticles coated with homopolymers of glycerol mono(meth)acrylate and their block copolymers. J. Mater. Chem. 2005, 15, 3424–3430.

26. Guo, M.; Yan, Y.; Zhang, H.; Yan, H.; Cao, Y.; Liu, K.; Wan, S.; Huang, J.; Yue, W. Magnetic and pH-responsive nanocarriers with multilayer core-shell architecture for anticancer drug delivery. J. Mater. Chem. 2008, 18, 5104–5112.

27. Lee, H.; Lee, E.; Kim, D.K.; Jang, N.K.; Jeong, Y.Y.; Jon, S. Antibiofouling polymer-coated superparamagnetic iron oxide nanoparticles as potential

magnetic resonance contrast agents for in vivo cancer imaging. J. Am. Chem. Soc. 2006,128, 7383–7389.

28. Wormuth, K. Superparamagnetic latex via inverse emulsion polymerization. J. Colloid Interface Sci. 2001, 241, 366–377.

29. Kumagai, M.; Imai, Y.; Nakamura, T.; Yamasaki, Y.; Sekino, M.; Ueno, S.; Hanaoka, K.; Kikuchi, K.; Nagano, T.; Kaneko, E.; et al. Iron hydroxide nanoparticles coated with poly(ethylene glycol)-poly(aspartic acid) block copolymer as novel magnetic resonance contrast agents for in vivo cancer imaging. Colloids Surfaces B 2007, 56, 174–181.

30. Sondjaja, R.; Alan Hatton, T.; Tam, M.K.C. Clustering of magnetic nanoparticles using a double hydrophilic block copolymer, poly(ethylene oxide)-b-poly(acrylic acid). J. Magn. Magn. Mater. 2009, 321, 2393–2397.

31. Akcora, P.; Zhang, X.; Varughese, B.; Briber, R.M.; Kofinas, P. Structural and magnetic characterization of norbornene–deuterated norbornene dicarboxylic acid diblock copolymers doped with iron oxide nanoparticles. Polymer 2005, 46, 5194–5201.

32. Akcora, P.; Briber, R.M.; Kofinas, P. TEM characterization of diblock copolymer templated iron oxide nanoparticles: Bulk solution and thin film surface doping approach. Polymer 2006, 47, 2018–2022.

33. Yoon, K.Y.; Kotsmar, C.; Ingram, D.R.; Huh, C.; Bryant, S.L.; Milner, T.E.; Johnston, K.P. Stabilization of superparamagnetic iron oxide nanoclusters in concentrated brine with cross-linked polymer shells. Langmuir 2011, 27, 10962–10969.

34. Lutz, J.-F.; Stiller, S.; Hoth, A.; Kaufner, L.; Pison, U.; Cartier, R. One-pot synthesis of PEGylated ultrasmall iron-oxide nanoparticles and their in vivo evaluation as magnetic resonance imaging contrast agents. Biomacromolecules 2006, 7, 3132–3138.

35. Basuki, J.S.; Jacquemin, A.; Esser, L.; Li, Y.; Boyer, C.; Davis, T.P. A block copolymer-stabilized co-precipitation approach to magnetic iron oxide nanoparticles for potential use as MRI contrast agents. Polym. Chem. 2014, 5, 2611–2620.

36. Papaphilippou, P.; Loizou, L.; Popa, N.C.; Han, A.; Vekas, L.; Odysseos, A.; Krasia-Christoforou, T. Superparamagnetic hybrid micelles, based on iron oxide nanoparticles and well-defined diblock copolymers possessing β-ketoester functionalities. Biomacromolecules 2009, 10, 2662–2671.

37. Moeser, G.D.; Green, W.H.; Laibinis, P.E.; Linse, P.; Hatton, T.A. Structure of polymer-stabilized magnetic fluids: Small-angle neutron scattering and mean-field lattice modeling. Langmuir 2004, 20, 5223–5234.

38. Zhi, J.; Wang, Y.; Lu, Y.; Ma, J.; Luo, G. In situ preparation of magnetic chitosan/Fe3O4 composite nanoparticles in tiny pools of water-in-oil microemulsion. React. Funct. Polym. 2006, 66, 1552–1558.

39. Zhang, Z.; Rondinone, A.J.; Ma, J.X.; Shen, J.; Dai, S. Morphologically templated growth of aligned spinel CoFe2O4nanorods. Adv. Mater. 2005, 17, 1415–1419.

40. Underhill, R.S.; Liu, G. Triblock nanospheres and their use as templates for inorganic nanoparticle preparation. Chem. Mater. 2000, 12, 2082–2091.

41. Ujiie, K.; Kanayama, N.; Asai, K.; Kishimoto, M.; Ohara, Y.; Akashi, Y.; Yamada, K.; Hashimoto, S.; Oda, T.; Ohkohchi, N.; et al. Preparation of highly dispersible and tumor-accumulative, iron oxide nanoparticles: Multi-point anchoring of PEG-b-poly(4-vinylbenzylphosphonate) improves performance significantly. Colloids Surfaces B Biointerfaces 2011,88, 771–778.

42. Pang, X.; Zhao, L.; Han, W.; Xin, X.; Lin, Z. A general and robust strategy for the synthesis of nearly monodisperse colloidal nanocrystals. Nat. Nanotechnol. 2013, 8, 426–431.

43. Yang, Y.; Jiang, J.-S. Gradual phase and morphology transformation of Fe3O4 nanoparticles to α-FeOOH nanorods in alcohol/water media in the presence of surfactant F127. J. Mater. Sci. 2008, 43, 4340–4343.

44. Zhang, M.; Müller, A.H.E.; Teissier, P.; Cabuil, V.; Krekhova, M. Polychelates of Amphiphilic Cylindrical Core–Shell Polymer Brushes with Iron Cations. In Trends in Colloid and Interface Science XVII; Springer: Berlin/Heidelberg, Germany, 2004; Volume 126, pp. 35–39.

45. Wan, S.; Huang, J.; Yan, H.; Liu, K. Size-controlled preparation of magnetite nanoparticles in the presence of graft copolymers. J. Mater. Chem. 2006, 16, 298–303.

46. Li, P.; Huang, J. Preparation of poly (ethylene oxide)-graft-poly (acrylic acid) copolymer stabilized iron oxide nanoparticles via an in situ templated process. J. Appl. Polym. Sci. 2008, 109, 501–507.

47. Abes, J.I.; Cohen, R.E.; Ross, C.A. Block-copolymer-templated synthesis of iron, iron-cobalt, and cobalt-nickel alloy nanoparticles. Mater. Sci. Eng. C 2003, 23, 641–650.

48. Yun, S.-H.; Sohn, B.-H.; Jung, J.C.; Zin, W.-C.; Lee, J.-K.; Song, O. Tunable magnetic arrangement of iron oxide nanoparticles in situ synthesized on the solid substrate from diblock copolymer micelles. Langmuir 2005, 21, 6548–6552.

49. Ghoshal, T.; Shaw, M.T.; Bolger, C.T.; Holmes, J.D.; Morris, M.A. A general method for controlled nanopatterning of oxide dots: A microphase separated block copolymer platform. J. Mater. Chem. 2012, 22, 12083–12089.

50. Uekawa, N.; Kaneko, K. Non stoichiometric properties of nanoporous iron oxide films. J. Phys. Chem. B 1998, 102, 8719–8724.

51. Yoon, M.; Kim, Y.M.; Kim, Y.; Volkov, V.; Song, H.J.; Park, Y.J.; Vasilyak, S.L.; Park, I.W. Magnetic properties of iron nanoparticles in a polymer film. J. Magn. Magn. Mater. 2003, 265, 357–362.

52. Yoon, T.; Chae, C.; Sun, Y.-K.; Zhao, X.; Kung, H.H.; Lee, J.K. Bottom-up in situ formation of Fe3O4 nanocrystals in a porous carbon foam for lithium-ion battery anodes. J. Mater. Chem. 2011, 21, 17325–17330.

53. Breulmann, M.; Cölfen, H.; Hentze, H.-P.; Antonietti, M.; Walsh, D.; Mann, S. Elastic magnets: Template-controlled mineralization of iron oxide colloids in a sponge-like gel matrix. Adv. Mater. 1998, 10, 237–241.

54. Sepúlveda-Guzmán, S.; Lara, L.; Pérez-Camacho, O.; Rodríguez-Fernández, O.; Olivas, A.; Escudero, R. Synthesis and characterization of an iron oxide poly(styrene-co-carboxybutylmaleimide) ferrimagnetic composite. Polymer 2007, 48, 720–727.

55. Hernández, R.; Sacristán, J.; Nogales, A.; Ezquerra, T.A.; Mijangos, C. Structural organization of iron oxide nanoparticles synthesized inside hybrid polymer gels derived from alginate studied with small-angle X-ray scattering. Langmuir 2009, 25, 13212–13218.

56. Xiong, Z.; Sun, Z.-B.; Zheng, M.-L.; Cao, Y.-Y.; Jin, F.; Chen, W.-Q.; Zhao, Z.-S.; Duan, X.-M. A facile method for the room-temperature synthesis of water-soluble magnetic Fe3O4 nanoparticles: Combination of in situ synthesis and decomposition of polymer hydrogel. Mater. Chem. Phys. 2011, 130, 72–78.

57. Ozay, O.; Ekici, S.; Baran, Y.; Aktas, N.; Sahiner, N. Removal of toxic metal ions with magnetic hydrogels. Water Res.2009, 43, 4403–4411.

58. Ozay, O.; Ekici, S.; Aktas, N.; Sahiner, N. P(4-vinyl pyridine) hydrogel use for the removal of UO22+ and Th4+ from aqueous environments. J. Environ. Manag. 2011, 92, 3121–3129.

59. Sivudu, K.S.; Rhee, K.Y. Preparation and characterization of pH-responsive hydrogel magnetite nanocomposite.Colloids Surfaces A Physicochem. Eng. Asp. 2009, 349, 29–34.

60. Zhang, J.; Xu, S.; Kumacheva, E. Polymer microgels: Reactors for semiconductor, metal, and magnetic nanoparticles. J. Am. Chem. Soc. 2004, 126, 7908–7914.

61. Pich, A.; Bhattacharya, S.; Lu, Y.; Boyko, V.; Adler, H.-J.P. Temperature-sensitive hybrid microgels with magnetic properties. Langmuir 2004, 20, 10706–10711.

62. García-Cerda, L.A.; Chapa-Rodríguez, R.; Bonilla-Ríos, J. In situ synthesis of iron oxide nanoparticles in a styrene-divinylbenzene copolymer. Polym. Bull. 2007, 58, 989–994.

63. Suh, S.K.; Yuet, K.; Hwang, D.K.; Bong, K.W.; Doyle, P.S.; Hatton, T.A. Synthesis of nonspherical superparamagnetic particles: In situ coprecipitation of magnetic nanoparticles in microgels prepared by stop-flow lithography. J. Am. Chem. Soc. 2012, 134, 7337–7343.

64. Bhattacharya, S.; Eckert, F.; Boyko, V.; Pich, A. Temperature-, pH-, and magnetic-field-sensitive hybrid microgels.Small 2007, 3, 650–657.

65. Yang, C.; Shao, Q.; He, J.; Jiang, B. Preparation of monodisperse magnetic polymer microspheres by swelling and thermolysis technique. Langmuir 2009, 26, 5179–5183.

Nanocrystals - Synthesis, Characterization and Applications

AF677928